APPLIED STATISTICS
using the computer

APPLIED STATISTICS
using the computer

RONALD S. KING
Baylor University

BRYANT JULSTROM
University of Iowa

Alfred Publishing Co., Inc.

8287

Production Management
and Art Direction by:
Michael Bass & Associates
Covered Design by:
Lorena Laforest Bass

     Alfred Publishing Co., Inc.
     15335 Morrison Street
     Sherman Oaks, California 91403

Printing (last digit)
10 9 8 7 6 5 4 3 2 1

T85-01876

Library of Congress Cataloging in Publication Data
King, Ronald S., 1943–

    Bibliography: p.
    Includes index.
    1. Mathematical statistics—Data processing.
I. Julstrom, Bryant, 1950–   . II. Title.
QA276.4.K55   519.5'028'54   81–22782
ISBN 0-88284-174-2   AACR2

To our parents:

Dr. and Mrs. Byron B. King
and
Dr. and Mrs. Clifford A. Julstrom

CONTENTS

PREFACE

Applied Statistics Using Computers is intended to be a first text in applied statistics. It requires no previous training in either statistics or computing. The text introduces fundamental ideas in both areas, and develops the classical statistical analyses while examining the corresponding routines from a statistical software package. At the conclusion of a course using this text, the student should be able to use a statistical software package to perform and interpret statistical analyses of a raw data file.

A minimum mathematical ability equivalent to college algebra is presumed, and some knowledge of calculus and linear algebra is required to understand certain derivations and demonstrations.

This book is designed to meet two main objectives. The first is to present the fundamental techniques of classical statistical analysis, both univariate and multivariate. The second is to illustrate the use of packaged statistical programs to perform these analyses.

Chapter 1 introduces the ideas of statistics and statistical inference, and describes the types of data analyzed with these techniques. Chapter 2 discusses computers and their use in statistical analysis, and describes the data file from which the majority of the examples and many exercises in the text are taken. In Chapter 3, the Statistical Package for the Social Sciences (SPSS), the package used in the examples in the text, is introduced by creating and entering the data file. This file is described in Chapter 4, which defines the usual measures of central tendency, variability, and distributional shape.

Fundamental concepts of probability—events, random variables, probability distributions, expected value and variance of a probability distribution—are covered in Chapter 5, while Chapter 6 examines the most important discrete and continuous distributions, and the use of one distribution to approximate another.

This discussion leads into Chapter 7, which considers the sampling distributions of the sample mean and other sample statistics. The Central Limit Theorem is stated and discussed.

Chapter 8 describes point estimation, interval estimation, and hypothesis testing. Investigation of hypothesis testing with one sample is continued in Chapter 9, with discussion of the Type II error, one- and two-tail tests, small-sample tests for the mean using the t distribution, and chi-square tests of variance. One-sample tests provide an opportunity to present a packaged program designed for a microcomputer, the Radio Shack TRS-80. Chapter 10 considers two-sample tests, beginning with the F test for the ratio of two variances, since the outcome of that test determines how to test the difference of two means when using independent samples. Matched-pair tests are also considered.

Chapter 11 describes the use of the chi-square statistic to perform tests of multinominal data, independence, goodness-of-fit, and missing data. Chapter 12 explores correlation, defining not only the Pearson correlation coefficient but other coefficients which apply to situations in which the Pearson R may not be appropriate. Chapter 13 discusses linear regression, both simple and multiple. Development of the multiple linear regression equation is accomplished through an iterative matrix technique rather than the cumbersome normal equations; and the final section of the chapter describes transformations which can be used to develop nonlinear regression equations with the techniques of linear regression. An example based on judgment analysis (JAN) is included. Chapter 14 discusses one-way and two-way analysis of variance, using completely randomized and randomized-block designs, and concludes with a discussion of the analysis of covariance. Chapter 15 is a short discussion of the design of experiments. Chapter 16 explores nonparametric hypothesis tests, beginning with tests of randomness and goodness-of-fit, and concluding with nonparametric substitutes for parametric tests.

In the final two chapters, the tools of statistical analysis are described and discussed. Chapter 17 examines hardware for statistical applications—calculators, small- and large-scale computer systems—and Chapter 18 considers the possible forms and desirable features of a statistical software package.

Statistical tables and data files for examples and exercises are contained in the appendices.

In each chapter, the discussion of a statistical procedure is accompanied by an example of statistical package software to perform that procedure. For example, cross-tabulations, introduced in Chapter 5 in conjunction with conditional probabilities, also appear in Chapters 11 and 12 (the chi-square statistic and correlation). These examples are drawn from SPSS, but this book can be used successfully with any statistical software package.

This book differs from other applied statistics texts in several important ways:

a. Except in simple examples, computations are left to the computer system. This frees the student to concentrate on selecting the correct procedure and interpreting the results. The student learns statistics by doing statistical analyses.

b. The use of a particular program package is not mandatory; analyses can be performed with any statistical software package. The package used in the text (SPSS) is present primarily for illustration.

c. To begin performing statistical analyses immediately, one data file is used in most discussions in the text.

d. Surveys of currently available software and hardware and guidelines for the evaluation of both are included, as is a short discussion of future trends.

e. Complicated concepts and relationships are explained verbally as well as mathematically.

f. The key to thinking like a statistician is the ability to visualize sampling distributions. These are explored and illustrated in detail.

Applied Statistics Using Computers is designed to be used in a one-year undergraduate course in applied statistics, or in a one-semester graduate course. The first nine chapters, which cover the basic ideas of descriptive statistics, probability, and statistical inference, and introduce SPSS, should always be included. The instructor may then select material from Chapters 10, 11, 12, and 13, 14, and 16 without disturbing the continuity of the course. Note that Chapter 15 depends on Chapters 13 and 14. Chapters 17 and 18 stand by themselves.

APPLIED STATISTICS
using the computer

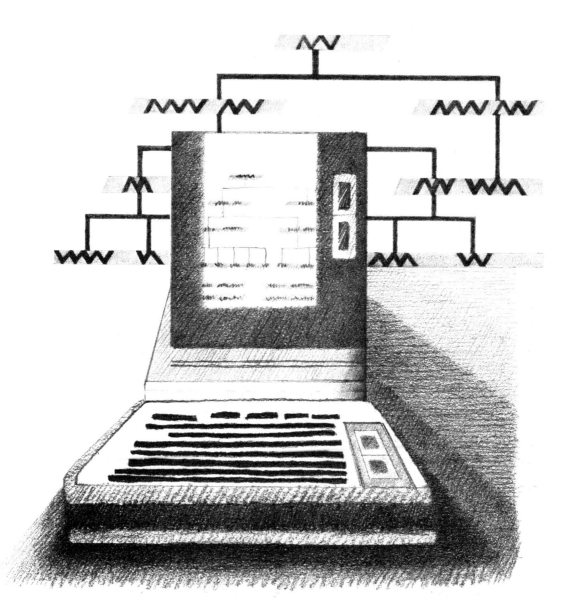

Introduction to Statistics

INTRODUCTION

What is statistics? Why should a person study statistics? How does one perform a statistical study? In this chapter, we consider these questions.

What Is Statistics?

According to Webster's *New Collegiate Dictionary*, statistics is

DEFINITION

the science of the collection and classification of facts on the basis of relative numbers of occurrence as a ground for induction; systematic compilation of instances for the inference of general truth.

Several aspects of statistics are brought out by the second part of this definition. The first one is the need for a well-defined method of summarizing the observations in an experiment, in order to make the information contained in the observations easier to understand. For example, a businessperson might summarize the sales per month of a particular item by calculating the "average" sales per month, thus reducing a group of numbers to a single value which tells him or her about a particular characteristic of the original values. The businessperson will learn more about the monthly sales figures if he or she also calculates a measure of dispersion, which would describe how sales of this item are distributed across the months. Are sales of the item steady, with all monthly sales figures near the "average," or are sales unusually high in certain months, low in others? Particularly in the latter case, the businessperson might also rank the months in order of decreasing sales to obtain each month's position with respect to the others. Finally, he or she could summarize monthly sales of the item by constructing a graph or table.

Each of these techniques facilitates the communication and interpretation of a large mass of data (a *population*), producing from that data information that can be used. This step leads us to the following definition:

DEFINITION

Descriptive statistics is a collection of methodologies used to describe a population's central tendencies, dispersion or variability, distribution, and the relative positions of the data in the population. Included in these methodologies are quantitative, graphical, and tabular techniques.

Our businessperson might also be concerned with more complicated questions, such as the interdepartmental allocation of resources within a large organization. Is it a major objective of the organization to maximize product output for a given level of resource input? Or is some other goal more important to the organization? What do the various resources (personnel, money, raw materials) contribute to the outputs and products of the organization? What are the economic implications of changing the system of resource allocation? Answers to these and similar questions involve huge amounts of data; practical limitations—time, personnel required, cost, government restrictions—will prevent their being answered simply by applying descriptive statistical techniques to entire populations of values.

Instead, the businessperson will examine subsets of the population called *samples* to which descriptive statistics can be reasonably applied. From these samples, he or she can make inferences and generalizations about the population from which they came. Studies of this kind are the second major aspect of statistics (though the first referred to in Mr. Webster's definition), *statistical inference*.

DEFINITION

Statistical inference is a collection of methodologies in which decisions about a population are made by examining a sample of the population.

In statistical inference, a *parameter* represents some numerical property of a population. A *statistic* is a numerical property of a sample, and is generally used to *estimate* the value of the corresponding population parameter. For example, the businessperson might take a sample from the population of all monthly sales figures of the item in which he or she is interested, and use the "average" of this sample to estimate the "average" monthly sales of the item since its introduction.

The Role of Probability in Statistical Inference

When the businessperson of our example is using a sample statistic as an estimate of a population parameter, he or she needs to know how accurate an estimate he or she is obtaining. That is, the businessperson needs a measure of the confidence that he or she may have in the results. *Probability* is the link between the characteristics of the sample and those of the population; it is the key to that measure of confidence.

DEFINITION

Probability is reasoning used to deduce characteristics of a sample from the characteristics of the population from which the sample was taken.

For example, if a researcher is familiar with the properties of several different populations, he or she could take the results of a given sample and determine from which of the populations, if any, the sample was taken. In this book we will detour into the study of probability before pursuing statistical inference.

Note that statistical inference, in which conclusions are drawn about a population based on a sample from the population, is *inductive reasoning*. This is the reverse of probability, in which we make statements about a sample based on the properties of the population from which it came.

TYPES OF DATA

The general area of statistics may also be divided according to the types of data being examined, and data can be classified according to two general schemes. The first scheme classifies data by *measurement scales*, the second by the number of values which the data may have. We usually think of measurement as the assignment of numbers to objects or observations, as when we measure the length of a piece of lumber. Such measurements, however, constitute just one in a range of *levels* or *scales* of measurements.

The lowest of these levels is *nominal measurement*, the classification of observations into categories. These categories must be mutually exclusive and collectively exhaustive; that is, they must not overlap, and must include all possible categories of the characteristic being observed. Examples of nominal variables are sex, type of automobile, and job classification.

An *ordinal scale* is distinguished from a nominal scale by the property of order among the categories, as in the rank of a contestant in a track event. We know that "first" is above "second," but we do not know how far above.

An *interval scale* is distinguished from an ordinal scale by having equal intervals between the units of measure. Scores on an exam are an example of values on an interval scale. However, though a person may score zero on an exam, this does not demonstrate that the person has none of the knowledge or traits that the exam intended to measure. *Ratio scales* have the properties of interval scales, with a true zero. Age, height, and weight are all measured on ratio scales.

This classification of scales of measurement has historically divided statisticians into two camps. The first holds that using the common arithmetic operations on nominal or ordinal data will distort the meaning of the data. Members of this first camp have developed procedures intended to be used with nominal and ordinal data, and these procedures constitute the field of *nonparametric statistics*. (The distinction between nonparametric and parametric statistics presented here is an oversimplification; further discussion occurs in Chapter 16.) The second camp believes that, since statistical analyses are performed on the numbers yielded by the measures rather than on the measures themselves, they may apply the same *parametric* statistical procedures to all the above types of data. They hold that common arithmetic operations performed on values produced by any measurement scale will not affect the validity of the results. Mathematical and empirical studies are available in the literature to support both contentions.

Data may also be classified according to the number of values over which the variable of interest ranges. If the variable can take on only a finite or countable number of values, it is said to be *discrete*. Political affiliation, sex, and number of cars sold in a month are examples of discrete variables. (Note that all nominal variables are discrete.) If a quantitative (numerically valued) variable can take on any value over a range of values, it is called *continuous*. Height, weight, distance, and temperature are thought of as continuous variables. Note, however, that while the underlying concept, the variable itself, is continuous in these cases, measurements of the values of the variable must be discrete, since we are limited by the accuracy of the tools with which we measure. We will often use discrete approximations of continuously valued quantities. Similarly, we will construct continuous approximations of discrete quantities.

The classification schemes discussed so far have dealt with quantitative variables. **Quantitative data** require that we describe the characteristics of the objects being studied numerically. **Qualitative data** are concerned with traits that are not numerically coded. A qualitative variable is called an *attribute*: in a data file of information on licensed drivers, "blue" would be a value of the attribute "eye color." Figure 1.1 shows the data classification schemes discussed above.

How to Perform a Statistical Study

A statistician, like a scientist, deals in probabilities. It is a common misconception that science deals in certainties. In the language of statistics, the best a scientist can hope to do is:

a. to specify the levels of the independent quantities in the research study,

b. to control the effects of extraneous quantities, and

c. to determine the probable effects to be observed on the dependent quantity being examined.

Once the scientist has formulated (c) from (a) and (b), he or she states the hypotheses to be tested, then states a decision rule to which the sample data, the results of the study, can be compared. Based on this comparison, the scientist decides whether to accept the hypotheses, or to reject them and begin the process again. Those accepted hypotheses become theories — not proven beyond all doubt, but accepted as indicating and describing the probable behavior of the system under study.

Throughout this process of theory development, scientists strive for theories which:

a. are consistent with known facts,
b. do not contradict one another,
c. can be tested experimentally,
d. generate explanations for a wider variety of phenomena, and
e. generate useful predictions.

FIGURE 1.1 Data classification schemes.

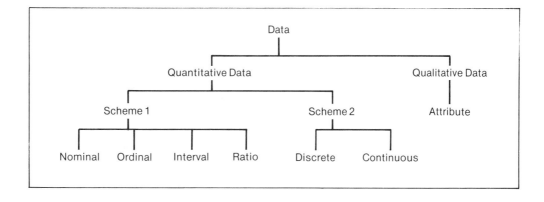

The same guidelines determine statistical methods for decision making under conditions of uncertainty. The decision maker must choose among alternative actions. The decision maker is uncertain about the possible results of these actions, since they also depend on conditions beyond the control of a person. Usually, he or she knows only the probabilities with which these other conditions will occur. Applying statistics to sample values of the independent and control variables, the decision maker calculates values of the dependent variables. These values are used in a previously formulated decision rule to decide which of the alternative actions to take. This approach to problem solving is called **decision analysis**. Most statistics texts devote a chapter to this topic, apart from the rest of statistics. The topic is called **Bayesian statistics**, as opposed to **classical statistics**. Here, since decision analysis is a philosophical approach to the question of decision making under uncertainty, it will be included throughout the book.

EXERCISES

1. Classify the following measures as nominal, ordinal, interval, and/or ratio:

    a. Plant maintenance expenditure

    b. Racial classification

    c. Level of educational attainment

    d. Percentage of workers with M.A. degrees

    e. Building age

    f. Rating the flavor of a soft drink on a scale of 1 to 9

    g. Religious affiliation

2. Repeat Exercise 1 with respect to whether the variables are continuous or discrete.

3. Discuss which of the fundamental arithmetic operations ($+$, $*$, $/$, $-$, exponentation) might invalidate the results of a study in which they are included.

4. Distinguish between the terms:

    a. Parametric and nonparametric statistics.

    b. Descriptive and inferential statistics.

    c. Quantitative and qualitative data.

5. Indicate which of the following terms are associated with a sample, S, or the population, P.

    a. Parameter

    b. Statistic

    c. Inductive reasoning is applied to the _____ to draw inferences about the _____.

6. State (in English) a question of interest related to a given file of data.

    a. Formalize a set of hypotheses which could be researched to resolve the answer.

    b. State the dependent variables of interest.

    c. State the independent variables of interest.

    d. State the control variables of interest.

    e. Discuss how you might draw the sample and describe a decision rule you might use.

2

Computing and Statistics

INTRODUCTION

The goal of statistics is to convert raw data into useful information. To do this, the statistician must determine the nature of the system that produced the data. Understanding this system is achieved by constructing mathematical models of all or part of it that mimic important aspects of its behavior, which can be used to relate the inputs and outputs of the system, and to predict its future behavior within some margin of error.

Consider this simple example. You have two lengths of rope which you plan to tie together, and you want to predict the length of the resulting rope. You measure the two lengths, add them together, subtract the length you think will be required to make the knot, and use the result as an estimate of the length of the entire (new) rope. The calculations you perform are a mathematical model of the system, the process, of tying the two original ropes together. This model allows you to relate the input of the system (the lengths of the original ropes) to the output (the resulting total length) and thereby predict the total length.

In situations more complicated than this, the search for such models is usually an iterative process: If one model fails to describe the system being studied in a useful way, another is tried, until satisfactory results are achieved. Often the researcher is forced to exhaust many possibilities before finding a model that "works." It is here that computing and statistics become linked. Computers and/or calculators can perform complex calculations rapidly and display the results in a variety of ways, including tables and graphs, allowing the researcher to consider many possible models in a reasonable length of time.

But what exactly is a computer system, and how can one be used as a statistical tool? In this chapter, we will answer this question, describe a set of data to be used throughout the book in discussions, examples, and exercises, and give an example of an applications package —programs already written and commercially available — which performs statistical analyses.

The Benefits of Using a Computer

Many tools are available today to make the calculations involved in statistical analysis faster and less tedious. One major step in this direction is the desk calculator, an electromechanical (or simply mechanical) device which performs the usual arithmetic operations ($+$, $-$, $\times$, $\div$, usually), and often more complex functions as well. These devices range from the most primitive four-function calculators (from \$5.00) to sophisticated programmable calculators, which will save a user-composed set of instructions (a program) for repeated use (from \$40 to \$400, depending on capabilities).

Calculators are inefficient for large-scale statistical analyses that involve many variables and many cases. Such analyses, unfeasible before the development of digital computers, are now common practice and a primary use of these powerful and versatile machines. Unlike nonprogrammable calculators, the digital computer and programmable calculator allow:

a. mass storage of and immediate access to the data which will be analyzed.

b. the set of instructions to perform the analytical procedure (the program) to be stored in the machine,

c. the performance of more complex computations,

d. the faster performance of computations,

e. the calculations to be performed without constant human intervention,

f. management (create, store, modify, rearrange, destroy) of data and sets of instructions, and

g. the production of more than one result.

By using a computer, the researcher can manipulate the data in a variety of ways, perform complex computations involving all or part of the data quickly and easily, and thereby arrive at a model which satisfactorily describes the important aspects of the system from which the data were collected.

The Components of a Computer

To understand how to use a computer as a statistical tool, we must first investigate the components of a computer system. Just as the structure of an automobile dictates how it will be used, so the anatomy of a computer directs our use of it. The schematic diagram in Figure 2.1 shows the fundamental parts of a computer system.

Data are entered into the system through an input device, then placed by the central processing unit (CPU) in a mass storage device where the data are stored as a pattern of electric charges or magnetic states. The program of instructions is also data, and this, too, is placed in memory. Then the CPU carries out (executes) the instructions in the program. These instructions may cause calculations to be performed, data to be generated or manipulated,

FIGURE 2.1 The fundamental parts of a computer system.

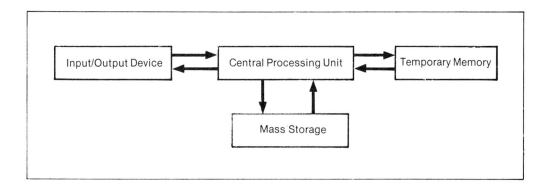

decisions made, and information read in or out. During execution, the program itself and the data on which it is operating are kept in temporary memory, from which data can be retrieved more rapidly than from mass storage.

The physical objects which make up the computer system are called *hardware*, while data and programs are called *software*. Both are necessary for the system to function.

Data Preparation and Storage

To use a computer for statistical analysis, data should be prepared in a manner compatible with the *input medium*, the mechanism by which information is transmitted from the outside world into the computer. Generally, when data are first placed in the memory of the computer, they are read in from cards on which each character is coded as a pattern of punches. The standard IBM (International Business Machines) 80-column card is shown in Figure 2.2.

Before the data can be punched onto cards, you must specify the format of the cards, the pattern which the data will follow, on a *coding sheet*. You must specify the assignment of the 80 columns of the card to the variables in the study in such a way that:

a. each individual or experimental unit (each case) has a unique identifying number.

b. each observation can be coded to the accuracy with which it was made.

c. all observations can be included.

d. a missing value (an observation not obtained, or lost) can be coded with a special value.

e. all variables, including those with nominal or ordinal values, are assigned numerical values.

FIGURE 2.2 Standard IBM 80-column card.

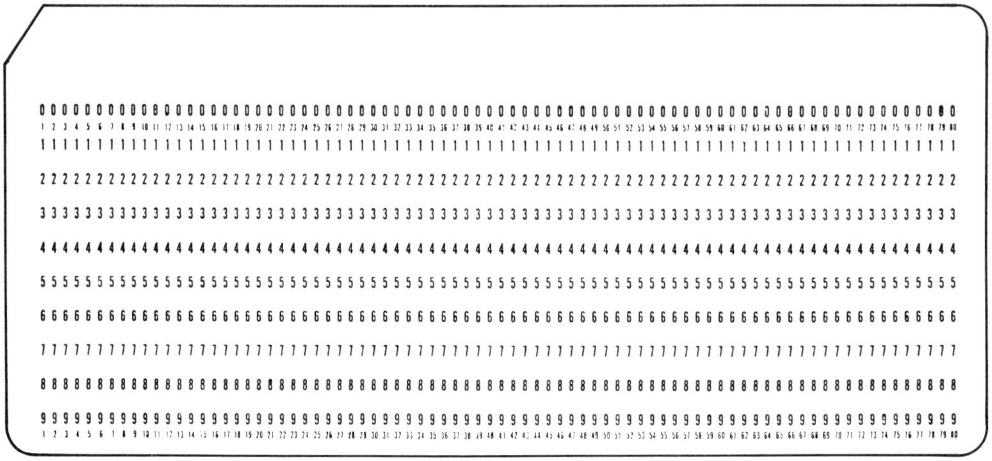

f. related variables are grouped together.

g. the number of variables is kept to a minimum.

If necessary, it is usually acceptable to use more than one card per case.

When the coding scheme has been fully described, punch the data onto cards using a key punch machine, then enter the data file into the computer system using the system commands appropriate at your installation. (These commands vary from one computer system to another, so check with your instructor or your computer center staff.) Your data file is now stored on a mass storage medium, generally magnetic tape or disk, and can be accessed in order to perform statistical analyses.

A magnetic tape drive may be used for both reading data from a tape and writing data on a tape, at a rate of about 150 inches of tape per second. The tape itself, similar to that used for audio recording, may contain as many as 15 million characters.

The magnetic disk, which physically resembles a stereo record, also stores data magnetically, operates at much faster speeds, and can store more data than tape, though it is more expensive.

DATA BASE CONCEPTS

When performing a statistical study, the researcher must have words with which to talk about all or part of the data being analyzed. Thus we have the following definitions.

DEFINITION

Data base: The set of all data files used in the study.

File: A subset of the data base containing related records, usually all with the same format.

Case or record: As an element of a file, a set of observations or measurements taken on one individual or experimental unit.

Field: That part of a record which contains the value of one variable.

The values that we collect from an experimental subject go into the fields of the record associated with that individual or set of measurements. All related records make up a file, and all the files are our data base.

Also, in most statistical studies, we will use some or all of these *data management techniques*, methods of manipulating files and the records within them:

DEFINITION

Data Base Creation: Building and storing the data base on the computer system.

Updating: Adding, deleting, or modifying records.

Searching: Finding and retrieving particular records.

Sorting: Arranging records in a prescribed order.

Subfile creation: Building subfiles of the original files by selecting particular records and excluding others.

Copying: Reproducing all or part of the data base.

Merging: Combining two or more files or subfiles.

All of these techniques allow us to arrange and modify the data base.

Processing Modes

Finally, there are two ways in which the researcher can process the data base. These are the **batch**- and the **interactive**-processing modes.

Batch processing occurs when several prepared programs, most often on cards, are run in a group at one time. Running statistical programs in the batch mode is depicted in Figure 2.3.

In the interactive mode, operations and processes are initiated throughout the analysis. Rather than submitting a program and waiting for the resulting printout, the researcher interacts with the processing, as it occurs. Most statistical analysis is done interactively, but the researcher should have the option of batch processing, which is less expensive, for very complex or long tasks. Interactive processing is more expensive because many users can perform interactive operations at essentially the same time, placing greater demand on the computer's

FIGURE 2.3 Batch processing.

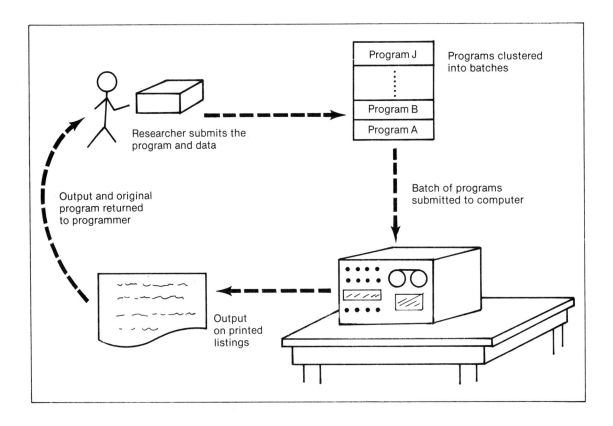

facilities. Each user communicates with the computer through a terminal, most commonly a typewriter-like keyboard and a screen, and interacts with the computer as processing occurs. Figure 2.4 depicts operations performed in the interactive mode.

The Case Study To Be Used Here

Since the computer frees us from the drudgery of computations, we can start "doing statistics" in a meaningful way almost immediately. We will pose questions, and you will be asked to make decisions and answer those questions. As various statistical procedures are introduced in the chapters that follow, you will use them to make those decisions, and use the computer as a tool to perform the procedures.

To illustrate the questions and processes that arise in a statistical study, we will perform such a study in this text. Most of our discussions and examples will be based on the following data base in Figure 2.5, which consists of economic and business data about 74 American

FIGURE 2.4 Interactive processing.

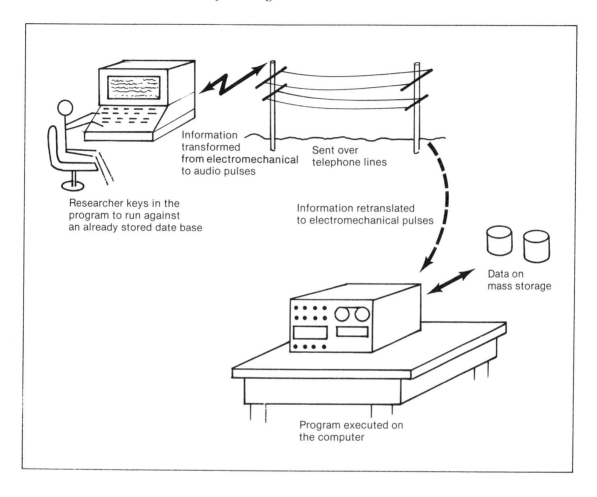

Information transformed from electromechanical to audio pulses

Sent over telephone lines

Researcher keys in the program to run against an already stored date base

Information retranslated to electromechanical pulses

Data on mass storage

Program executed on the computer

cities in 1980. Note that the file as given is simply a large collection of numbers: raw data. We will use descriptive statistics to summarize these data, thereby making them more understandable, and with statistical inference we will draw interesting and useful conclusions about the economic climate which produced the data.

Statistical Software

Statistical software is a package of programs to perform statistical analyses and an accompanying manual available to users at their computer center. The programs which make up the

FIGURE 2.5 Data base of 74 American cities in 1980. From *U.S. News & World Report*, April 28, 1980, pp. 77, 79. Basic data from U.S. Departments of Labor and Commerce.

A Look at Business Across the Country

	DEPARTMENT-STORE SALES COMPARED WITH A YEAR AGO ‡	EXTENT OF UNEMPLOYMENT	NONFARM EMPLOYMENT COMPARED WITH A YEAR AGO	INCOME OF AVERAGE FACTORY WORKER*	FACTORY WORKERS' INCOME COMPARED WITH A YEAR AGO	CONSTRUCTION ACTIVITY COMPARED WITH A YEAR AGO †
IN EASTERN CITIES						
Atlanta	Up 10.7%	4.7%	Up 3.2%	$13,432	Up 12.4%	Up 0.5%
Baltimore	Up 9.8%	Not available	Up 1.1%	$16,178	Up 9.0%	Up 5.2%
Boston	Up 10.9%	6.2%	Up 4.1%	$14,132	Up 5.7%	Up 21.5%
Bridgeport	Up 12.7%	5.5%	Up 2.2%	$15,784	Up 9.5%	Up 13.6%
Buffalo	Up 7.1%	9.7%	Up 1.1%	$18,362	Up 2.5%	Up 21.4%
Burlington, Vt.	Not available	6.4%**	Up 4.9%	$13,979	Up 11.6%	Up 27.4%**
Charleston, S.C.	Up 3.3%	6.1%	Up 4.0%	$13,002	Up 8.7%	Up 11.8%
Charleston, W.Va.	Up 11.9%	6.3%	Up 0.2%	$18,740	Up 13.8%	Up 4.7%
Charlotte, N.C.	Up 13.3%	3.9%	Up 1.5%	$10,905	Up 7.6%	Up 4.2%
Greensboro	Up 12.8%	4.4%	Up 1.8%	$11,299	Up 10.6%	Down 6.1%
Hartford	Up 13.1%	4.8%	Up 10.3%	$16,547	Up 7.9%	Up 20.2%
Jacksonville	Down 7.3%	5.1%	Up 1.5%	$13,790	Up 7.7%	Up 3.3%
Manchester, N.H.	Not available	3.5%‡	Up 1.9%	$10,337	Up 8.9%	Up 3.4%
Miami	Up 11.0%	4.8%	Up 5.1%	$10,718	Up 9.2%	Up 22.5%
Newark	Up 11.1%	5.8%	Up 1.6%	$15,456	Up 7.6%	Up 7.6%
New York	Up 7.9%	8.8%	Up 1.2%	$12,285	Up 8.7%	Up 4.9%
Norfolk	Down 1.2%	5.3%	Up 1.0%	$13,658	Up 3.1%	Down 5.1%
Philadelphia	Up 10.7%	6.6%	Up 3.6%	$15,073	Up 6.9%	Up 17.5%
Pittsburgh	Up 12.0%	7.1%	Up 0.4%	$18,483	Up 6.1%	Down 1.5%
Portland, Me.	Up 10.1%	5.4%	Up 5.1%	$11,619	Up 9.5%	Up 3.0%
Providence	Up 4.5%	8.1%	Down 0.4%	$11,012	Up 7.6%	Up 2.8%
Richmond	Down 3.9%	3.1%	Up 1.8%	$15,848	Up 19.9%	Down 2.9%
Rochester	Up 5.5%	6.5%	Up 1.6%	$17,890	Up 5.0%	Up 1.1%
Savannah	Up 8.2%	5.5%	Down 1.3%	$16,024	Up 13.5%	Down 10.0%
Tampa–St. Petersburg	Up 7.5%	4.9%	Up 3.4%	$12,333	Up 5.6%	Up 2.2%
Washington, D.C.	Up 7.3%	Not available	Up 2.7%	$17,139	Up 13.4%	Up 6.4%
Wilmington	Up 19.0%	7.4%	Up 0.2%	$17,606	Up 11.4%	Down 4.3%
IN CENTRAL CITIES						
Birmingham	Up 2.4%	6.8%	Up 2.4%	$15,681	Up 8.7%	Up 3.0%
Chicago	Up 19.8%	Not available	Up 0.4%	$15,226	Up 10.6%	Down 1.2%
Cincinnati	Up 9.8%	6.0%‡	Up 1.5%	$16,622	Up 6.7%	Up 9.4%
Cleveland	Down 1.7%	5.6%	Up 0.6%	$17,782	Up 0.3%	Up 5.3%
Columbus	Up 6.3%	4.6%	Up 2.1%	$15,261	Up 6.5%	Up 8.6%
Dallas–Ft. Worth	Up 23.1%	4.2%	Up 5.6%	$13,590	Up 10.9%	Up 13.4%
Des Moines	Up 22.4%	5.5%	Up 1.1%	$17,819	Up 13.4%	Down 2.9%
Detroit	Up 3.3%	12.0%	Down 4.4%	$20,378	Down 3.1%	Down 1.3%
Fargo	Up 13.4%	5.0%	Up 3.6%	$13,412	Up 11.4%	Down 2.9%
Grand Rapids	Up 14.3%	6.7%	Down 0.1%	$15,917	Up 3.2%	Up 4.4%

City-by-city figures are based on official reports for February, latest available, except department-store sales. Most statistics are for metropolitan areas. *Factory workers' weekly earnings at annual rates. †Based on number of workers in construction. **Rate for the state—roughly the same as for major cities in that state. ‡January data.

U.S.NEWS & WORLD REPORT, April 28, 1980

	DEPARTMENT-STORE SALES COMPARED WITH A YEAR AGO†	EXTENT OF UNEMPLOYMENT	NONFARM EMPLOYMENT COMPARED WITH A YEAR AGO	INCOME OF AVERAGE FACTORY WORKER*	FACTORY WORKERS' INCOME COMPARED WITH A YEAR AGO	CONSTRUCTION ACTIVITY COMPARED WITH A YEAR AGO‡
Houston	Up 16.2%	3.9%	Up 4.2%	$18,274	Up 7.9%	Up 1.4%
Indianapolis	Up 9.5%	6.9%	Up 1.9%	$17,122†	Up 5.0%†	Up 11.5%
Jackson, Miss.	Up 5.3%	4.2%	Up 2.5%	$12,387	Up 9.2%	No change
Kansas City, Mo.-Kans.	Up 13.5%	5.4%	No change	$16,577	Up 9.6%	Down 6.0%
Knoxville	Up 8.4%	5.5%	Down 0.7%	$14,298	Up 9.6%	Down 3.6%
Little Rock	Up 2.5%	4.3%	Up 1.9%	$12,846	Up 8.8%	Down 2.4%
Louisville	Up 13.1%	7.4%	Down 2.1%	$16,224	Up 4.1%	Down 1.9%
Memphis	Up 14.2%	5.9%	Up 0.5%	$13,351	Up 7.2%	Up 8.5%
Milwaukee	Up 18.8%	4.3%	Up 2.9%	$18,062	Up 7.3%	Up 5.9%
Minneapolis–St. Paul	Up 7.7%	4.2%	Up 3.7%	$16,298	Up 8.6%	Up 13.1%
Mobile	Up 24.3%	7.2%	Up 2.7%	$17,171†	Up 13.3%†	Up 16.3%
New Orleans	Up 11.3%	6.6%	Up 1.9%	$16,337	Up 12.9%	Up 2.4%
Oklahoma City	Up 20.4%	3.0%	Up 8.7%	$15,918	Up 28.3%	Up 10.3%
Omaha	Up 14.1%	5.2%	Up 1.8%	$15,053	Up 6.0%	Down 2.1%
San Antonio	Up 4.3%	6.4%	Up 2.8%	$10,525	Up 12.8%	Up 12.0%
St. Louis	Up 19.1%	Not available	Down 1.5%	$16,349	Up 5.1%	Up 7.2%
Sioux Falls, S.D.	Up 20.0%	3.8%	Down 0.2%	$17,372	Up 19.2%	Down 8.3%
Topeka	Up 6.9%	5.3%	Up 1.8%	$14,702	Down 5.7%	Up 22.2%
Tulsa	Up 21.0%	3.3%	Up 3.8%	$15,238	Up 7.3%	Up 5.6%
Wichita	Up 16.9%	3.3%	Up 5.3%	$15,918	Up 7.6%	Up 12.5%
IN WESTERN CITIES						
Albuquerque	Up 6.6%	7.1%	Up 5.0%	$12,871	Up 19.1%	Up 4.1%
Boise	Up 3.1%	5.7%	Down 0.9%	$11,275	Up 0.5%	Down 3.8%
Casper	Not available	4.1%**	Up 13.9%	$14,768	Down 10.9%	Up 40.6%
Denver–Boulder	Up 4.9%	4.0%	Up 5.1%	$14,818	Up 11.3%	Up 9.5%
Fresno	Up 10.8%	10.2%	Up 5.4%	$14,410	Up 15.2%	Up 14.5%
Great Falls	Up 1.4%	7.5%	Down 0.3%	$19,307**	Up 9.7%**	Down 15.4%
Honolulu	Up 4.3%	4.4%	Up 3.9%	$14,524	Up 19.7%	Down 1.1%
Las Vegas	Up 22.1%	5.9%	Up 7.0%	$17,985†	Up 4.5%†	Up 4.1%
Los Angeles–Long Beach	Up 10.9%	5.6%	Up 3.6%	$14,291	Up 7.7%	Up 1.3%
Phoenix	Up 6.8%	4.9%	Up 4.8%	$14,083	Up 7.6%	Down 4.7%
Portland, Oreg.	Not available	5.3%	Up 4.2%	$14,142†	Up 6.9%†	Up 13.9%
Sacramento	Up 12.8%	7.7%	Up 5.4%	$15,953	Up 10.0%	Up 10.8%
Salt Lake City–Ogden	Down 3.0%	5.3%	Up 5.8%	$12,708	Up 10.0%	Up 5.1%
San Diego	Up 13.3%	6.4%	Up 4.5%	$14,228	Up 8.2%	Down 1.8%
San Francisco–Oakland	Up 11.1%	5.4%	Up 2.0%	$17,639	Up 7.4%	Up 3.3%
Seattle	Up 8.1%	5.8%	Up 6.1%	Not available	Not available	Up 9.9%
Tucson	Up 10.8%	4.7%	Up 5.7%	$13,459	Up 10.1%	Up 2.8%
U.S. AVERAGE	**Up 11.6%**	**6.0%**	**Up 2.3%**	**$14,410**	**Up 5.7%**	**Up 8.3%**

package are usually written in a high-level, problem-oriented computer language such as FORTRAN, PL/1, COBOL, or BASIC. (More is said on this topic in Chapter 17.)

Most packaged programs are stored on magnetic tapes or magnetic disks. When a user accesses one of these programs by entering the appropriate control statements, the program is called up and loaded into the computer's main memory under the supervision of the central processing unit. The control statements that cause this to happen can be entered on cards in the batch-processing mode, or through a terminal keyboard in the interactive mode. The manual accompanying the program package explains the syntax of the control statements (what they should look like) and the results to expect when you are using them.

Several of the popular statistical program packages require the user to prepare an input deck of cards consisting of:

a. System control cards, which inform the computer of the requirements of the task: what program is being used, how data are to be input/output.

b. Program control cards, which give the details necessary to execute the desired program: how many variables, records, and subfiles, and their names, and which statistical tasks to perform.

c. Special cards, which give instructions for labeling variables and values, modifying data, marking the end of the run, etc.

d. Data cards, though the data may also be input from tape, disk, or some other medium.

Examples of packages organized in this way are:

Package	Developer
SPSS—Statistical Package for the Social Sciences	Norman H. Nie National Opinion Research Center University of Chicago Chicago, IL 60637
BMD—Biomedical Computer Programs	W.J. Dixon Health Sciences Computing Facility University of California Los Angeles, CA 90000
SAS—Statistical Analysis System	SAS Institute, Inc. Institute of Statistics North Carolina State University at Raleigh Raleigh, NC 27607

Some other popular packages are sets of subprograms. The subprograms are called by a main program in a problem-oriented language (like FORTRAN) that defines all variables and input/output procedures, and which the user must write. The manuals accompanying these packages explain the subprograms and describe writing the main program which calls them. An example of such a package is:

Package	Developer
SSP—Scientific Subroutine Package	IBM 112 112 East Post Road White Plains, NY 10601

In the third approach, the package is itself a specialized programming language, which is generally used interactively. An example is:

Package	Developer
FOSOL	Anton Florian
	Sangamon State University
	Springfield, IL

These packages are mentioned here because of their popularity or historical significance, and one or more of them is probably available to you. SAS and SPSS have been highly rated in *Datamation* magazine's survey of systems and applications software packages for several years. BMD and SSP were among the first packages to be widely used in education, government, and industry.

For uniformity and clarity, SPSS will be used throughout this text for illustrations and examples. The authors feel that SPSS offers great flexibility in data formatting, labeling of variables and values, generation of new variables (called *process variables*) during processing, analysis of missing data, and selection of statistical procedures. Both batch and interactive modes are available, and versions of SPSS can be implemented on IBM, CDC (Control Data Corporation), Honeywell, Univac, HP (Hewlett-Packard), and Digital Equipment Corporation computers.

While we will be using SPSS to construct examples, the statistical analyses we will examine can be performed using any statistical applications package. In particular, you can use this text, and perform the examples and exercises given here, with whatever statistical package is used at your computer center.

EXERCISES

1. Describe the construction of a data base to evaluate the following research question: "Are male and female teachers treated equally at your school?"

   a. Identify the files.

   b. Identify the cases (records).

   c. Identify the fields of interest in each case.

2. Describe the construction of a data base to aid the concession stand at the football stadium to maintain a high degree of cost-effectiveness:

   a. Identify the files.

   b. Identify the cases (records).

   c. Identify the fields of interest per case.

3. For Exercises 1 and 2, record some data observations on coding sheets and, under the direction of your computer center staff or instructor, store the data base on the computer.

4. Store the data bases in Appendix B on your computer system.

5. Refer to your computer center staff or instructor to

    a. determine which statistical program package is to be used in your course.

    b. investigate the manual.

    c. try running some of the simpler programs in the package on a data base.

6. Ask your computer center if there exist any computer-assisted instruction (CAI) packages illustrating how to use your local facilities. Look for a package which

    a. defines basic system commands to create, access, save, and kill files, and

    b. creates and manipulates data files.

    If such a package exists at your school, sit down and use it!

7. Repeat Exercise 6 with a CAI session on the statistical package to be used in this course at your computer site.

8. Consult your instructor or other advisor about journals in your field of interest which apply statistical analyses to data bases to extract informational content. Begin reading articles of interest from these sources.

READINGS

1. "User Ratings of Software Packages," *Datamation* (each December issue), John L. Kirkley, ed.

2. "Computers as an Instrument for Data Analysis," M.E. Muller, *Technometrics 12* (May 1970): 259.

3. "A Survey of Statistical Packages," W.R. Schucancy, Paul D. Minton, and B. Stanley Shannon, Jr., *ACM Computing Surveys*, Vol. 4, no. 2 (June 1972): 65–81.

Any good introductory text in computer science; ask your instructor for suggestions. The text should define and discuss the following terms: *time-sharing, batch processing, computer hardware components* (*I/O* devices, central processing unit, memory, mass storage), *types of software*, and *survey I/O devices.*

Creating a Data File and Using a Program Package

INTRODUCTION

We now begin to describe and analyze the data contained in the example file which we will call USNEWS. Before asking specific questions about the data in the USNEWS file, we should first become familiar with some fundamental characteristics of the file. In particular, we would like to describe the central tendency *of each of the variables in the study (e.g., what seems to be a middle or central value for each variable), and the* variation *in the values of the variables around their respective centers. Also, we might investigate the* shape *of the distribution of values for each variable, and the* location *of particular values with respect to that distribution. These descriptions of the variables in the data file will often suggest lines of inquiry into the meaning of the data. We must also know exactly what the values in the data file represent.*

Operational Definitions

The **operational definition** of a variable is a precise description of the origin of the values recorded for that variable. For example, there are many definitions of the word *intelligence*; a researcher wishing to include a measurement of this elusive quantity in a study might operationally define "intelligence" as an individual's score on the Stanford-Binet test. While the concept of "intelligence" might be vague, this operational definition is precise, and tells the source of the values.

The variables in the file USNEWS have the following operational definitions:

NUM   Case number   Number of the city, in the order of the file as given in Chapter 2.

X1   City region   EASTERN = eastern seaboard
                                 CENTRAL = from Ohio, Kentucky, Tennessee, Alabama west to the Dakotas, Kansas, Nebraska, Oklahoma, Texas
                                 WESTERN = west of CENTRAL

X2   Change in depart-   Percent change in department store sales from 1979
        ment store sales   to 1980.*

X3   Unemployment   Percent of work force currently unemployed and seeking work.*

X4   Change in nonfarm   Percent change in nonfarm employment from 1979
        employment   to 1980.*

X5   Average income of   Average factory income; weekly earnings at annual
        factory workers   rate.*

X6   Change in factory   Percent change in factory workers' income from
        workers' income   1979 to 1980.*

X7   Change in con-   Percent change in construction activity from 1979
        struction activity   to 1980, based on the number of workers employed.*

Data Preparation and Entry

Having formulated operational definitions of the variables in the data file, the researcher faces the task of preparing the data for machine processing. To avoid problems during analysis, it is wise to be aware of data-processing considerations during the planning of the

---

*As compiled by the U.S. Departments of Labor and Commerce.

study and the collection of data. The forms used for data collection should be checked for accuracy and completeness. Also, it is better to gather more data than are required to answer the research questions than to fail to collect data which may be needed later in the study.

Likewise, be familiar with the processing capabilities of your system; it is frustrating to discover that your experiment requires a procedure which, for hardware or software reasons, your system cannot perform. In particular, check limitations on the processes available, the number of variables, the number of cases, and possible input and output formats.

Unless the data file is very small, or the system on which the analysis is to be performed is a mini- or microcomputer which does not read cards, the data will be entered into the computer from punched cards in the batch mode. The data are punched onto cards using a *card punch machine*, a device superficially resembling an electric typewriter. Cards are loaded into the hopper on the top right of the machine. They travel through the punching station directly above the keyboard, across the machine through the read station to the left where they are held in order in another hopper. Letters are arranged on the keyboard in the same pattern as on a typewriter, but numbers are "above" certain letters on the right side of the keyboard, and are reached by using the *shift* key. Other keys whose functions you should notice are: DUP, which causes the card in the read station to be *duplicated*; REG, which *registers* another card in the punching station, FEED, which moves a card from the input hopper to the "on-deck" position; and the CLEAR switch, which clears all cards from both stations into the output hopper.

The punched cards should be checked for accuracy. This can be done by hand, or with a device called a *verifier*, into which a copy of the data file is keyed. The verifier reads the card deck, and notches the offending column of any card on which a discrepancy is found. These cards can then be repunched.

Backup copies of card decks should always be kept, in case cards in the original deck get folded, spindled, mutilated, or simply lost. These can be created using a *duplicating punch*, which reads one deck and punches another just like it. Alternately, if your computer center has the capability, you can write a simple program to duplicate the deck.

Also available for manipulating cards are *card sorters and collaters*. A card sorter arranges a deck of cards into 12 stacks, corresponding to the 12 punch positions in a selected column, while a collater will merge two decks in such a way as to pair cards with matching columns or fields of columns.

An alternative to punched cards is *mark-sense* cards or sheets. On these, data are recorded by blackening small spaces with an electrographic or no. 2 lead pencil. These *marks* can be *sensed* by a reader which causes a corresponding deck of cards to be punched. The mark-sense process has disadvantages; the medium is delicate, and it may be inefficient to transcribe data from their source to mark-sense cards or sheets. Also, the equipment is more expensive and less reliable than card-manipulation devices.

When the data have been recorded on cards, they can be transferred by the computer to magnetic tape or disk for processing and storage, or the cards themselves can be used directly. Keyboard-to-tape and keyboard-to-disk direct devices are available, but they are very expensive and limited in their use. On a time sharing system data could also be entered directly onto disk from a terminal.

When the researcher is using a mini- or microcomputer system, he or she will probably not use cards. The researcher will enter data through the keyboard into memory, then load the data file from memory onto a cassette tape or diskette. Diskettes are much faster and more convenient than tapes, but, again, are more expensive.

Obtaining Descriptive Statistics—
An Introduction to SPSS

We will now write a program using the statistical software package SPSS to find descriptive statistical characteristics of the variables in the data file. We will find measures of various characteristics of the variables in the file USNEWS. You should perform these analyses yourself using the statistical package implemented on your computer system (the manual that accompanies the system will show example sessions and programs); we will show here how to use SPSS to find these descriptive statistics.

An SPSS program is a sequence of statements or *commands*, each of which causes the package to perform one of these operations:

1. *File definition*—specifying the variables in a file, input medium and format, the name of the file.

2. *Data transformation* — calculating new values from the data, making decisions based on data values.

3. *Task definition* — specifying the statistical processes to be performed on the data.

4. *Execution control* — determining the repetition of processes, stopping execution.

5. *File management* — subfile creation and deletion, sorting, selecting cases, adding and deleting cases.

Each command is composed of two *fields*. The *control field*, columns 1 through 15, always contains a *control word*, beginning in column 1, which gives the command that is to be performed. The *specification field*, columns 16 through 80, contains the details — specifications — of the operation named in the control field.

We construct an SPSS program which reads in the data from the USNEWS study, names and labels the variables, calculates a number of descriptive statistics, and saves the data file on a mass storage device (generally a disk) under the name USNEWS. (See page 30 for the entire program.)

The data deck follows a *read input data* card (Figure 3.1).

Note that the data were coded and arranged on the cards according to these conventions:

Variable	Columns	Coding
NUM, CASE NUMBER	1, 2	One- or two-digit integer
X1, CITY REGION	3	1 = EASTERN 2 = CENTRAL 3 = WESTERN
X2, CHANGE IN DEPART- MENT STORE SALES	4–7	Column 4 for + or −; 5–7 for a three-decimal number with one position to the right of the decimal point.
X3, UNEMPLOYMENT	8–11	Same as X2, beginning in column 8.

FIGURE 3.1 A read input data card.

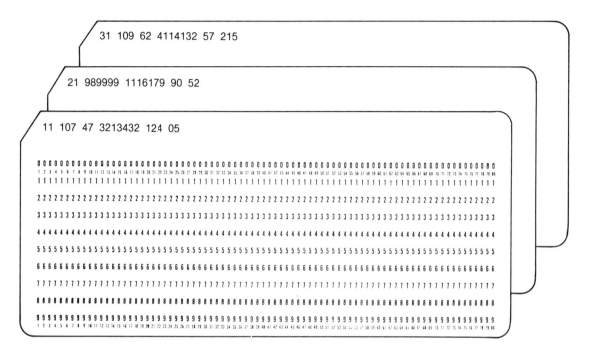

X4, CHANGE IN NONFARM EMPLOYMENT	12–15	Same as X2, beginning in column 12
X5, AVERAGE INCOME OF FACTORY WORKERS	16–20	Right-justified five-digit number; no decimal place.
X6, CHANGE IN FACTORY WORKERS' INCOME	21–24	Same as X2, beginning in column 21
X7, CHANGE IN CON-STRUCTION ACTIVITY	25–28	Same as X2, beginning in column 25.

We will now examine the statements in the SPSS program.

The program itself must have a name, and this is supplied by the RUN NAME command. The control field contains the control word RUN NAME (though RUN NAME is two English words, it is one SPSS control word) and the specification field contains the name itself, which may be any collection of characters up to 64 characters long. This name will be printed at the top of each page of output produced by the program, and is a convenient way to include in the printout a general description of what the program does.

Similarly, the FILE NAME command attaches a name to the data file. This name may contain up to eight alphanumeric characters, must begin with a letter, and may not include embedded blanks. Thus, XFILE1 and MYDATA are acceptable file names, while 2NDFILE and SURVEY DATA are not. After the file name, and separated from it by a comma or blank, is a *label* for the file. (To SPSS, commas and blanks are identical, and wherever one of these *common delimiters* may be used, any number may be included.) This label may be up to 64 characters long, and is documentation which will be printed on any output generated from the file.

The variables are given names in the VARIABLE LIST command; the rules for variable names are the same as those for file names. The INPUT MEDIUM command specifies the source of the data (in our case, CARD), and the N OF CASES statement specifies the number of cases (records) in the data file. In the file USNEWS, each city is a case, so there are 74 cases.

There are several ways to specify the *input format*, or arrangement of data on the data deck. *Fixed* specifies that all data are real numbers in decimal form, and the contents of the parentheses indicate just how the data are punched on the data deck. *Fm.n* specifies a field containing real numbers *m* columns wide, with *n* places to the right of the decimal. The decimal point itself is implied, not physically punched. *kFm.n* means that the forma. specification *Fm.n* is repeated *k* times. Note how this format specification corresponds to the arrangement of the data deck, represented by its first card (Figure 3.2).

These statements specify all the characteristics of the data in the file. To perform statistical analyses on these data, we specify a *task*. Each set of task cards contains a task command stating which subprogram in SPSS is to be run on the data, and may also include a *statistics* card, indicating which statistics are to be printed in the output, and an *options* card, which causes certain special procedures to be used. Most SPSS subprograms have sets of statistics and options from which we may choose.

The SPSS subprogram CONDESCRIPTIVE calculates a variety of common descriptive statistics—measures of central tendency, variation, and distributional shape. Using the word ALL in the specification field of the CONDESCRIPTIVE command causes these calculations to be performed for *all* the variables named in the VARIABLE LIST statement. Had we wanted these calculations only for X4 and X7, the appropriate statement would be

FIGURE 3.2 First card of the data file USNEWS.

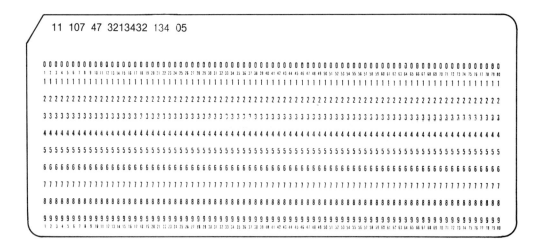

CONDESCRIPTIVE X4, X7

Similarly, we request that all statistics be printed by using the statement STATISTICS ALL. The statements

CONDESCRIPTIVE   ALL
STATISTICS   ALL

are the task statements for the first task.

FREQUENCIES is a subprogram which not only finds the descriptive statistics produced by CONDESCRIPTIVE, but calculates several others, and creates frequency and cumulative distributions, and graphs of frequency distributions (called *histograms*). The FREQUENCIES subprogram may be used in either of two modes. In the *general mode*, the variables listed may take on any decimal values, and this mode is used in the example program. In the *integer mode*, the variables must take on integer values, and the smallest and largest values must be specified following the variable name. X1 takes on only integer values, so it would be acceptable to use this command:

FREQUENCIES   INTEGER = X1(1,3)

Statistics of X1, city region, would be meaningless, so we omit the STATISTICS command, and the card OPTIONS 8 causes the printing of the histogram. Thus the task cards for the second task are

FREQUENCIES   GENERAL = X1
OPTIONS   8

Between the two tasks in the program is a card containing the control word READ INPUT DATA. This card and the data deck must always follow the *first* task in an SPSS program. All other tasks will follow the data deck. The command FINISH signals the end of the SPSS program.

The commands discussed so far are sufficient to write and run a program in SPSS, but we can include additional information in the output to make it more readable. This is done by attaching *labels* to variable names and values. These labels appear on the printed output whenever the corresponding variable name or value appears.

Variable labels are attached with the VAR LABELS command. In the specification field, a variable name is followed by its label of up to 40 characters, and is separated from it by at least one comma or blank. If you wish to attach labels to more than one variable with a single VAR LABELS command, the label of each variable name except the last is followed by a slash (/), as in the example program.

The format of the VALUE LABELS command is similar. The variable name is given, then a value in parentheses, then the label for that value. Other values of the same variable may also be named. To label the values of another variable, separate it from the preceding specifications with a slash.

The MISSING VALUES card is used to identify values which represent data that could not be obtained. In the specification field, a variable name or list is given, followed in parentheses by a list of up to three missing values. Again, to identify missing values for other variables, separate that information from the rest with a slash:

MISSING VALUES X1 TO X4,X6,X7(999.9)/ X5(99999)

In the data deck of the USNEWS file, we coded missing values of X5 as 99999, and missing values of the other variables as 999.9. (In most SPSS subprograms, it is possible to include in the calculations those values specified as missing by a MISSING VALUES command by choosing the appropriate OPTION.)

The preceding example also illustrates the "TO convention for variable lists." Where a list of variables is called for, in the specification field of any command, that list can be *implied* by giving the first and last variables in the list, in the order in which they appear on the VARIABLE LIST card. Thus, "X1 TO X4" is equivalent to "X1,X2,X3,X4" in the MISSING VALUES command. Similarly, variables can be named on the VARIABLE LIST card using the TO convention if the names fit the pattern

alpha*xxx* TO alpha*yyy*

where alpha is an alphabetic prefix, and *xxx* and *yyy* are integers with *yyy* greater than *xxx*. The VARIABLE LIST command in the example program below could be replaced by the simpler statement

VARIABLE LIST   NUM, X1 TO X7

The SAVE FILE command causes the data file, with its associated names, labels, and missing values information, to be saved on a mass storage device for later use, under the file name USNEWS. (This name need not be the same as that given in the FILE NAME command.) The data file and all its associated information can be retrieved for later runs with a GET FILE command, using the name USNEWS given in the SAVE FILE statement.

The entire program is this:

```
RUN NAME          STATISTICS AND FILE CREATION
FILE NAME         USNEWS
VARIABLE LIST     NUM, X1,X2,X3,X4,X5,X6,X7
INPUT FORMAT      FIXED (F1.0,3F4.1,F5.0,2F4.1)
INPUT MEDIUM      CARD
N OF CASES        74
VAR LABELS        X1 CITY REGION/
                  X2 CHANGE IN DEPARTMENT STORE SALES/
                  X3 UNEMPLOYMENT/
                  X4 CHANGE IN NONFARM EMPLOYMENT/
                  X5 AVERAGE INCOME OF FACTORY WORKERS/
                  X6 CHANGE IN FACTORY WORKER INCOME/
                  X7 CHANGE IN CONSTRUCTION ACTIVITY

VALUE LABELS      X1   (1) EASTERN   (2) CENTRAL   (3)WESTERN

MISSING VALUES    X1 to X4,X6,,X7(999.9)/X5(99999)

CONDESCRIPTIVE    ALL
STATISTICS        ALL
READ INPUT DATA
  (data deck)

FREQUENCIES       GENERAL = X1
OPTIONS           8
SAVE FILE         USNEWS
FINISH
```

You should notice several things about the printout: the appearance at the top of each page of the run name and file name; the output produced by the subprograms CON-DESCRIPTIVE and FREQUENCIES; the appearances of variable and value labels; and the mention of missing values. In Chapter 4 we will discuss the meanings of the statistics calculated by this run.

Job Control Language

At every computer facility, all programs will be preceded and followed by *job control* statements which communicate to the computer's *operating system*, the set of programs that supervise the operation of the computer, information about the job being submitted. The program you have just seen was run on a Control Data Corporation Cyber 72, with these JCL (*job control language*) statements surrounding the SPSS program:

```
FILECR,T20,CM70000.
ACCOUNT,MISC037,KING.
header card
DEFINE(SVFILE = USNEWS)
CALL (SPSS)
7/8/9 (multipunched)

SPSS program

6/7/8/9 (multipunched)
```

On an IBM SYSTEM/370 computer, the JCL might look like this:

```
// JOB FILECR R1101001 NAME = (JULSTROM)
disposition card
PROC SPSS,(NOTAPE)

SPSS program

/*
/& END OF JOB
```

The required JCL statements will vary from one installation to another, even among those using similar hardware. Check with your instructor or computer center about appropriate JCL to run statistical package programs.

EXERCISES

1. Become familiar with the operation of a key punch machine.

2. This is a typical data card from a study involving 5 variables.

FIGURE 3.3.

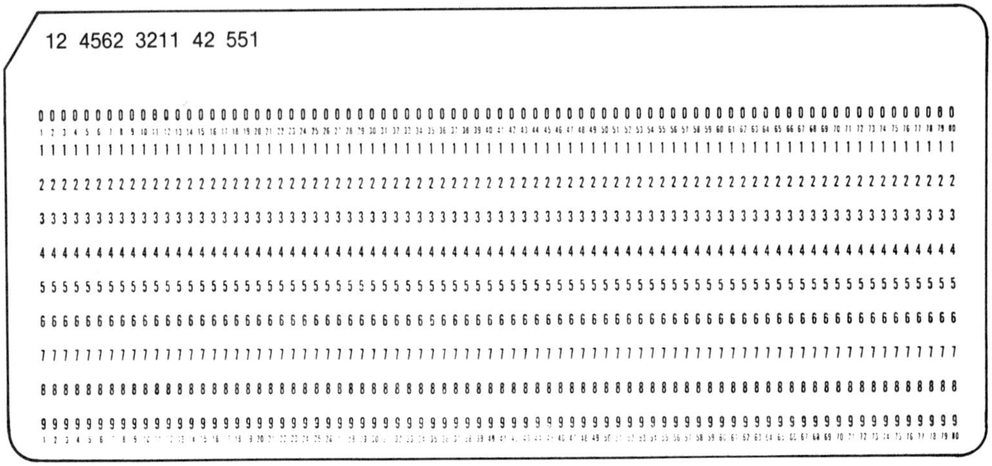

Develop a VARIABLE LIST and an INPUT format card for this data file.

3. Corresponding to these statements, draw a typical data card.

VARIABLE LIST   NUM, EDLEVEL, INCOME

       .

       .

       .

INPUT FORMAT   FIXED(F3.0,F3.1,F7.0)

4. For one of the data files in Appendix C, or one of your own choice, give the operational definitions of the variables.

5. For the data file you chose in Exercise 4, create and describe a coding scheme for the data. Give VARIABLE LIST and INPUT FORMAT cards corresponding to your coding scheme.

6. For the data file you chose in Exercise 4, write statements to attach labels to variable names and values where they would be useful.

7. For the data file you chose in Exercise 4, write a MISSING VALUES statement corresponding to missing data.

8. Rewrite the following using the "TO convention":

   a. VARIABLE LIST   Y1,Y2,Y3,Y4,Y5,Y6,Y7,V8,Y9,Y10

   b. VARIABLE LIST   NUM, EDLEVEL, INCOME, STATUS
      .
      .
      .
      FREQUENCIES   GENERAL = EDLEVEL, INCOME, STATUS

   c. VARIABLE LIST   CASENUM, XY10,XY11,XY12,XY13,Z1,Z2
      .
      .
      .
      MISSING VALUES   XY10,XY11,XY12(99)/
      .   XY13,Z1,Z2(9999)
      .
      .
      CONDESCRIPTIVE   XY11,XY12,XY13,Z1,Z2

9. Write and run a program using SPSS or whatever package is available at your installation to find descriptive statistics for all the variables in the data file you selected in Exercise 4. Save the file on a mass storage device for convenient use in the future.

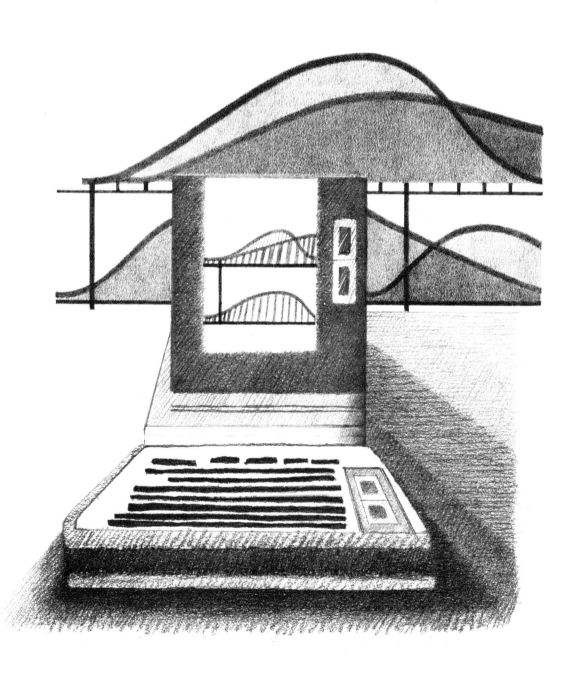

Descriptive Statistics

INTRODUCTION

In descriptive statistics we are concerned with finding measures of central tendency, variation, distributional shape, and relative location of individual values for the values of a variable. From a large group of values, we want to extract numbers which will characterize certain qualities of the group; from the original mass of data, we want to distill information that we can more easily understand and communicate.

Tables and Graphs

We begin by considering tabular and graphical representations of the last two qualities in the list above, distributional shape and relative location of individual values. The output shown below was produced by these SPSS task commands.

```
            FREQUENCIES     GENERAL = X1
            OPTIONS         8

   X1       CITY REGION
```

CATEGORY LABEL	CODE	ABSOLUTE FREQ	RELATIVE FREQ (PCT)	ADJUSTED FREQ (PCT)	CUM FREQ (PCT)
EASTERN	1.	27	36.5	36.5	36.5
CENTRAL	2.	30	40.5	40.5	77.0
WESTERN	3.	17	23.0	23.0	100.0
	TOTAL	74	100.0	100.0	

```
   X1       CITY REGION

      CODE
         |
       1.|**************************  (        27)
         | EASTERN
         |
       2.|****************************  (      30)
         | CENTRAL
         |
       3.|****************  (      17)
         | WESTERN
         |
         |.........|.........|.........|.........|.........|
         0        10        20        30        40        50
         FREQUENCY

   VALID CASES   74          MISSING CASES   0
```

The primary function of the FREQUENCIES subprogram is to produce *frequency distributions* of the values of the variables named in the specification field of the FREQUEN-CIES command; that is, a table for each variable listing its values in order, with the *absolute frequency* of each value, the number of times it occurs. From our printout-table above, we see that 27 of the cities in our survey are in the eastern region.

The next column of the table forms, with the first, a *relative frequency distribution*, whose entries give the *proportion* of all the values equal to each particular value. In the table, we see that 36.5%, $(27/74) \times 100\%$, of the cities in our survey are in the East. The fourth column gives the same information as the third, but excludes from construction of the table all values classified as missing. Since there are no missing values of X1, these *adjusted frequencies* are identical to the relative frequencies.

We should be cautious about giving percentages for particular values when the total number of cases is small. Percentages carry a ring of precision, yet in our example the addition to or removal from a category of only one value would change the corresponding relative frequency by 1.35%. In a study with only 10 cases, shifting one entry would change the corresponding relative frequency by 10%.

The first and last columns of the table form a *cumulative relative frequency distribution*, in which the entries give the percentage of all the values less than or equal to each value of the variable. We see that $(27 + 30)/74 \times 100\% = 77.0\%$ of the values of X1 are less than or equal to 2; that is, 77% of the cities are central or eastern.

Use of the statement OPTIONS 8 with the FREQUENCIES command caused the printing of the graphical display, called a *histogram*, of the absolute frequency distribution. This graph and the table summarize the regional information of the variable X1 and display it in a way that is clear and easily understood.

Statistics of the Data

Both FREQUENCIES and CONDESCRIPTIVE calculate a variety of descriptive statistics, but the selection provided by FREQUENCIES is larger. Therefore, consider this simple SPSS program and its output, in which we use the GET FILE command to retrieve the file USNEWS. The file, with its associated labels, was saved at the end of the file creation program of Chapter 3, so we do not need to repeat any of the statements that described the file, labeled variables and values, or identified missing values. OPTION 7 suppresses the printing of the tables, so we get only the statistics. Note that we need no READ INPUT DATA command.

```
RUN NAME        FINDING DESCRIPTIVE STATISTICS WITH
                FREQUENCIES
GET FILE        USNEWS
FREQUENCIES     GENERAL = X5
STATISTICS      ALL
OPTIONS         7
FINISH
```

X5   AVERAGE INCOME OF FACTORY WORKERS

MEAN	15065.918	STD ERR	270.903	MEDIAN	15226.000
MODE	15918.000	STD DEV	2314.596	VARIANCE	5357352.382
KURTOSIS	− .550	SKEWNESS	− .126	RANGE	10041.000
MINIMUM	10337.000	MAXIMUM	20378.000	SUM	1099812.000
C.V. PCT	15.363	.95 C.I.	14525.883	TO	15605.953
VALID CASES	73			MISSING CASES	1

We now consider the meanings of some of these statistics.

Measures of Central Tendency

When we are confronted by a long list of values, we often ask, "What is a typical value?" or "What is an average value?" We are asking for a measure of the *center* of the values, for which there are several commonly used statistics.

MODE

The most simpleminded measure of the central tendency of a group of values is the *mode*, the most common value.

DEFINITION

The **mode** is the value with the largest frequency.

In the group of values 5, 7, 6, 7, and 10, the mode is 7, since it appears more often than any of the others. The mode of the variable X5, average income of factory workers, is given as 15918.00. However, if several values share the largest frequency, the subprogram FREQUENCIES reports only the smallest of these values, all of which are, in fact, modes.

As a measure of the center of a distribution of values, the mode has several characteristics which should be noted.

1. It is the fastest and roughest measure of central tendency.

2. It is not necessarily unique, since a given group of values may have more than one mode.

3. It does not necessarily exist; if all values occur only once, there is no mode.

4. It is determined by the most common value or values, and does not consider any others.

When giving the mode, it is wise to state the frequency of the mode and the total number of values.

MEDIAN

A more useful measure of a "typical" value in a group of values is the *median*, which is defined in this way:

The **median** is that value which divides all the values of the variable so that half are larger than, and half smaller than, the median.

To find the median, arrange the values in order (they are then called *order statistics*), and select the one in the middle. If the number of values n is odd, then the $[(n + 1)/2]$th largest (or smallest) value is the median. For example, the median of the values 2, 7, 5, 4, 6, 9, 6 is 6:

$$2, 4, 5, 6, 6, 7, 9$$
$$\uparrow$$
$$\text{median}$$

If the number of values is even, the median is halfway between the $(n/2)$th and $[(n/2) + 1]$st values:

$$10, 12, 16, 17, 20, 23, 25, 26$$
$$\uparrow$$
$$\text{median} = 18.5 = \frac{20 + 17}{2}$$

(The FREQUENCIES subprogram finds the median in a slightly different but related way, which we will discuss on page 53-54.)

Relevant characteristics of the median are these:

1. It always exists; for any group of values, the median can be computed.

2. It is unique; any group of values has only one median.

3. It is not greatly affected by extreme values; in the last example above, if we include the value 1643 in the list, the median moves up to 20.

Note that the median does take into account the *relative positions* of all the values, and is more sensitive to *all* the values than is the mode.

MEAN

The measure of the middle of a group of values which considers the *magnitudes* of all the values is the *mean* or *arithmetic mean*, the quantity most often associated with the word "average."

The **mean** is the sum of all the values of the variable divided by the number of values.

Stated more precisely, if $X_1, X_2, \ldots, X_N$ are a *population* of values, their *mean* is

$$\frac{X_1 + X_2 + \ldots + X_N}{N} = \frac{\sum\limits_{i=1}^{N} X_i}{N}, \text{ and is denoted } \mu \text{ (mu)}$$

For example, the mean of the values 9, 15, 23, 21, 12, and 16 is

$$\frac{9 + 15 + 23 + 21 + 12 + 16}{6} = 16$$

If the values are elements of a *sample*, the notation is different, but the process is the same:

If $X_1, X_2, \ldots, X_n$ are a *sample* of values, their *mean* is

$$\frac{X_1 + X_2 + \ldots + X_n}{n} = \frac{1}{n} \sum\limits_{i=1}^{n} X_i, \text{ and is denoted } \bar{X} \text{ (}X\text{-bar)}$$

The mean has several important characteristics:

1. Like the median, it always exists, and is unique.

2. It is a good estimator; if we take repeated samples from the same population, the means of the samples will tend to cluster around the population mean.

3. It is sensitive to extreme values; consider the previous example, with the value 1920 included. The mean is then:

$$\frac{9 + 15 + 23 + 21 + 12 + 16 + 1920}{7} = 288.$$

Measures of the Middle and Distributional Shape

When a distribution of values is symmetrical and unimodal (has one mode), the mean is the best measure of central tendency. In fact, in such a case, the mean, median, and mode will be approximately equal, as shown in Figure 4.1.

If a distribution of values is unimodal but not symmetrical, we say that it is *skewed*, and since the mean, median, and mode differ in sensitivity to extreme values, they will be different, as shown by Figure 4.2.

On the left side of Figure 4.2, we see a distribution which is *skewed to the left*, or *negatively skewed*. The mean is most affected by the low extreme values, and is less than the median, which is less than the mode. In a distribution that is *skewed to the right (positively skewed)*, this relationship is reversed. Such distributions arise in situations where all the extreme values lie on the same side of the bulk of the data, as with salaries in a business, where a few of the values are larger than the majority.

FIGURE 4.1   Measures of central tendency of a symmetrical distribution.

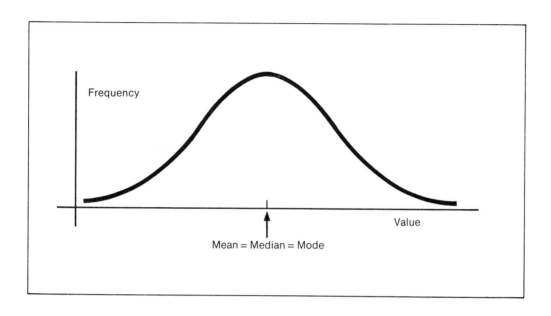

FIGURE 4.2   Measures of central tendency of skewed distributions.

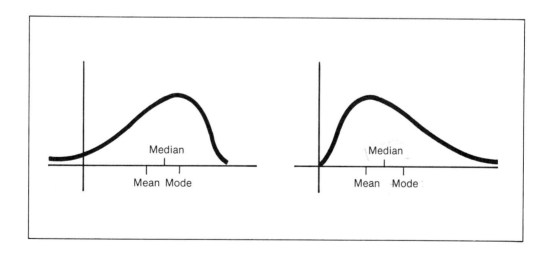

Measures of Variation

After graduation, you are offered two jobs. Employer A tells you that the median salary at his plant is $18,500, while employer B's organization pays a mean salary of $19,000. If the salary were your only consideration, would you accept B's offer over A's? Do you have enough information here to make an informed decision?

Probably not. You know nothing of the variability of the salaries in the two businesses, whether they are bunched tightly around their respective centers, suggesting a good starting salary but little chance for increase, or widely varying, suggesting that you might work toward pleasantly high levels. What is needed is a measure of the *variation* or *dispersion* of a group of values.

RANGE

The simplest and quickest measure of dispersion is the *range*; as its name implies, it is the difference between the maximum and minimum values of the variable.

DEFINITION

range = maximum value − minimum value.

The maximum and minimum values are found by the subprogram CONDESCRIPTIVE, while FREQUENCIES gives these and the range itself. For the variable X5 in our example study, the range is

$$20378 - 10337 = 10041.$$

Clearly the range will tend to be larger if the values are more varied, smaller if they are more uniform, but the range is an unreliable measure of variation since it is entirely determined by only two values out of the entire sample or population.

VARIANCE AND STANDARD DEVIATION

We would like a value that indicates variability and takes into account *all* the values in the sample or population.

Suppose that $X_1, X_2, \ldots, X_N$ are a population of values. We begin our search for a better measure of variability by considering the *deviations from the mean*, $(X_i - \mu)$, for $i = 1, 2, \ldots, N$. These will be larger if the values, the X_i's, are more varied, so we might consider the mean of these deviations. However,

$$\frac{1}{N} \sum_{i=1}^{N} (X_i - \mu) = \frac{1}{N} \sum_{i=1}^{N} X_i - \frac{1}{N} \sum_{i=1}^{N} \mu = \mu - \frac{N\mu}{N} = 0,$$

no matter what the values might be, so this quantity is not a measure of anything.

The above result is due to the canceling out of positive and negative deviations from the mean. This can be prevented by using the *absolute values* of the deviations. Thus we can define the

$$mean\ deviation\ =\ \frac{1}{N}\ \sum_{i=1}^{N}\ |X_i - \mu|.$$

This is a valid measure of variability, which increases as the values of the population become more dispersed, but it is rarely used. Instead, statisticians have taken advantage of the fact that the square of any real number is nonnegative. That is, $(X_i - \mu)^2$ is a positive number whose magnitude depends on the difference between X_i and μ. From this, we make the following definition.

DEFINITION

The **variance** of a *population* of values is the mean of the squared deviations from the mean, and is denoted σ^2 (read "sigma squared"). That is,

$$\sigma^2 = \frac{1}{N}\ \sum_{i=1}^{N}\ (X_i - \mu)^2.$$

Using calculus, it can be shown that for any value d, the sum of the squared deviations from d of the values in a population,

$$\sum_{i=1}^{N}\ (X_i - d)^2,$$

achieves its minimum value when $d = \mu$. (The proof of this is left as an excercise to that most treasured member of our audience, the interested reader.) This lends a geometric as well as an arithmetic meaning to the idea of the variance.

Note that every value in the population is included in the calculation of the variance, which is widely used as a measure of dispersion. Both CONDESCRIPTIVE and FREQUEN-CIES calculate it.

There are, however, two difficulties with the variance. The first can be seen by consulting the printout of descriptive statistics for X5.

X5		AVERAGE INCOME OF FACTORY WORKERS			
MEAN	15065.918	STD ERR	270.903	MEDIAN	15226.000
MODE	15918.000	STD DEV	2314.596	VARIANCE	5357352.382
KURTOSIS	− .550	SKEWNESS	− .126	RANGE	10041.000
MINIMUM	10337.000	MAXIMUM	20378.000	SUM	1099812.000
C.V. PCT	15.363	.95 C.I.	14525.883	TO	15605.953
VALID CASES	73	MISSING CASES	1		

When the values themselves are large, as they are in this case, the variance can become huge, and it is difficult to relate the variance to the original values.

Also, the values of X5 are measured in "dollars." Therefore the units of the variance of X5 are "dollars squared," creatures never seen outside the laboratory.

To solve both these problems, we define a related measure of variability.

DEFINITION

The **standard deviation** of a population, denoted σ, is the square root of the variance:

$$\sigma = \sqrt{\sigma^2}.$$

The standard deviation is in the same units as the original values, and its magnitude can be more easily related to the dispersion of those values.

For example, if 9, 15, 23, 21, 12, and 16 are a population of values,

$$\mu = \frac{9+15+23+21+12+16}{6} = 16,$$

$$\sigma^2 = \frac{(9-16)^2+(15-16)^2+(23-16)^2+(21-16)^2+(12-16)^2+(16-16)^2}{6} = 23.33,$$

and

$$\sigma = \sqrt{23.33} = 4.83$$

In the definitions of variance and standard deviation, we specified that the values form a *population*. When the values are elements of a *sample*, we compute the variance in a slightly different way:

If $X_1, X_2, \ldots, X_n$ are a sample of values, their **variance**, denoted s^2, is

$$\frac{1}{n-1} \sum_{i=1}^{n} (X_i - \bar{X})^2.$$

We divide by $(n-1)$, one less than the number of values, to make s^2 a better estimator of the variance of the population from which the sample was taken. This procedure will be explained and justified in Chapter 7. Also,

---

DEFINITION

The **standard deviation** of a sample, denoted s, is the square root of the variance:

$$s = \sqrt{s^2}.$$

If the values 4, 7, 6, 3, 10 are a sample,

$$\bar{X} = \frac{4+7+6+3+10}{5} = 6,$$

$$s^2 = \frac{(4-6)^2+(7-6)^2+(6-6)^2+(3-6)^2+(10-6)^2}{5-1} = 7.5,$$

and

$$s = \sqrt{7.5} = 2.739.$$

Since SPSS is designed to analyze sample data in statistical studies, these latter definitions are those used in SPSS calculations (though the formulas employed are slightly different; see Exercise 7).

Tchebysheff's Inequality

We know that the variance and standard deviation of a population or sample of values measure the variability of the population or sample, but we can be more precise in our understanding of the relationship between these measures and the locations of the values through a result called

---

DEFINITION

Tchebysheff's Inequality: If a population of values has mean μ and variance σ^2, then for any number $k \geq 1$, at least $(1-(1/k^2)) \times 100\%$ of the values must lie within k standard deviations of μ; alternately, at most $(1/k^2) \times 100\%$ of the values lie above $\mu + k\sigma$ or below $\mu - k\sigma$.

This result also holds for a sample with mean $\bar{X}$ and variance s^2.

Looking again at X5, which has mean and standard deviation approximately 15066 and 2315, respectively, we can say that in at least $(1-(1/3^2)) \times 100\% = 88.89\%$ of the cities in our sample, the average income of factory workers is between $15066 - 3 \times 2315 = 8121$ and $15066 + 3 \times 2315 = 22011$. The proportion between these limits may be greater, but Tchebysheff's inequality assures us that it is at least 88.89%.

Alternately, we can say that at most $(1/3^2) \times 100\% = 11.11\%$ of the values lie below 8121 or above 22011. Again, the proportion may be less, but we are assured that it is no more than 11.11%.

Tchebysheff's inequality tells us nothing about the shape of the distribution of values, though if the distribution is unimodal and symmetrical, we can apply the *empirical rule*, a stronger statement than Tchebysheff's inequality.

DEFINITION

The **empirical rule:** For a unimodal, symmetrical distribution of values, Figure 4.3 shows the approximate percentage of all the values which lie within specific intervals centered on the mean.

FIGURE 4.3

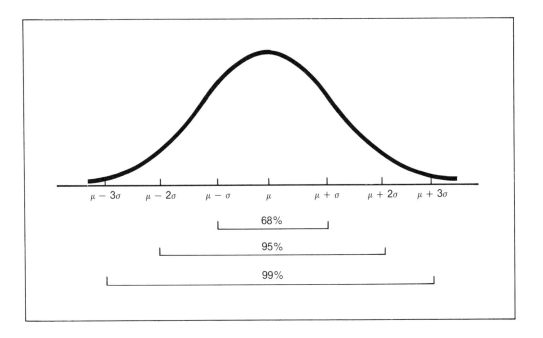

Therefore, if the 73 values of X5 were distributed in this way, we could say that in more than two-thirds of the cities in our survey, the average income of factory workers is between $15066 - 2315 = 12751$ and $15066 + 2315 = 17381$; more than 95% of these values lie between $15066 - 2 \times 2315 = 10436$ and $15066 + 2 \times 2315 = 19696$; and nearly all the values lie between 8121 and 22011.

Comparing Variabilities

Often, the researcher will want to compare the variability of one population with that of another. If the means of the two populations are different, it would be risky to compare their variances or standard deviations directly. We require a measure of variability that is independent of the magnitudes of the values in the sample or population. This definition provides such a measure:

DEFINITION

The **coefficient of variation** of a group of values is their standard deviation expressed as a percentage of their mean; that is,

$$\frac{s}{\bar{X}} \times 100\% \text{ or } \frac{\sigma}{\mu} \times 100\%.$$

On SPSS printouts, this value is identified as C.V.PCT, and we see that of the variables in the USNEWS file, the greatest variability is shown by X7, change in construction activity, while the variable whose values vary the least is X5, average income of factory workers.

Measures of Distributional Shape-Skewness

We have mentioned that a distribution of values may be *symmetrical* or *skewed*. A measure of this characteristic of distributional shape is defined in this way:

DEFINITION

The **skewness** of a sample of values $X_1, X_2, \ldots, X_n$ is:

$$\frac{\sum_{i=1}^{N} (X_i - \bar{X})^3}{ns^3} \ .$$

If the skewness is near 0, the distribution is symmetrical. Positive and negative values of skewness (see Figure 4.4) indicate positively and negatively skewed distributions, respectively, though skewness is not considered extreme unless it is less than -1 or greater than $+1$.

This statistic should be applied only to unimodal distributions composed of interval- or ratio-level data, and is referred to in some texts as *relative skewness*.

Again consulting the printout, we see that the values of X3, unemployment, are skewed to the right, while the values of X5 and X6 form relatively symmetrical distributions.

FIGURE 4.4   Skewness of unimodal distributions.

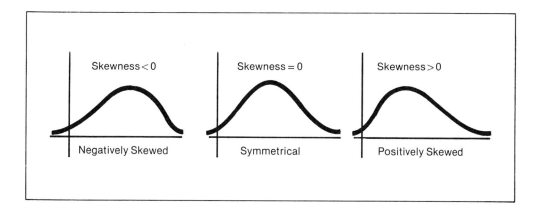

Measures of Distributional Shape-Kurtosis

The **kurtosis** of a sample of values $X_1, X_2, \ldots, X_n$ is defined to be

$$\frac{\displaystyle\sum_{i=1}^{N} (X_i - \bar{X})^4}{n^4} - 3.$$

This statistic is a measure of the relative variation of a symmetrical, unimodal distribution; kurtosis measures the "pointness" or "flatness" of a distribution of values, as shown in Figure 4.5. For example, the values of X7 form a slightly peaked distribution.

This concludes our discussion of the descriptive statistics provided by the subprograms FREQUENCIES and CONDESCRIPTIVE. (The meanings of *minimum, maximum,* and *sum* are obvious.) The two quantities we have not examined, STD ERR and .95 C.I., deal with estimating the population mean μ with a sample mean, and will be discussed in Chapters 7 and 8.

The frequency distributions examined earlier in this chapter gave frequencies, relative frequencies, and cumulative relative frequencies for every value (or code) of the variable being considered. Such distributions are called **ungrouped.**

These distributions and their associated histograms will often produce confusing or unintelligible pictures of the arrangement of the data values, particularly when the number of different values is large. For an example of this we use FREQUENCIES to produce a frequency distribution and histogram of the values of X5 from the USNEWS file in the next section.

FIGURE 4.5   Kurtosis measurements.

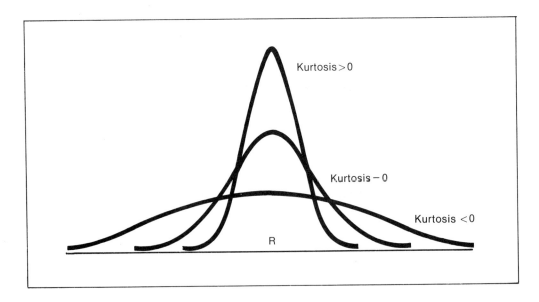

Interactive Processing

Before we examine the production of more useful frequency distributions and histograms, we describe another way in which SPSS and many other statistical program packages can be used. This is the *time-sharing* or *interactive* mode, in which the user is in contact with the computer through a terminal throughout processing. In an interactive session, there are four major steps.

1. Log on the computer system.
2. Enter the statistical software system.
3. Write and execute statements to perform statistical tasks.
4. Leave the statistical system and log off the computer.

The following printout illustrates this process in producing the output described above, a frequency distribution and histogram of the values of X5, in an interactive session on a CDC Cyber 72. All user-supplied material is lowercase, while the computer produced the capitals.

```
        ┌─  80/07/15.   10.17.11.
        │   EDUCATIONAL COMPUTING NETWORK              NOS 1.3–485/485.
        │   USER NUMBER: misc037
  1.    │   PASSWORD
        │   TERMINAL:        52,   TTY
        │   RECOVER /SYSTEM:    attach,usnews
        └─  READY.
        ┌─ -spss

        │   SPSS/ONLINE 8.0

  2.    │   USE AN SPSS SYSTEM FILE THIS RUN
        │   ? yes
        │   ENTER FILE NAME
        │   ? usnews
        │   USE A RAW DATA FILE THIS RUN
        │   ? no
        └─  AUTO-MODE.
        ┌─  ? 10 set file;usnews
        │
  3.    │   ? 20 frequencies;general = x5
        │   ? 30 options;8
        └─  ? e

            ENTERING SPSS.
            — — — FREQUENCIES — — —

            FREQUENCIES — INITIAL CM ALLOWS FOR 1030 VALUES
                            MAXIMUM CM ALLOWS FOR 11547 VALUES


            X5        AVERAGE INCOME OF FACTORY WORKERS
```

CATEGORY LABEL	CODE	ABSOLUTE FREQ	RELATIVE FREQ (PCT)	ADJUSTED FREQ (PCT)	CUM FREQ (PCT)
	10337.	1	1.4	1.4	1.4
	10525.	1	1.4	1.4	2.7
	10718.	1	1.4	1.4	4.1
	10905.	1	1.4	1.4	5.5
	.	.	.	.	.
	.	.	.	.	.
	.	.	.	.	.
	18740.	1	1.4	1.4	97.3
	19307.	1	1.4	1.4	98.6
	20378.	1	1.4	1.4	100.0
	99999.	1	1.4	MISSING	
	TOTAL	74	100.0	100.0	

X5   AVERAGE INCOME OF FACTORY WORKERS

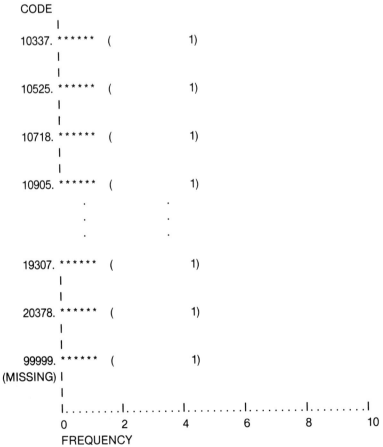

```
          VALID CASES     73        MISSING CASES      1
          CPU TIME REQUIRED..       .3780 SECONDS

          TOTAL CPU TIME USED..     .4180 SECONDS

         ┌ SPSS/ONLINE AUTO-MODE
         │ ? stop
         │ *TERMINATED*
         │ READY
   4.    │ bye
         │
         │ MISC037     LOG OFF    10.25.06.
         └ MISC037     SRU        2.706 UNTS.
```

As you can see from this example, interactive-mode processing of SPSS commands differs in several significant ways from the batch mode. First, the lines of the SPSS program are numbered; they are executed in order of ascending line number, rather than in the physical order of the statements. Second, the control and specification fields are separated by semicolons, rather than having the specification field begin in column 16. Third, a program which would be incomplete in the batch mode is acceptable here; note that there are no RUN NAME or FINISH statements in the example session.

Again, use of the interactive mode on your computer system will probably be different from the example shown here, so consult your instructor or computer center.

Grouped Frequency Distributions

We have seen in the previous example that simple frequency distributions or histograms provide very little information when the number of individual values is large. In situations like this, we can create more intelligible frequency distributions and graphs by grouping values into *intervals* or *classes*, then counting the number of values in each interval. Such a distribution is called a *grouped frequency distribution*.

To choose the number of intervals into which the values will be divided and the widths of those intervals, two conventions are useful:

1. Use not less than 5 nor more than 15 class intervals.

2. The interval widths should be 1, 2, 3, 5, or 10, or some multiple of 2, 3, 5, or 10, and should be equal.
   (An exception: the highest and lowest intervals may be unbounded.)

Violation of these rules will tend to produce tables and graphs which are difficult to read.

To create a grouped frequency distribution by hand, simply choose the class intervals, and count the number of values in each. With a large number of cases, this becomes tedious, so we again turn to the computer. In SPSS, the relevant command is RECODE.

RECODE is used to replace a value or list of values of a variable with some other value. For example:

RECODE   X1(1 = 4) (2 = 5)/X7(11.0 THRU 15.0 = 13)

This command changes X1 from 1 to 4 and from 2 to 5, and changes all the values of X7 between 11.0 and 15.0 (inclusive) to 13. We can use such a statement, which is placed after the data file is described or retrieved, and before the first task, to create intervals of values, and from these, grouped frequency distributions, as shown in this program in the batch mode, which groups the values of X5 into six classes.

```
RUN NAME        GROUPED FREQUENCY DISTRIBUTION OF X5
GET FILE        USNEWS
RECODE          X5 (LOWEST THRU 11000 = 10000)
                (11001 THRU 13000 = 12000)
                (13001 THRU 15000 = 14000)
                (15001 THRU 17000 = 16000)
                (17001 THRU 19000 = 18000)
                (19001 THRU HIGHEST = 20000)
FREQUENCIES     INTEGER = X5(10000,22000)
STATISTICS      ALL
OPTIONS         8
FINISH
```

In the interactive mode, an equivalent program would be this:

```
? 10 GET FILE;USNEWS
? 20 RECODE;X5(LOWEST THRU 11000 = 10000)
? 20.010 (11001 THRU 13000 = 12000)
? 20.015 (13001 THRU 15000 = 14000)
? 20.020 (15001 THRU 17000 = 16000)
? 20.025 (17001 THRU 19000 = 18000)
? 20.030 (19001 THRU 21000 = 20000)
? 30 FREQUENCIES;GENERAL = X5
? 40 STATISTICS;ALL
? 50 OPTIONS;8
```

We have chosen the new values of X5 to be the *class marks* or *midpoints* of the intervals. Any values might be selected, but these are the most representative of the values in each class. The output of the interactive program is this:

X5	AVERAGE INCOME OF FACTORY WORKERS				
CATEGORY LABEL	CODE	ABSOLUTE FREQ	RELATIVE FREQ (PCT)	ADJUSTED FREQ (PCT)	CUM FREQ (PCT)
	10000.	4	5.4	5.5	5.5
	12000.	10	13.5	13.7	19.2
	14000.	20	27.0	27.4	46.6
	16000.	22	29.7	30.1	76.7
	18000.	15	20.3	20.5	97.3
	20000.	2	2.7	2.7	100.0
	99999.	1	1.4	MISSING	
	TOTAL	74	100.0	100.0	

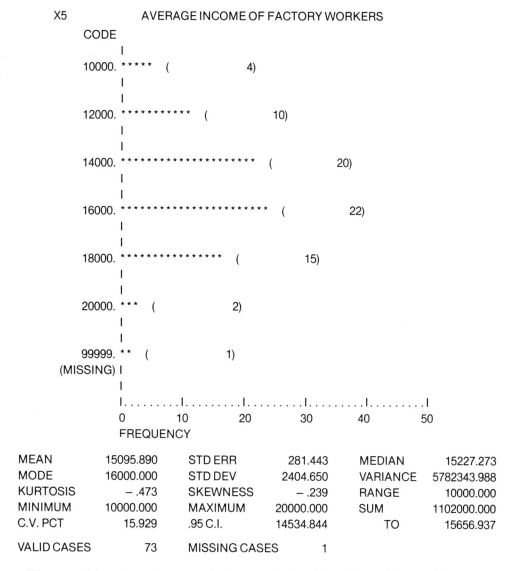

MEAN	15095.890	STD ERR	281.443	MEDIAN	15227.273
MODE	16000.000	STD DEV	2404.650	VARIANCE	5782343.988
KURTOSIS	− .473	SKEWNESS	− .239	RANGE	10000.000
MINIMUM	10000.000	MAXIMUM	20000.000	SUM	1102000.000
C.V. PCT	15.929	.95 C.I.	14534.844	TO	15656.937
VALID CASES	73	MISSING CASES	1		

We have achieved a great increase in the readability of the table and the graph by group-ing the values. We can easily see the number of values in each class, and the general shape of the distribution of the values of X5.

However, to obtain this readability of the frequency distribution and the histogram, we have sacrificed accuracy in the computation of the statistics. Notice that all of them differ from the values found by CONDESCRIPTIVE as reproduced on page 43. We grouped the values of X5 with the RECODE command, thereby changing them in this case to their respective class marks. The statistics given have been calculated from the class marks, not the original values.

The mean was calculated using this formula:

$$\bar{X} = \frac{\displaystyle\sum_{i=1}^{M} f_i \, cm_i}{M} \, ,$$

where M = number of classes, cm_i = class mark of ith class, and f_i = frequency of ith class.

The median was found using an interpolation procedure which assumes that the values in each class are evenly distributed through the class. To find the median, we proceed through the class in which the median must lie until we reach the hypothetical middle value:

$$\text{median} = L + \frac{j}{f} \times c,$$

where L = lower endpoint of the median class, f = frequency of median class, j = (number of values /2) $-$ (*number of values* $\leq L$), and c = width of median class.

The median is always calculated this way; the FREQUENCIES subprogram assumes that it is working with grouped values (see page 38-39).

Likewise, the mode is the class mark of the class with the largest frequency. This result would be more accurately reported by saying: "The modal class is 15,001 to 17,000, with a frequency of 22 out of 73."

Similar observations can be made about all statistics computed with recoded values. Therefore, to obtain maximum accuracy, use the original data values when computing statistics.

It is possible to produce results based on both grouped and ungrouped values in the same program if the grouping (or other value changing) is done with the *RECODE command. *RECODE is identical to RECODE, except that it may be placed immediately before any task, and the changes it causes are temporary, lasting only through that single task, while changes associated with RECODE persist throughout the SPSS program. The following program will find statistics for X5 based on the original values, then group the values to build the frequency distribution and histogram shown above.

```
? 10 GET FILE;USNEWS2
? 20 CONDESCRIPTIVE;X5
? 25 STATISTICS;ALL
? 30 *RECORDE;X5(LOWEST THRU 11000 = 1000)
? 30.010  (11001  THRU 13000 = 12000)
? 30.015  (13001  THRU 15000 = 14000)
? 30.020  (15001  THRU 17000 = 16000)
? 30.025  (17001  THRU 19000 = 18000)
? 30.030  (19001  THRU 21000 = 20000)
? 40 FREQUENCIES:GENERAL = X5
? 45 OPTIONS;8
```

Selecting Cases

Often, a researcher wants to find statistics of the values of one variable for all those cases in which a second variable has a particular value, or to separate the cases into groups by the values of one of the variables. In the USNEWS file, it would be interesting to find and compare the means and standard deviations of X3, unemployment, and X5, average income of factory workers, for the three regions of the United States (the three values of X1).

This can be done using the task command BREAKDOWN, which has this format:

BREAKDOWN   variable list BY variable name

The BREAKDOWN subprogram will provide means and standard deviations for the variables in the variable list, with the cases grouped by the values of the variable named after the word BY.

For example, this interactive program produces the statistics of X3 and X5 for the three regions described above:

?10   GET FILE; USNEWS
?20   BREAKDOWN; X3, X5 BY X1

Its output is this:

CRITERION VARIABLE	X3		UNEMPLOYMENT		
BROKEN DOWN BY	X1		CITY REGION		
VARIABLE	CODE	MEAN	STD DEV	N VALUE LABEL	
FOR ENTIRE POPULATION		5.691	1.647	70	
X1	1.	5.836	1.567	25 EASTERN	
X1	2.	5.446	1.792	28 CENTRAL	
X1	3.	5.882	1.554	17 WESTERN	

TOTAL CASES =   74
MISSING CASES =   4   OR 5.4 PTC.

CRITERION VARIABLE	X5		AVERAGE INCOME OF FACTORY WORKERS		
BROKEN DOWN BY	X1		CITY REGION		
VARIABLE	CODE	MEAN	STD DEV	N VALUE LABEL	
FOR ENTIRE POPULATION		15065.918	2314.596	73	
X1	1.	14504.852	2636.825	27 EASTERN	
X1	2.	15723.667	2019.017	30 CENTRAL	
X1	3.	14779.438	2064.741	16 WESTERN	

TOTAL CASES =   74
MISSING CASES =   1   OR 1.4 PCT

It would appear that unemployment in the *central* region is less than that in the rest of the country, and that workers' average income is higher there as well. However, the differences between these means are not large, particularly in relation to the standard deviations, so we cannot now be sure that the differences are meaningful. The mean values of unemployment and workers' income might be essentially equal across the country. In later chapters we will investigate ways to determine if these differences are significant.

Suppose that we are interested only in those cases for which a particular variable has a certain value or falls in a certain range of values. We can examine only those cases and ignore all others with this statement:

SELECT IF   (condition)

For each case, if the condition is true, that case is included in further analyses.

This program selects those cases for which X1 equals 1 (that is, which represent eastern cities), then finds statistics of X3 and X5 for just those cases:

```
? 10 GET FILE;USNEWS2
? 20 SELECT IF;(X1 EQ 1)
? 30 CONDESCRIPTIVE;X3,X5
? 40 STATISTICS;ALL
```

UNEMPLOYMENT   VARIABLE

MEAN	5.836	STD ERR	.313	STD DEV	1.567
VARIANCE	2.457	KURTOSIS	.531	SKEWNESS	.630
MINIMUM	3.100	MAXIMUM	9.700	SUM	145.900
C.V. PCT	26.856	.95 C.I.	5.189	TO	6.483
VALID CASES	25	MISSING CASES	2		

VARIABLE X5   AVERAGE INCOME OF FACTORY WORKERS

MEAN	14504.852	STD ERR	507.457	STD DEV	2636.825
VARIANCE	6952843.977	KURTOSIS	− 1.219	SKEWNESS	.013
MINIMUM	10337.000	MAXIMUM	18740.000	SUM	391631.000
C.V. PCT	18.179	.95 C.I.	13461.759	TO	15547.945
VALID CASES	27	MISSING CASES	0		

Note that the mean and standard deviation of X3 and X5 for the 27 eastern cities in our study are the same as the values given for those cases by the previous run using BREAKDOWN.

The SELECT IF command, like RECODE, must follow retrieval or description of the data file, and must precede the first task. Selections made by the SELECT IF command remain in effect throughout the program. Also like RECODE, we may *temporarily* select cases with the *SELECT IF command. It may precede any task, has the same form and function as the SELECT IF command, and is effective only for the one subsequent task. At the conclusion of that task, all the cases are once again included in the analysis.

PACKAGE CAPABILITIES USED IN THIS CHAPTER

In this chapter we have used SPSS to perform calculations and present results, but the processes used here can be carried out with most statistical software packages. It is desirable that a statistical package allow the following procedures, which were demonstrated in the preceding discussion:

1. The construction of histograms (FREQUENCIES).

2. The computation of descriptive statistics (FREQUENCIES and CONDESCRIPTIVE).

3. The changing of values of variables based on existing values (RECODE).

4. The selection of particular cases (BREAKDOWN and SELECT IF).

EXERCISES

1. Find the mean, median, and mode of this sample of values: 4, 6, 12, 6, 9, 5.

2. Find the mean, median, and mode of this population of values: 26, 35, 29, 27, 33, 35, 29.

3. Over the past 10 years, mutual fund A has paid a mean return of 12%, with a median return of 7%; mutual fund B has paid a mean return of 10%, with a median return of 9%. You plan to invest a sum of money for two years. Based on this information, would you choose to invest in fund A or fund B? Why?

4. Ten workers are paid the following hourly wages: $4.10, $4.75, $5.55, $7.50, $7.70, $4.80, $6.15, $7.75, $5.30, and $6.20. What is the mean wage paid to the workers? The median? Is this group of values skewed?

5. A fleet of 30 cars achieves mean mileage of 19.2 mpg, while another fleet of 50 cars gets 17.5 mpg. What is the mean mileage of both fleets together? Does this suggest any generalization about the mean of two groups of values whose means are known?

6. Use the SPSS subprogram FREQUENCIES, or its equivalent, on your system, to find the mean, median, and mode of the variable X7 (change in construction activity) from the USNEWS data file. What do these values tell you?

7. The formulas given by the definitions of population and sample variance are cumbersome, but we can develop shortcut formulas which involve fewer operations. Show that these statements are true:

   a. $\sigma^2 = \dfrac{1}{N} \sum_{i=1}^{n} (X_i - \mu)^2 = \left[\dfrac{1}{N} \sum_{i=1}^{n} X_i^2 \right] - \mu^2$.

   b. $s^2 = \dfrac{1}{n-1} \sum_{i=1}^{n} (X_i - \bar{X})^2 = \dfrac{\sum\limits_{i=1}^{n} (x_i^2) - n\bar{X}^2}{n-1}$.

8. Find the range, variance, and standard deviation of the sample of values given in Exercise 1.

9. Find the range, variance, and standard deviation of the population of values given in Exercise 2.

10. The values $X_1, X_2, \ldots, X_N$ are a population. Use calculus to show that the quantity

    $f(d) = \sum_{i=1}^{n} (X_i - d)^2$ achieves its minimum value when $d = \mu$.

11. Find the range, variance, and standard deviation of the values given in Exercise 4. Are these values most appropriately considered a population or a sample?

12. What is the smallest value a population variance can ever have? Under what conditions will this value occur?

13. Use FREQUENCIES or CONDESCRIPTIVE, or their equivalent on your system, to find the range, variance, and standard deviation of the values of the variable X7 in the USNEWS data file. What are the units of these statistics?

14. Use the statistical software system implemented on your computer system to find the range, variance, and standard deviation (and other statistics if you're interested) of the values of X3 in the USNEWS file. What are the units of these statistics?

15. Would it make sense to find the mean, median, variance, and range of the values of X1, city region, in the USNEWS file? Why or why not? What does this tell you about these measurements?

16. Consider the statistics of X3 found in Exercise 14 and the mean of X3, 5.691.

    a. What proportion of all the values of X3 must lie between 2.397 and 8.985?
    b. Between what values must lie 8/9 of the values of X3?

17. Consider the statistics of X7 found in Exercises 6 and 13.

    a. What proportion of the values of X7 must lie between -22.380 and 33.426?
    b. Between what values must lie 75% of the values of X7?

18. Are the values of X3 more or less variable than the values of X7? How do you know?

19. Find the skewness and kurtosis of the sample of values given in Exercise 1.

20. Find the skewness and kurtosis of the population of values given in Exercise 2.

21. Consider the variable X3 in the USNEWS file. Find and interpret the skewness and kurtosis of X3.

22. Consider the variable X7 in the USNEWS file. Find and interpret the skewness and kurtosis of X7.

23. Use your computer system interactively to generate all the statistics discussed in this chapter for all the variables in the USNEWS file. Keep this information; it will be useful later on.

24. Group the values of X3 from the USNEWS file into five classes using a RECODE command or its equivalent. Then generate a histogram of the values of X3, and all the usual statistics. Compare these statistics to those found with the ungrouped values of X3. Why are they different?

25. Group the values of X7 into six classes using a RECODE command or its equivalent. Generate a histogram of the grouped values, and their statistics. Compare the statistics to those of the ungrouped values.

26. Generate a histogram of the values of X3 for those cases in the USNEWS file representing central cities. Use a SELECT IF statement to choose only those cities.

27. Generate a histogram of the values of X7 for those cases in the USNEWS file representing central or western cities. Use a SELECT IF statement to choose the appropriate cases.

28. Use a BREAKDOWN command to find the mean and standard deviation of X7 for each of the three regions in the USNEWS file.

29. Write a VALUE LABELS statement for the program on p. 52 which will label the classes of X5 created by the RECODE command.

Elementary
Probability

INTRODUCTION

Having defined several descriptive statistics, our next objective is the development of the tools of statistical inference. To do this, we must first study probability, and this topic, useful and interesting in its own right, is pursued in the next three chapters.

Many intuitions you may have about what probability is and how it behaves are likely to be true. If we flip a coin, it seems reasonable to say that the probability that it will land with the "heads" side up is $\frac{1}{2}$. This simple example illustrates the major goal of this chapter: To develop a set of processes and rules for assigning to an event that might occur a number (its probability) which is proportional to its likelihood.

Random Experiments

We begin by defining the situation to which we apply our methods:

DEFINITION

A **random experiment** is any well-defined situation whose outcome is uncertain and in which we make an observation or take a measurement.

The word "random" indicates the element of chance; the experiment may result in any one of several possible outcomes, and we do not know which one will occur. Flipping a coin is a random experiment; it will result in either heads or tails, but we cannot know which. Other examples of random experiments are these:

1. Roll two dice. How many dots are on the two upper faces?

2. Deal a hand of five cards. Which particular group of five cards is dealt?

3. Choose one city at random from the 74 cities represented in the USNEWS file. Which city was chosen?

4. Observe the closing Dow Jones average.

5. Count the number of cars passing through an intersection in a 1-hour period.

Sample Spaces and Events

In each of these experiments, the precise result is unknown, but we can list all the possibilities. Each of these possibilities is called an **outcome** (or **elementary event**), and they cannot be subdivided. That is, outcomes are the smallest units of what might happen. For a given experiment, the set of all outcomes is called the **sample space** (sometimes *event space* or *outcome space*), and it is indicated by the letter S.

If we flip one coin, the two outcomes are "heads" and "tails," which we can represent as H and T, and we write $S = \{H, T\}$. If we roll two dice, the total number of dots showing on the upper faces might be any integer from 2 to 12; $S = \{2,3,4, \ldots ,12\}$. In Example 2 above, the sample space is the set of all 5-card hands that can be created from the usual deck of 52 cards. In Example 3, S is the set of all cities in our sample of 74; $S = \{$Atlanta, Baltimore, $\ldots$, Tucson$\}$. (You should describe sample spaces for Examples 4 and 5 as well.)

Note that the performance of a random experiment always results in one and only one outcome; when a coin is flipped, it must fall either heads or tails. Also, no two outcomes can ever occur simultaneously.

There are many experiments in which we are concerned with the occurrence or nonoccurrence of a set of outcomes, rather than just one. For example, we choose one city from the USNEWS file at random. Is it a western city? Here we are asking if the experiment resulted in one of a set outcomes, and from this idea we extract the following definition:

DEFINITION

An **event** is any subset of the sample space S.

Since the set of western cities is a subset of S, the set of all cities in our study, it is an event. Further:

DEFINITION

If A is an event in S (written $A \subseteq S$), we say that A *occurs* if the experiment results in an outcome which is in A.

If A is the event that we pick a western city, then when we choose one city at random, for instance, Sacramento, we say that A occurs. If Chicago is chosen, then A does not occur.

Two events deserve special attention: The sample space S is a subset of itself, so S is an event. Since S contains all the outcomes associated with the experiment, the experiment always results in an outcome which is in S; S *always* occurs. The *null set* (or *empty set*) $\varnothing$ is a subset of every other set, so it is a subset of S, and, therefore, an event, the *null event*. Since it contains no outcomes, it *never* occurs. Also, each outcome by itself is a subset of S, so each outcome is also an event.

What Probability Means

An experiment is a situation whose result is uncertain. We can list, in the sample space, all the possible outcomes of the experiment, but we do not know which will occur. We want to assign numbers—probabilities—to events in the sample space indicating how likely they are to occur. What will these numbers mean?

Consider the simple experiment of rolling a single die. The event A is the appearance of six dots on its upper face. We performed this experiment many times (actually, a computer was used to simulate the repetitions), and the number of times that A occurred (the *frequency* of A) was counted (Table 5.1).

TABLE 5.1

NUMBER OF TRIALS	FREQUENCY OF A	RELATIVE FREQUENCY OF A
50	7	0.14
100	19	0.19
500	77	0.154
1000	176	0.176
5000	871	0.1742
10000	1692	0.1692

Also calculated was the **relative frequency** of A, the proportion of all the trials in which A occurred. It is clear that as the number of trials increased the relative frequency of A appeared to stabilize around approximately 0.17. This seems to indicate that we can expect A to occur about 17 times out of every 100 repetitions of the experiment, and that the *probability* of A might then be near 0.17.

This is an example of the relative frequency definition of probability:

DEFINITION

As the number of repetitions of an experiment increases, the relative frequency of an event A will appear to stabilize around some value p. We call p the *probability* of A, written $P(A)$.

In our example, we would conclude that $P(A) \doteq 0.17$. (We will see that $P(A) = \frac{1}{6} = 0.1667$.)

Equally Likely Outcomes

We have a definition of the probability of an event; we must now develop ways to assign probabilities to events. The simplest case is that in which all the outcomes in the sample space of the experiment are assumed to be **equally likely**. In situations of this kind, we make the following definition:

DEFINITION

Let the sample space S contain k equally likely outcomes. Then each outcome in S is assigned probability $1/k$.

An example is the rolling of a single die. The die may come to rest with any face on top, and it seems reasonable to believe that the six faces are equally likely. Therefore, the probability that any one face will come to rest on top is $\frac{1}{6}$. In particular, the probability of rolling 6 is $\frac{1}{6} \doteq 0.1667$, verifying the experimental result on page 63.

Similarly, if we select one city from the USNEWS file at random (the phrase *at random* means "in such a way that all are equally likely"), the probability that Denver is chosen is $\frac{1}{74} = 0.0135$, since there are 74 cities in the file.

We are also interested in the probabilities of *events* in sample spaces in which the outcomes are equally likely. The definition of the probability of such an event is a reasonable extension of the previous definition:

DEFINITION

Let A be an event in S, a sample space in which the outcomes are equally likely. Then

$$P(A) = \frac{\text{number of outcomes in } A}{\text{number of outcomes in } S} \, .$$

If we roll a single die, the probability of rolling 5 or 6 is $\frac{2}{6} = \frac{1}{3}$; if we select a city from the USNEWS file, the probability that it is an eastern city is $\frac{27}{74}$ = 0.365, since there are 27 eastern cities in the file.

Putting Events Together—Union, Intersection, and Complement

We know that given several sets, we can apply the set operations of union, intersection, and complement to them to produce other sets. Events are subsets of the sample space S, so we can apply set operations to events. What kinds of things emerge?

The *union* of two sets is the set that contains all the elements which are in either or both of the original sets. The union of two events A and B contains all the outcomes which are in A, in B, or in both. Thus:

DEFINITION

If A and B are events in S, the event $A \cup B$, read "A union B" or "A or B," occurs when A occurs *or* B occurs, or both.

Similarly, the *intersection* of two sets contains the elements which are simultaneously in both. The intersection of two events contains the outcomes in both events:

DEFINITION

If A and B are events in S, the event $A \cap B$, read "A intersect B" or "A and B," occurs when both A *and* B occur.

Finally, the *complement* of a set is the set of all elements not in the set. The complement of an event is composed of all the outcomes in the sample space which are not in the event:

DEFINITION

If A is an event in S, the event A', read "A complement" or "not A," occurs when A does *not* occur.

Consider the experiment of drawing one card at random from an ordinary deck of 52 cards, with A the event that we draw a club, and B the event that we draw a face card (jack, queen, or king).

Then $A \cup B$ is the event that we draw either a club *or* a face card; $A \cap B$ occurs if the card drawn is both a club *and* a face card; and A' is the event that the card is a diamond, heart, or spade—*not* a club.

Consider A and the event C that a diamond is drawn. It is impossible for A and C to never occur simultaneously, since they have no outcomes in common. In such a case, we say that A and C are **mutually exclusive** and write $A \cap C = \varnothing$.

Two useful results from set theory are De Morgan's Laws, which relate the operations of union, intersection, and complement:

$$(A \cap B)' = A' \cup B'$$

$$(A \cup B)' = A' \cap B'$$

For events A and B in a sample space S, consider the interpretations of these expressions. For example, $(A \cup B)'$ is the event that neither A nor B occurs.

Venn Diagrams

Sets and their relationships can be depicted schematically with Venn diagrams, in which the interiors of circles represent the elements of the sets. Events and their interactions are often illustrated with Venn diagrams; circles represent events, and the enclosing rectangle corresponds to the sample space S.

The diagrams of Figure 5.1 illustrate (a) $A \cap B$, (b) $A \cup B$, and (c) A'.

FIGURE 5.1.   These Venn diagrams illustrate (a) $A \cap B$, (b) $A \cup B$, and (c) A'.

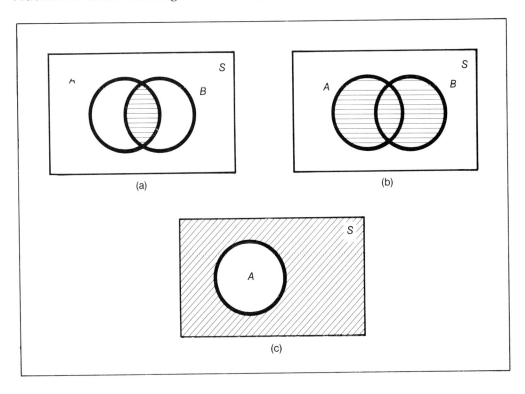

THE AXIOMS OF PROBABILITY

We have discussed the idea of probability, and have defined a method for assigning probabilities to events in experiments where the outcomes in the sample space are equally likely. We have also outlined the results of applying set theoretical operations to events. It is in investigating the probabilities of events made with the set operations that we begin to build the mathematical structure of probability theory. A', $A \cup B$, and $A \cap B$ are events; what are their probabilities?

The construction of a mathematical system begins with a statement of **axioms**, assumptions from which the structure will be derived. In probability, we state three axioms.

Let A be an event in the sample space S. Then:

Axiom 1. $P(A) \geq 0$.
Axiom 2. $P(S) = 1$.
Axiom 3. If B is another event in S, with $A \cap B = \varnothing$,
then $P(A \cup B) = P(A) + P(B)$.

These are reasonable assumptions. It would not make sense to assign an event a negative probability, and since S contains all the outcomes for the experiment and therefore must occur, it has probability 1. To see the validity of Axiom 3, consider again selecting a city at random from the USNEWS file, with C the event that we choose a central city, and W the event that the city is western. $P(C) = \frac{30}{74} = .405$, $P(W) = \frac{17}{74} = 0.230$, and $C \cap W = \varnothing$. Then $P(C \cup W) = \frac{47}{74} = 0.635 = P(C) + P(W)$.

Theorems Derived from the Axioms ⎯⎯⎯⎯⎯⎯⎯⎯⎯⎯

From the axioms of probability, we can develop theorems that tell us more about the probabilities of events in S.

THEOREM 5.1

For any event $A \subseteq S$, $P(A) = 1 - P(A')$.

Proof

$S = A \cup A'$, and $A \cap A' = \varnothing$.

By Axiom 3,

$$P(S) = P(A) + P(A').$$

But $P(S) = 1$, so $1 = P(A) + P(A')$;

Then $P(A) = 1 - P(A')$.

This result will be useful in finding the probabilities of complicated events whose complements can be more easily examined.

THEOREM 5.2

For any event $A \subseteq S$, $P(A) \leq 1$.

Proof

From Theorem 5.1,

$$P(A) = 1 - P(A').$$

By Axiom 1, $P(A') \geq 0$, so $P(A) \leq 1$.

This result verifies a statement which is intuitively reasonable—that the probability of an event cannot exceed 1. For any event A, $0 \leq P(A) \leq 1$.

THEOREM 5.3

$P(\emptyset) = 0$.

Proof

Let $A = \emptyset$ in the statement of Theorem 5.1.

Then $P(\emptyset) = 1 - P(\emptyset')$.

But $\emptyset' = S$, so $P(\emptyset) = 1 - P(S) = 1 - 1 = 0$.

This again is a reasonable result; the null event contains no outcomes, therefore it cannot occur, and has probability 0.

THEOREM 5.4

For any events A and B in S, $P(A \cup B) = P(A) + P(B) - P(A \cap B)$.

Proof

$A \cup B = A \cup (A' \cap B)$, and $A \cap (A' \cap B) = \emptyset$, so by Axiom 3,

$P(A \cup B) = P(A) + P(A' \cap B)$.   (*)

Also, $B = (A \cap B) \cup (A' \cap B)$ with $(A \cap B) \cap (A' \cap B) = \emptyset$,

so $P(B) = P(A \cap B) + P(A' \cap B)$.

Subtracting $P(A \cap B)$ from both sides of this equation,

$P(B) - P(A \cap B) = P(A' \cap B)$.

Substituting into (*) above, we obtain

$P(A \cup B) = P(A) + P(B) - P(A \cap B)$.

Theorem 5.4 is an important one, and more subtle than the previous three. To visualize what is happening, it is useful to turn to Venn diagrams in which the *area* of the part of the diagram that represents an event corresponds to the *probability* of the event. The entire area $A \cup B$ is almost the area of A plus the area of B (Figure 5.2), but this would include the area $A \cap B$ *twice*. To avoid this situation, we subtract that area once. Area corresponds to probability, so $P(A \cup B) = P(A) + P(B) - P(A \cap B)$.

(If A and B are mutually exclusive, $A \cap B = \varnothing$ and $P(A \cap B) = 0$; we get a restatement of Axiom 3.)

As an example of this result, consider again choosing one city from the USNEWS file at random. Let W be the event that the city chosen is in the West, and let U be the event that unemployment in that city is 5.0% or greater. Then $P(W) = \frac{17}{74} \doteq 0.230$, and $P(U) = \frac{46}{74} \doteq 0.622$. There are 12 western cities with unemployment 5% or greater, so $P(W \cap U) = \frac{12}{74} \doteq 0.162$. Then the probability that the city chosen is either western *or* has unemployment greater than 5.0% is:

$$P(W \cup U) = P(W) + P(U) - P(W \cap U)$$
$$= 0.230 + 0.622 - 0.162$$
$$= 0.690.$$

FIGURE 5.2.   **Venn diagram illustrating Theorem 5.4.**

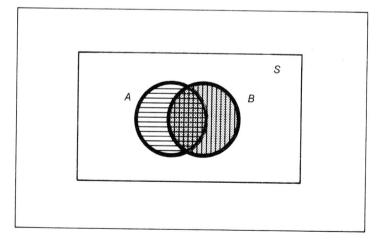

Counting Techniques—Permutations and Combinations

In the examples so far, we have used experiments whose sample spaces contained equally likely outcomes, and we have assigned probabilities by counting outcomes in events and in sample spaces. It might seem that this technique is limited to simple situations, but we can increase its usefulness by extending our ability to count.

Suppose we decide to classify the cities in the USNEWS file into 3 regions, 3 levels of unemployment (low to high), and 5 levels of growth, as measured by construction activity (very low to very high). If by "different" we mean "different in any detail," in how many different ways can a city be classified?

Given the number of alternatives at each step of the classification, it seems reasonable that there are $3 \times 3 \times 5 = 45$ possible city classifications.

This is an example of the **multiplication principle:**

DEFINITION

If a selection consists of k steps, each with n_i alternatives ($i = 1,2,\ldots,k$), then the entire selection can be made in $n_1 \times \ldots \times n_k$ different ways.

If we extend the classification scheme of the example above to include three categories of change in nonfarm employment and four categories of workers' income, then a city could be classified in $3 \times 3 \times 5 \times 3 \times 4 = 540$ different ways.

Similarly, consider how many different nonsense sequences of four letters can be made from the letters A, B, C, D, and E, if letters can be repeated. (These nonsense words would be things like ABAD, CACB, etc.) We assemble these words by making a four-step selection, and at each step, we have five alternatives. Thus, there are

$$5 \times 5 \times 5 \times 5 = 625$$

possible nonsense words. We can think of this as selecting a letter four times from a hat containing five letters, each time replacing the letter chosen.

If we are not allowed to repeat letters, and if we select without replacement, the number of remaining alternatives decreases by 1 at each step of the selection, so there are only

$$5 \times 4 \times 3 \times 2 = 120$$

possible words.

In general, consider the number of ways r objects might be selected *in order* from n objects. For the first selection, we have n alternatives, for the second selection, since one object has been used, $n-1$, and so on. At the last selection, there remain $n-r+1$ objects from which to choose (after all r objects have been selected and lined up, there remain $n-r$ unchosen objects), so the number of such arrangements is

$$n \times (n-1) \times (n-2) \times \ldots \times (n-r+1),$$

and they are given a name.

Each of these ordered arrangements is called a **permutation** of n objects taken r at a time.

For example, of the 17 cities in the western region of the United States listed in the USNEWS file, you plan to visit 5. If the order in which you visit the cities matters, then each possible trip is a permutation of the 17 cities taken 5 at a time, and there are

$$17 \times 16 \times 15 \times 14 \times 13 = 742,560$$

such permutations.

Expressions and formulas involving products like these can be written more efficiently using this notation:

For any positive integer n, the product $n \times (n-1) \times (n-2) \times \ldots \times 2 \times 1$ is called n **factorial**, and is written $n!$. Also, $0! = 1$.

For example, $5! = 5 \times 4 \times 3 \times 2 \times 1 = 120$. This notation lets us write the number of permutations of n objects taken r at a time as

$$\frac{n!}{(n-r)!} \quad \text{(see Exercise 20).}$$

At this point, reconsider the trip described above, and assume that the order in which the cities are visited does *not* matter. When order *does* matter, there are

$$\frac{17!}{(17-5)!}$$

possible trips. Each unordered group of 5 cities could be ordered in $5! = 120$ ways, and these have all been counted separately by the number

$$\frac{17!}{(17-5)!}$$

The number of permutations has counted each group of 5 cities $5!$ times, so that the number of different *unordered* groups of 5 cities is

$$\frac{17!}{(17-5)!} \Big/ 5! = \frac{17!}{(17-5)!5!} = 6188.$$

This brings us to the following definition:

DEFINITION

The number of *unordered* groups of r objects that can be selected out of n objects is

$$\frac{n!}{(n-r)!r!} \text{ , often written } \binom{n}{r}$$

and sometimes read "n choose r." Each of these unordered groups is called a **combination** of n objects taken r at a time.

A classic example of this concept comes from card playing: How many 5-card poker hands are possible from an ordinary deck of 52 cards? The order of cards in a hand does not matter, so each hand is a combination of 5 cards selected from 52 cards. The total number of such hands is

$$\binom{52}{5} = \frac{52!}{(52-5)!5!} = \frac{52 \times 51 \times 50 \times 49 \times 48}{5 \times 4 \times 3 \times 2 \times 1} = 2{,}598{,}960.$$

Counting Techniques and Probability

We can use these counting techniques to expand the range of events to which we can assign probabilities, as in this example.

We select 3 cities at random from those in the USNEWS file. What is the probability that all three are eastern?

Each possible group of 3 cities is one outcome in the sample space of this experiment, and there are

$$\binom{74}{3} = \frac{74!}{(74-3)!3!} = 64{,}824 \text{ such groups.}$$

Of these,

$$\binom{27}{3} = \frac{27!}{(27-3)!3!} = 2{,}925$$

are composed entirely of eastern cities, and these are the outcomes in the event whose probability we seek. Therefore, the probability that all 3 cities are in the East is:

$$\frac{\binom{27}{3}}{\binom{74}{3}} = \frac{2{,}925}{64{,}825} = 0.0451.$$

In later chapters we will see other applications of counting techniques to questions of probability.

CONDITIONAL PROBABILITIES

In many random experiments, knowledge of the occurrence or nonoccurrence of one event may change our estimate of the probability of another event. We construct an example of such a situation from the USNEWS file, and introduce a new SPSS task.

Our experiment will be the random selection of one city from the 74 in our study, and we will consider the interaction of events based on the variables $X1$ and $X3$, region and unemployment. There are many distinct values for $X3$, however, and our illustration will be clearer if we group these values into *low, medium,* and *high* classes. This is done with the RECODE command, as shown in the following program, which produces frequency distributions and a histogram of the grouped values of $X3$.

```
? 10 GET FILE;USNEWS2
? 20 RECODE;X3(LO THRU 4.99 = 4)
? 20.010 (5.00 THRU 6.99 = 6)(7.00 THRU 15.00 = 8)
? 30 VALUE LABELS;X3 (4) LOW (6) MEDIUM (8) HIGH
? 40 FREQUENCIES;GENERAL = X3
? 50 OPTIONS;8
```

Execution of the program produces this output; note the appearances of the value labels specified in statement 30 above.

X3   UNEMPLOYMENT

CATEGORY LABEL	CODE	ABSOLUTE FREQ	RELATIVE FREQ (PCT)	ADJUSTED FREQ (PCT)	CUM FREQ (PCT)
LOW	4.	24	32.4	34.3	34.3
MEDIUM	6.	34	45.9	48.6	82.9
HIGH	8.	12	16.2	17.1	100.0
	1000.	4	5.4	MISSING	
	TOTAL	74	100.0	100.0	

X3   UNEMPLOYMENT

```
CODE
     |
  4. ************************ (            24)
     |   LOW
     |
  6. ********************************* (            34)
     |   MEDIUM
     |
  8. ************* (            12)
     |   HIGH
```

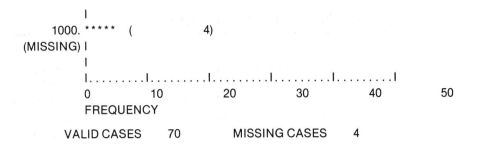

```
           |
      1000. *****   (              4)
 (MISSING) |
           |
           |........|........|........|........|........|
           0        10       20       30       40       50
           FREQUENCY

         VALID CASES    70      MISSING CASES    4
```

If L is the event that the city chosen has *low* unemployment, we can see that $P(L) = \frac{24}{74} \doteq 32.4\% = 0.324$.

Suppose that C is the event that the chosen city is in the central region, and we know that C occurred. We can use this information to revise the probability of L by finding how many central cities have low unemployment.

This kind of analysis—finding how many cases in a category related to one variable fall into a category based on another variable—is called *cross-tabulation*, and is performed by the SPSS subprogram CROSSTABS, whose task statement is this:

CROSSTABS   TABLES = variable list BY variable list.

This command will cause a cross-tabulation to be produced pairing each variable in the first list with each variable in the second list. In the following program, the values of $X3$ are grouped as described above, and a cross-tabulation of $X3$ by $X1$ is produced.

```
10.     GET FILE
10.005  USNEWS2
20.     RECODE
20.005  X3(LO THRU 4.99 = 4)
20.010  (5.00 THRU 6.99 = 6)(7.00 THRU 15.00 = 8)
30.     VALUE LABELS
30.005  X3 (4) LOW (6) MEDIUM (8) HIGH
40.     CROSSTABS
40.005  TABLES = X3 BY X1
```

This is Table 5.2 created by the program. Again, note the value labels.

TABLE 5.2

X3 By X1	UNEMPLOYMENT CITY REGION			
	X1			
COUNT ROW PCT COL PCT TOT PCT	EASTERN 1.	CENTRAL 2.	WESTERN 3.	ROW TOTAL

X3

	4.	8	11	5	24
LOW		33.3	45.8	20.8	34.3
		32.0	39.3	29.4	
		11.4	15.7	7.1	
	6.	12	14	8	34
MEDIUM		35.3	41.2	23.5	48.6
		48.0	50.0	47.1	
		17.1	20.0	11.4	
	8.	5	3	4	12
HIGH		41.7	25.0	33.3	17.1
		20.0	10.7	23.5	
		7.1	4.3	5.7	
COLUMN	25	28	17	70
TOTAL	35.7	40.0	24.3	100.0

MISSING OBSERVATIONS —   4

The key printed in the upper left describes the contents of each cell of Table 5.2. For example, in the upper-left cell we see that 8 cities in our file are in the East and have low unemployment; 33.3% of the low-unemployment cities are in the East; 32.0% of eastern cities have low unemployment, and the 8 cases in this cell of the table are 11.4% of all the cases in the file with non-missing values of $X1$ and $X3$. Row and column total counts and percentages are printed in the rightmost column and at the bottom of the table, respectively. (There are statistics available with CROSSTABS which will be discussed in later chapters.)

We can now reconsider the interaction of the events L and C. We have seen that $P(L) = \frac{24}{70} = 0.343$, but if we know that C occurred, we know that the experiment resulted in one of the 28 outcomes in C. Of these, 11 are also in L (that is, 11 outcomes are in $C \cap L$), so, if C has occurred, the probability of L is:

$$\frac{\text{the number of outcomes in } C \cap L}{\text{the number of outcomes in } C} = \frac{11}{28} \doteq 0.393.$$

The probability of L given that C has occurred is different from $P(L)$. Thus the following definition:

DEFINITION

Let A and B be events in S, with $P(B) > 0$. The **conditional probability** of A given B, written $P(A|B)$, is

$$\frac{P(A \cap B)}{P(B)}.$$

In the example, we found $P(L|C)$ the conditional probability of L given C. Note that

$$P(L|C) = \frac{P(L \cap C)}{P(C)} = \frac{0.157}{0.400} = 0.393,$$

and that all these values can be found, as percentages, in the cross-tabulation table.

Probabilities of the form $P(L \cap C)$ are called **joint probabilities**, and these are given in Table 5.2. For example, $P(M \cap W) = 0.114 = 11.4\%$. And in the margins of the table are **marginal probabilities**, of the form $P(L)$ and $P(C)$. For example, $P(M) = 0.486 = 48.6\%$. A conditional probability is a joint probability divided by a marginal probability.

Another example is to suppose that A and B are events in S with $P(A) = 0.46$ and $P(A \cap B) = 0.14$. Then

$$P(B|A) = \frac{P(A \cap B)}{P(A)} = \frac{0.14}{0.46} \doteq 0.304.$$

Independent Events

Consider the simple experiment of rolling two dice. Let A be the event that 1 or 2 dots appear on the first die, and let B be the event that an even number of dots appear on the second die. Then

$$P(A) = \frac{12}{36} = \frac{1}{3}; \ P(B) = \frac{18}{36} = \frac{1}{2}; \ \text{and} \ P(A \cap B) = \frac{6}{36} = \frac{1}{6}.$$

(See Exercise 4 for more discussion of this experiment.)

With this information, we can compute conditional probabilities:

$$P(A|B) = \frac{P(A \cap B)}{P(B)} = \frac{1/6}{1/2} = \frac{1}{3} = P(A).$$

$$P(B|A) = \frac{P(A \cap B)}{P(A)} = \frac{1/6}{1/3} = \frac{1}{2} = P(B).$$

Note that the conditional and unconditional probabilities are the same; knowing that one of these events has occurred does not affect the probability of the other. In fact, the above two statements are equivalent:

$$P(A|B) = P(A) \leftrightarrow$$

$$\frac{P(A \cap B)}{P(B)} = P(A) \leftrightarrow$$

$$P(A \cap B) = P(A) \times P(B) \leftrightarrow$$

$$\frac{P(A \cap B)}{P(A)} = P(B) \leftrightarrow$$

$$P(B|A) = P(B)$$

Both statements are equivalent to the third statement above, from which we obtain this definition:

DEFINITION

Let A and B be events in S. A and B are **independent** if $P\ (A \cap B) = P(A) \times P(B)$.

Independent events do not interact in the sense that the occurrence of one has no effect on the probability of the other. This is true of the events in the dice example, where

$$P(A \cap B) = \frac{1}{6} = \frac{1}{3} \times \frac{1}{2} = P(A) \times P(B).$$

Again, an example: C and D are events in S, with $P(C)=0.6$, $P(D)=0.5$, and $P(C \cup D)=0.8$. Are C and D independent?

We must find $P(C \cap D)$, and we know that $P(C \cup D) = P(C) + P(D) - P(C \cap D)$. Therefore, $0.8 = 0.6 + 0.5 - P(C \cap D)$, and $P(C \cap D) = 0.3$. Then $P(C \cap D) = 0.3 = 0.6 \times 0.5 = P(C) \times P(D)$, so the events C and D are independent.

Random Variables

Consider the experiment of flipping three coins and counting the number of heads that appear. Equally likely outcomes for this experiment are triples of H's and T's, and from them we can see that these probabilities hold:

$$S = \{HHH,\ HHT, \qquad P(\text{no heads}) = \frac{1}{8} \qquad P(1\ \text{head}) = \frac{3}{8}$$
$$HTH,\ HTT,$$
$$THH,\ THT, \qquad P(2\ \text{heads}) = \frac{3}{8} \qquad P(3\ \text{heads}) = \frac{1}{8}$$
$$TTH,\ TTT\}$$

What we have done is associate a number (0, 1, 2, or 3) with each outcome in the sample space (HHH has "3" attached to it, and so on); when we perform the experiment, an outcome will occur, and with it, the number of that outcome. This is an example of a **random variable**:

DEFINITION

A **random variable** is a numerical quantity whose value is determined by a random experiment.

That is, a random variable is a function from S to the real numbers that attaches a number to every outcome. When an outcome occurs, the random variable takes on the value attached to that outcome. Here are some examples.

1. In the above experiment, let the random variable X be the number of heads.
2. Roll 2 dice; let the random variable Y be the total number of dots on the upper faces.
3. Inspect 20 items out of a production lot of 1000; let Z be the number that fail the inspection.
4. Select a city at random from the USNEWS file; let X be its unemployment.
5. Flip a coin until the first head appears; let Y be the number of trials before the first head.

In each of these examples, the random variable can take on only certain exact values. Such random variables are called *discrete*, and are defined in this way:

DEFINITION

A random variable is **discrete** if the set of its possible values (called the *range* or *space* of the random variable) is finite or countably infinite.

In the Example 5, the range of the random variable is 0,1,2,3, . . . ,and is countably infinite.

DEFINITION

A random variable is **continuous** if it can assume any real value in an interval (that is, if its range is an interval).

These are examples of continuous random variables.

1. Let X be the height of a randomly chosen person.
2. Let Y be the time required to perform some task.
3. Let Z be the distance required to stop a car from 50 mph.

In these experiments, the random variables can take on any value within an interval of values (though in practice we might approximate them with discrete random variables).

For now, we will consider discrete random variables, then extend the concepts we develop to the continuous case.

Probability Distributions

We have seen that frequency distributions are tabular displays of the values of a variable which help to illustrate the shape of the distribution, its central tendency, variability, and other characteristics. Since probabilities can be interpreted as relative frequencies when the number of trials is large, it is natural to use the idea of a distribution to describe the allocation of probability among the possible values of a random variable.

Return to the experiment in which we flip three coins (page 77), and let the random variable X be the number of heads which appear. The range of X is $\{0,1,2,3\}$, and we can give the probabilities associated with each possible value of X:

$$P(X=0) = \frac{1}{8} \qquad P(X=2) = \frac{3}{8}$$

$$P(X=1) = \frac{3}{8} \qquad P(X=3) = \frac{1}{8}$$

This display above resembles a relative frequency distribution, and is one way of describing the **probability distribution** of X. It is also an example of this definition:

DEFINITION

Let X be a discrete random variable. The **probability function** f associated with X is defined to be:

$$f(x) = P(X=x).*$$

In the example

$$f(0) = P(X=0) = \frac{1}{8}, \ f(1) = \frac{3}{8}, \ f(2) = \frac{3}{8}, \text{ and } f(3) = \frac{1}{8};$$

$f(x) = 0$ for all other values of x. This can be most economically written as

$$f(x) = \begin{cases} \frac{1}{8}, & x=0,3, \\[2mm] \frac{3}{8}, & x=1,2, \\[2mm] 0, & \text{all other } x, \end{cases}$$

and is another way of describing the probability distribution of X.

Two observations should be made about discrete probability functions:

1. The value of a probability function is a probability ($f(x) = P(X=x)$), so for any value of x, $0 \leq f(x) \leq 1$.

2. A random variable divides the sample space into distinct events, so the sum of the values of $f(x)$ over all possible values of x must be one; $\sum_{x} f(x) = 1$.

---

*f is also called the *probability density function* (PDF), or *probability mass function* (PMF) of X.

FIGURE 5.3. The random variable X divides the sample space into separate events.

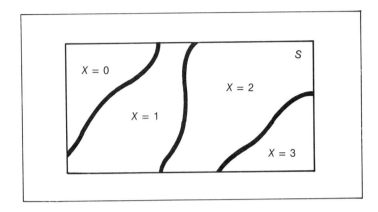

Note that both of these conditions are met by the probability function in the example (Figure 5.3).

Just as histogram can be constructed corresponding to a frequency distribution, so we can draw the graph of a probability function. The probability associated with each value in the range of the random variable is represented by a vertical line above that value, as shown in Figure 5.4.

FIGURE 5.4. The probability function $f(x)$ of X.

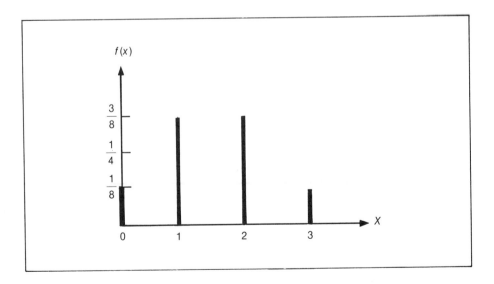

Finally, another function can be defined which corresponds to the *cumulative* relative frequency distribution:

DEFINITION

Let X be a discrete random variable. The *cumulative density function F* associated with X is defined to be $F(x) = P(X \leq x) = \sum_{y \leq x} P(X = y)$.

That is, while $f(x)$ gives the probability that the random variable X will take on the value x, $F(x)$ gives the probability that X will take on a value less than or equal to x.

This function can be graphed as in Figure 5.5, which shows the cumulative density function and its graph for our example. Note that each step in the value of F is equal to the value of f at that step.

Expected Value

Just as we can indicate the center of a distribution of values by its mean, so we can indicate the center of the probability distribution of a random variable. This center is an average of the possible values of the random variable, weighted by their respective probabilities, and corresponds to the center of the graph of the probability function. More precisely,

---

FIGURE 5.5.   The cumulative density function.

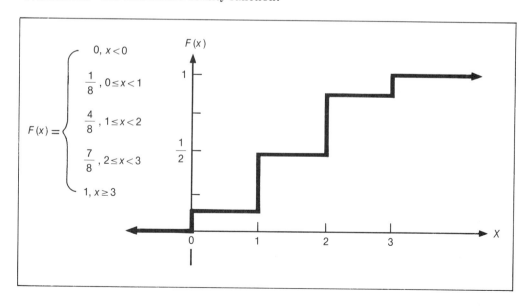

DEFINITION

If X is a random variable with probability function f, the **expected value** or *mean* of X, indicated by $E(X)$ or μ, is defined to be

$$\sum_x (x \cdot f(x)) = \sum_x (x \cdot P(X=x)),$$

where the sum is taken over all possible values of X.

In the example developed in the previous section, we found a probability function and graph. (Refer to Figure 5.4.) The expected value or mean of X is

$$E(X) = \mu = \sum_x x \cdot f(x) = \left(0 \times \frac{1}{8}\right) + \left(1 \times \frac{3}{8}\right) + \left(2 \times \frac{3}{8}\right) + \left(3 \times \frac{1}{8}\right) = \frac{3}{2}.$$

Note that this value corresponds to the center of the distribution as shown by Figure 5.4, and that there is one term in the sum for each value in the range of X; each term is the product of a possible value of the random variable with the probability that X takes on that value.

The random variable of the example can never take on the value $\frac{3}{2}$; we can never "expect" this to happen. Rather, expected value can be thought of as the *average* (mean) value of the random variable over many repetitions of the experiment. If we repeated the tossing of the three coins 1,000 or 10,000 times, the mean of the values of X from all the trials would be near $\frac{3}{2}$.

Several results about expected value should be established for later use.

1. For any constant c, $E(c) = c$.

   That is, if a random variable always takes on the same value, that value is the expected value of the random variable. Though this result may seem trivial, it will be useful in simplifying complicated expressions which involve expected values.

2. For any constant c, $E(cX) = cE(X)$.

   If we create a new random variable from another by multiplying by a constant, the mean of the new random variable is the constant times the mean of the old.

3. For any random variables X and Y, $E(X+Y) = E(X) + E(Y)$.

   The expected value of the sum of two (or more) random variables is the sum of their expected values.

VARIANCE

Having established a measure of the middle of a probability distribution, we then inquire into its *variability*. Are the possible values of the random variable concentrated around the mean, or are they widely scattered? We answer this question as we did when we investigated the variability of a frequency distribution, by using the squared deviations of the values from their mean:

Let X be a random variable with probability function f and mean μ. The *variance* of X, indicated by Var(X) or σ^2, is

$$E\left[(X-\mu)^2\right] = \sum_x (x-\mu)^2 f(x).$$

Notice how similar this is to the definition of the variance of a sample or population. The definition of variance leads to another familiar definition:

The *standard deviation* of X, indicated by σ, is $\sqrt{\sigma^2} = \sqrt{\text{Var}(X)}$.

Again, using the example of the three coins, with X the number of heads appearing when the coins are tossed,

$$\sigma^2 = \text{Var}(X) = \left(0-\frac{3}{2}\right)^2 \times \frac{1}{8} + \left(1-\frac{3}{2}\right)^2 \times \frac{3}{8} + \left(2-\frac{3}{2}\right)^2 \times \frac{3}{8} + \left(3-\frac{3}{2}\right)^2 \times \frac{1}{8}$$

$$= \frac{24}{32} = \frac{3}{4} = 0.75, \text{ and}$$

$$\sigma = \sqrt{\sigma^2} = \sqrt{0.75} = 0.866.$$

As with population and sample variances, we can develop a simpler formula for Var (X), using the properties of expected value given in the last section:

$$\begin{aligned}
\sigma^2 = \text{Var}(x) &= E[(X-\mu)^2] = E[X^2 - 2X\mu + \mu^2] \\
&= E(X^2) + E(-2X\mu) + E(\mu^2) \\
&= E(X^2) - 2\mu E(X) + \mu^2 \\
&= E(X^2) - 2\mu^2 + \mu^2 \\
&= E(X^2) - \mu^2 \\
&= \sum_x (x^2 f(x)) - \mu^2
\end{aligned}$$

This formula is very similar to that developed in Exercise 7 of Chapter 4. We use it to find the variance of the random variable in the example of the three coins:

$$\sigma^2 = \text{Var}(X) = \left(0^2 \times \frac{1}{8}\right) + \left(1^2 \times \frac{3}{8}\right) + \left(2^2 \times \frac{3}{8}\right) + \left(3^2 \times \frac{1}{8}\right) - \left(\frac{3}{2}\right)^2$$

$$= \frac{24}{8} - \frac{9}{4} = \frac{3}{4} = 0.75 .$$

We can state two results about the variance of a random variable:

1. For any constant c, $\text{Var}(cX) = c^2 \text{Var}(X)$. If we multiply a random variable by a constant, the variance of the new random variable is the constant squared times the variance of the old.

2. For any constant c, $\text{Var}(c + X) = \text{Var}(X)$. Adding a constant to a random variable does not change its variance.

Independent Random Variables

We have seen that when two events A and B are independent, $P(A \cap B) = P(A) \times P(B)$, and the occurrence of one does not change the probability of the other. We can extend this concept to include random variables:

DEFINITION

The random variables X and Y are **independent** if, for all possible values of X and Y,

$$P(X = x \cap Y = y) = P(X = x) \times P(Y = y).$$

That is, any pair of events $(X = x)$ and $(Y = y)$ is independent; the value taken on by one of the random variables has no effect on the probability distribution of the other.

Rolling two dice provides us with an example of a pair of independent random variables. Let X be the number of dots on the upper face of the first die, and let Y be the number on the second. Since there is no physical connection between the two dice, it seems reasonable that these random variables should be independent, and a check of the 36 pairs of nontrivial events of the form $(X = x) \cap (Y = y)$ will show that this is so.

You may have noticed that in the previous section we said nothing about $\text{Var}(X + Y)$. We will make no statement about $\text{Var}(X + Y)$ in general, unless X and Y are independent. To do that, we must first prove this result:

THEOREM 5.5

If X and Y are independent random variables, then $E(XY) = E(X)E(Y)$.

Proof

$$E(XY) = \sum_{x,y} xy \, P(X = x \cap Y = y)$$

$$= \sum_{x,y} xy \, P(X = x) \, P(Y = y)$$

$$= \sum_{x,y} x \cdot P(X = x) \cdot y \cdot P(Y = y)$$

$$= \sum_{x,y} x \cdot f(x) \cdot y \cdot g(y)$$

$$= \sum_{x} \left[\sum_{y} x \cdot f(x) \cdot y \cdot g(y) \right]$$

$$= \sum_{x} \left[x \cdot f(x) \sum_{y} y \cdot g(y) \right]$$

$$= \left[\sum_{y} y \cdot g(y) \right] \left[\sum_{x} x \cdot f(x) \right]$$

$$= E(Y) \cdot E(X)$$

Using this result, we can prove the following theorem.

THEOREM 5.6

If X and Y are independent random variables, then $\text{Var}(X + Y) = \text{Var}(X) + \text{Var}(Y)$.

Proof

$$\text{Var}(X + Y) = E[(X + Y)^2] - [E(X + Y)]^2$$

$$= E(X^2 + 2XY + Y^2) - [E(X) + E(Y)]^2$$

$$= E(X^2) + E(2 \times Y) + E(Y^2) - [E(X)^2 + 2E(X)E(Y) + E(Y)^2]$$

$$= [E(X^2) - E(X)^2] + [E(Y^2) - E(Y)^2]$$

$$\quad + [2E(X \times Y) - 2E(X)E(Y)]$$

$$= \text{Var}(X) + \text{Var}(Y) + 0.$$

It does not follow from this result that $\text{Var}(X - Y) = \text{Var}(X) - \text{Var}(Y)$. In fact, if X and Y are independent,

$$\text{Var}(X - Y) = \text{Var}(X + (-1Y))$$

$$= \text{Var}(X) + \text{Var}(-1Y)$$

$$= \text{Var}(X) + (-1)^2 \text{Var}(Y)$$

$$= \text{Var}(X) + \text{Var}(Y).$$

Also,

$$\sigma(X + Y) = [\text{Var}(X) + \text{Var}(Y)]^{1/2} \neq \sigma(X) + \sigma(Y), \text{ and}$$

$$\sigma(X - Y) = [\text{Var}(X) + \text{Var}(Y)]^{1/2}.$$

EXERCISES

1. Describe sample spaces for these random experiments.

    a. Roll one die and observe the number of dots on its upper face.

    b. Measure the mileage of a car at a test track.

    c. Flip a coin until the first head appears.

    d. Select a person at random and measure his or her height.

    e. Choose a group of 5 people from your class.

    f. Administer an IQ test to a randomly chosen person.

2. a. In 550 repetitions of an experiment, the event A occurred 380 times. What does this suggest about $P(A)$?

    b. If $P(B) = 0.22$, how many times would you expect B to occur in 250 repetitions of the experiment? Must B occur this often?

3. Choose one card from a well-shuffled deck of 52 cards. Find the probability that the card chosen is:

    a. The 3 of diamonds.

    b. A club.

    c. A face card (jack, queen, or king)

    d. A face card and a club.

    e. A face card or a club.

4. We roll two dice. Describe a sample space for this experiment in which the outcomes are equally likely. (*Hint:* Assume that you can tell the two dice apart, and represent the outcomes as ordered pairs.)

5. Using the sample space described in Exercise 4, find the probabilities of these events.

    a. The two dice show a total of 7.

    b. The two dice show a total of 12.

    c. An even number of dots on the first die.

    d. 5 or 6 dots on either die.

6. We choose one city at random from the USNEWS file. A is the event that a western city is chosen; B is the event that unemployment in the city is less than 5%. Describe the following events.

    a. $A \cap B$

    b. $A \cup B$

    c. A'

    d. $A' \cup B'$

    e. $A \cap B'$

    f. $(A \cap B)'$

7. Let C be the event that tomorrow's weather is hot, while D is the event that it rains. Describe these events.

    a. $C \cap D$

    b. $C \cup D$

    c. D'

    d. $C \cap D'$

    e. $C' \cup D'$

    f. $(C \cup D)'$

8. Choose a city from the USNEWS file. E is the event that the city is eastern, A that it is western, and B that unemployment is less than 5% there. Which of these pairs of events are mutually exclusive?

    a. A and E

    b. B and E

    c. A and B

9. A and B are events in S. Indicate the portions of Venn diagrams corresponding to:

    a. A'

    b. $A \cup B$

    c. $A' \cap B$

    d. $(A \cap B)'$

    e. $A' \cup B$

10. Use Venn diagrams to verify De Morgan's laws.

11. A and B are events in S, with $P(A) = 0.25$, $P(B) = 0.55$, and $P(A \cap B) = 0.10$. Find the following probabilities.

    a. $P(A')$

    b. $P(A \cup B)$

    c. $P(A' \cap B')$ (*Hint:* Apply De Morgan's laws.)

    d. $P(A \cap B')$

    e. $P((A \cap B)')$

12. C and D are events in S, with $P(C) = 0.42$, $P(D) = 0.66$, and $P(C \cup D) = 0.84$. Find these probabilities.

    a. $P(C \cap D)$

    b. $P(C' \cup D')$

    c. $P(C' \cap D)$

13. E and F are mutually exclusive events in S, with $P(E) = 0.33$ and $P(F) = 0.45$. Find these probabilities.

    a. $P(E \cap F)$

    b. $P(E \cup F)$

    c. $P(E \cap F')$

    d. $P(E \cup F')$

14. Prove this theorem. If A and B are events in S, and $A \subseteq B$, then $P(A) \leq P(B)$.

15. We flip 4 coins. What is the probability that at least one head appears? (*Hint*: Use Theorem 5.1.)

16. In a marketing survey, persons interviewed are classified according to 5 education categories, 7 job categories, 4 income categories, and 3 marital status categories. In how many ways can a person be classified?

17. How many 4-letter patterns can be made from the letters a,b,c,d,e,f, and g

   a. if letters can be repeated?

   b. if letters cannot be repeated?

18. Five brands of beer are ranked in order by a taste tester.

   a. How many different orders are possible?

   b. In how many of these is Splatz beer ranked last?

19. How many different ways can a group of 10 people be divided into

   a. two committees of 5 each?

   b. a committee of 4 and another of 6?

20. How many poker hands (5 cards) contain no hearts?

21. You plan a trip to visit 7 cities; in how many orders might this be done?

22. Show that $n \times (n-1) \times (n-2) \times . \times (n-r+1) = \dfrac{n!}{(n-r)!}$! (*Hint:* Write out the factorials, and see what simplifies.)

23. How many poker hands contain two hearts? (*Hint*: Choose the hearts, then the other three cards. Use the multiplication principle to connect the numbers of alternatives of the two steps.)

24. A group of 30 people consists of 18 women and 12 men. If 4 people are chosen at random from this group, what is the probability that

   a. all 4 chosen are women?

   b. at least one of those chosen is a woman?

25. Refer again to the cross-tabulation of $X3$ by $X1$ on p. 000. Use Table 5.2 to find these probabilities.

   a. $P(H)$

   b. $P(M)$

   c. $P(H \cap M)$

   d. $P(M|H)$

   e. $P(L|W)$

   f. $P(L \cup M|C)$

26. Choose a pair of variables from another data set, and generate a cross-tabulation with them. Group the values of one or both with a RECODE command or its equivalent (if necessary). Report some conditional probabilities based on the cross-tabulation.

27. We can restate the axioms of probability for conditional probability:
   Let A and B be events in S with $P(B) > 0$.
   Then:
   (i) $P(A \mid B) \geq 0$
   (ii) $P(B \mid B) = 1$
   (iii) If C is an event in S with $A \cap C = \varnothing$, then $P(A \cup C|B) = P(A|B) + P(C|B)$.

    a. Show that these restatements hold.

    b. Restate the four theorems of probability (Theorems 5.1–5.4) for conditional probability. If the axioms hold for conditional probability, what do we know about the restated theorems?

28. A and B are events in S, with $P(A) = 0.28$, $P(B) = 0.54$, and $P(A \cap B) = 0.14$. Find these probabilities:

    a. $P(A|B)$

    b. $P(B|A)$

    c. $P(A|B')$

    Use the conditional probability axioms and theorems stated in Exercise 27 to find

    d. $P(A'|B)$

    e. $P(C \cup A|B)$, if $C \cap A = \varnothing$ and $P(C|B) = 0.10$

29. At Acme Mail Order Specialties, 80% of all orders are shipped within 24 hours, and 50% are shipped within 24 hours and delivered within a week.

    a. What is the probability that an order shipped within 24 hours will arrive within a week?

    b. If 60% of all orders are delivered in a week, what is the probability that an order delivered in a week was shipped within 24 hours?

30. In August in a certain city, the probability of temperatures above 80° on any day is 0.85, while the probability of both high temperatures and thunderstorms is 0.25. If, on a given day, the temperature is certain to exceed 80°, what is the probability of thunderstorms?

31. E and F are mutually exclusive events in S, with $P(E) = 0.4$ and $P(F) = 0.3$. Find these probabilities.

    a. $P(E|F)$

    b. $P(E|F')$

    c. $P(E'|F)$

32. C and D are events in S, with $P(C) = 0.4$, $P(D) = 0.5$, and $P(C \cap D') = 0.2$. Show that C and D are or are not independent.

33. We roll 2 dice. A is the event that the first die shows 4 dots, and B is the event that the total of the two dice is 7.

    a. Are A and B independent? (Refer to Exercise 4.)

    b. If C is the event that the two dice show a total of 6, are A and C independent?

34. Consider the events $L, M, H, E, C,$ and W associated with the cross-tabulation of Table 5.2 on p. 74-75. Are any pairs of events among these independent?

35. Repeat Exercise 34 for the cross-tabulation generated in answer to Exercise 26.

36. Are there any circumstances in which mutually exclusive events will be independent?

37. If events A and B are independent, are A' and B? A and B'? A' and B'? Why?

38. Identify these random variables as discrete or continuous, and describe their ranges:

    a. A baseball game will be played. Let X be the total number of runs scored.

    b. Let Y be the time required for a sprinter to run 100 yards.

    c. Let Z be the number of jobs submitted to a computer center in a week.

39. Why, in practice, must we generally approximate continuous random variables with discrete ones?

40. We select a city at random from those listed in the USNEWS file. X is the value of $X3$, unemployment, as changed by the RECODE statement in the program on p. 73. Use the output of this program to construct the probability function of X and draw its graph. Construct the cumulative density function associated with X and draw its graph.

41. Repeat Exercise 40 for a variable chosen from another data file.

42. It is possible to define more than one random variable on an experiment. For example, if we roll a single die, let X be the number of dots on the upper face of the die; let Y be twice the number of dots; and let Z be 0 unless a 6 is rolled, in which case Z will be 1.

    a. Give the probability functions and their graphs for X, Y, and Z.

    b. Give the cumulative density functions and their graphs for X, Y, and Z.

43. Let X be a random variable with this probability function: $f(x) = x/6$, $x = 1,2,3$.

    a. Show that f is acceptable as a probability function.

    b. Find $E(X)$ and Var(x) (μ and σ^2).

44. Prove the following results about expected value for discrete random variables.

    a. $E(c) = c$, where c is a constant.

    b. $E(cX) = cE(X)$.

    c. $E(X+c) = E(X) + c$

45. Find the mean and variance of the random variable in Exercise 41.

46. Find the mean and variance of the random variable in Exercise 40.

47. Find the mean and variance of the random variables in Exercise 42. Find $E(X + Y + Z)$.

48. Let X be as in Exercise 42. If $Y = 3X - 4$, find $E(Y)$ and $\text{Var}(Y)$.

49. Y and Z are random variables with $E(Y) = 10$, $\text{Var}(Y) = 9$, $E(Z) = 15$, $\text{Var}(Z) = 16$. Find the following.

    a. $E(2Y + Z)$

    b. $\text{Var}(2Y)$ and $\sigma(2Y)$

    c. $E(Z/4)$

    d. $\text{Var}(Z/4)$ and $\sigma(Z/4)$

    e. $E(Y - 3Z)$

    f. $\text{Var}(Y + 5)$ and $\sigma(Y + 5)$

50. Consider the random variables in Exercise 49. Can we say anything about $\text{Var}(Y + Z)$? Why or why not?

51. We flip two coins. Let X be 1 if the first coin is heads, 0 otherwise; let Y be 1 if the second coin is heads, 0 otherwise.

    a. State the probability functions of X and Y.

    b. Find the means, variances, and standard deviations of X and Y.

    c. Show that X and Y are independent.

    d. Find $\text{Var}(X + Y)$ and $\sigma(X + Y)$.

    e. Find $\text{Var}(X - Y)$ and $\sigma(X - Y)$.

6

Probability
Distributions

INTRODUCTION

Probability distributions arise in a variety of situations in the real world, and many of these distributions fall into a relatively small number of categories. Investigations of these categories of distributions will provide powerful tools for analyzing the natural systems to which they correspond. We will be able to construct mathematical models of real-world systems incorporating these probability distributions, and to describe and predict the behavior of the systems through analysis of the models.

Problem Solving through Modeling

In this chapter we will use *deductive* logic, reasoning from the whole to a part, rather than the *inductive* logic we will apply in statistical inference. Probability distributions will be constructed as models of populations, and we will find probabilities of events within those populations. Corresponding to each population, the model will consist of a type of probability distribution with the mean and variance of the associated random variable.

Figure 6.1 illustrates the process of solving a real-world problem through the application of such a probability model. This process is composed of these steps.

1. From the real-world problem, identify the appropriate type of probability model.

2. Using values from the real-world problem, describe the parameters of the probability model.

3. Find a solution to the model, and examine its plausibility in terms of the model. If the solution is unreasonable, repeat this step.

4. From the solution of the model, find the corresponding solution to the original problem, and consider its plausibility. If the real-world solution is unreasonable, reconsider step 2 and continue.

There are two ways in which this process can fail. First, the solution of the model may be incorrect, an error which generally is easily found and easily corrected. Second, the model itself may not correspond sufficiently to the real-world problem to provide an adequate representation of its relevant features. This will render incorrect any solution based on the model. This type of error is more subtle than the first, and requires modifying or replacing the model.

FIGURE 6.1. Process of solving a real-world problem.

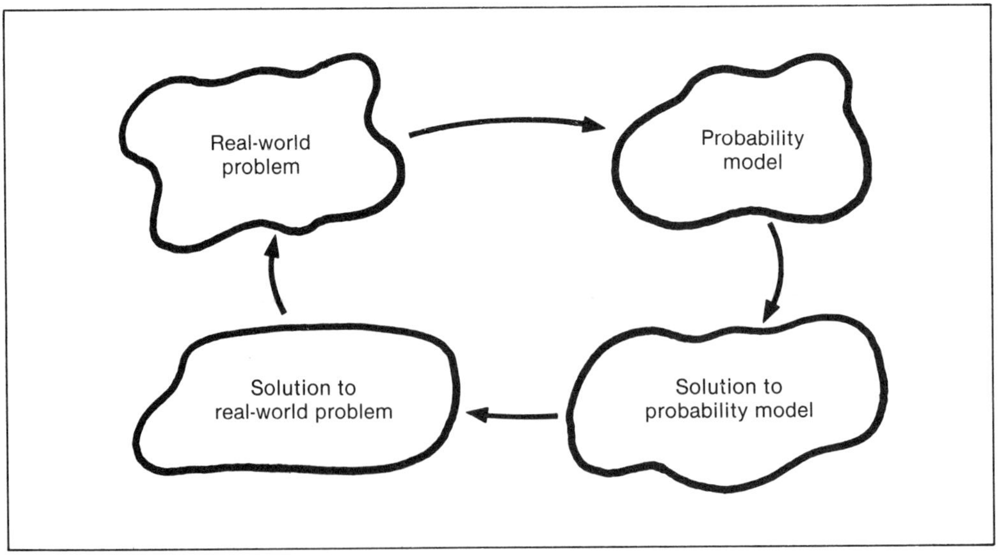

Important Discrete Distributions

As observed above, probability distributions arise in a variety of real-world situations, and many of these distributions fall into a small number of categories. We now examine several important discrete distributions.

THE UNIFORM DISCRETE DISTRIBUTION

We have seen that a probability distribution, however represented (probability function, graph, cumulative density function, etc.), consists of an experiment, a sample space, a random variable, and the associated functions. A *uniform discrete distribution* arises when every one of the k values in the range of the discrete random variable has associated with it probability $1/k$. That is,

---

DEFINITION

Let X be a discrete random variable. X has a **uniform discrete distribution** if, for each of the k evenly spaced values in the range of X, $f(x) = P(X = x) = 1/k$.

For example, let X be the number of dots on the upper face of a single die. (See Exercise 41, Chapter 5.) X may take on the values 1, 2, 3, 4, 5, and 6, each with probability $\frac{1}{6}$, and has the uniform discrete distribution shown in Figure 6.2.

FIGURE 6.2.   A uniform discrete distribution.

$$f(x) = \begin{cases} \dfrac{1}{6}, & x = 1,2,3,4,5,6 \\ 0, & \text{otherwise} \end{cases}$$

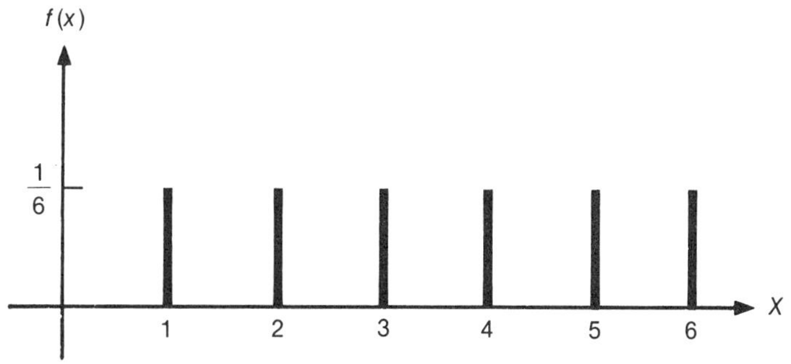

The graph of a uniform discrete distribution will always have this appearance: an evenly spaced set of lines of equal height.

From the probability function we construct the cumulative density function F and its graph (Figure 6.3), recalling that

$$f(x_0) = P(X \leq x_0) = \sum_{x \leq x_o} f(x).$$

The mean of example distribution is

$$\mu = E(X) = \sum_X x \cdot f(x) = \left(1 \times \frac{1}{6}\right) + \left(2 \times \frac{1}{6}\right) + \left(3 \times \frac{1}{6}\right) + \left(4 \times \frac{1}{6}\right) +$$

$$\left(5 \times \frac{1}{6}\right) + \left(6 \times \frac{1}{6}\right)$$

$$= \frac{21}{6} = 3.5.$$

FIGURE 6.3.   The cumulative density function F and its graph.

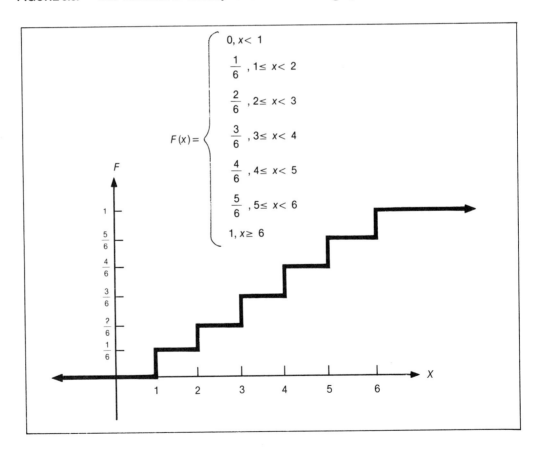

$$F(x) = \begin{cases} 0, & x < 1 \\ \dfrac{1}{6}, & 1 \leq x < 2 \\ \dfrac{2}{6}, & 2 \leq x < 3 \\ \dfrac{3}{6}, & 3 \leq x < 4 \\ \dfrac{4}{6}, & 4 \leq x < 5 \\ \dfrac{5}{6}, & 5 \leq x < 6 \\ 1, & x \geq 6 \end{cases}$$

This value is at the middle of the graph of the probability function, horizontally its geometric center.

Using the simplified procedure for finding variance,

$$\sigma^2 = \text{Var}(X) = \sum_x x^2 f(x) - \mu^2$$

$$= \left(1^2 \times \frac{1}{6}\right) + \left(2^2 \times \frac{1}{6}\right) + \left(3^2 \times \frac{1}{6}\right) + \left(4^2 \times \frac{1}{6}\right) + \left(5^2 \times \frac{1}{6}\right) +$$

$$\left(6^2 \times \frac{1}{6}\right) - 3.5^2$$

$$= \frac{91}{6} - 3.5^2 = \frac{105}{6} \doteq 2.917,$$

and the standard deviation is

$$\sigma = \sqrt{\sigma} \doteq \sqrt{2.917} \doteq 1.708.$$

THE BINOMIAL DISTRIBUTION

The distribution we consider next arises in a wide variety of real-world situations, but rather than approach it directly, we first define a very simple experiment and its associated distribution.

DEFINITION

A **Bernoulli experiment*** is one in which all the outcomes are in one of two mutually exclusive events, usually called "success" and "failure," with probability of *success p* and probability of *failure q* = 1 − *p*.

We perform a Bernoulli experiment when we roll a single die with success occurring if 6 dots appear on the upper face (Figure 6.4).

FIGURE 6.4.   A Bernoulli experiment.

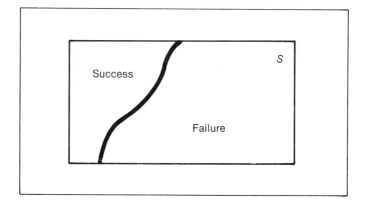

---

* Named for the seventeenth-century mathematician Jackob Bernoulli.

Associated with this simple experiment, we define an equally simple random variable: X is 1 if success occurs, 0 if failure; X is the number of successes in the Bernoulli experiment. The probability function and graph associated with X are shown in Figure 6.5.

The mean and variance of X are also easily found:

$$\mu = E(x) = \sum_{x} x \cdot f(x) = (1 \times p) + (0 \times q) = p$$

$$\sigma^2 = \mathrm{Var}(X) = \sum_{x} x^2 f(x) - \mu^2 = (1^2 \times p) + (0^2 \times q) - p^2$$

$$= p - p^2 = p(1-p) = pq.$$

If success is the occurrence of a 6 on a single roll of a die, and X is the number of successes on that single roll, then

$$p = \frac{1}{6}, \; q = \frac{5}{6}$$

$$\mu \;\; = E(X) = p = \frac{1}{6}$$

$$\sigma^2 \;\; = \mathrm{Var}(X) = pq = \frac{1}{6} \times \frac{5}{6} = \frac{5}{36}, \text{ and}$$

$$\sigma \;\; = \sqrt{5/36} \doteq 0.373.$$

Such a simple random variable is by itself of limited utility, but we can put several of them together:

FIGURE 6.5. **Probability function and graph of X.**

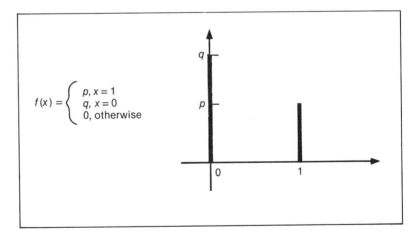

The random variable associated with this larger experiment is the number of successes in the n trials. If we roll a die 10 times, and success is the appearance of 6 dots, the random variable X is the number of times this occurs in the 10 trials. (Note that the successive rolls are independent, and the probability of success is always $\frac{1}{6}$.)

To find the probability function of X first consider the probability of one particular sequence of, say, 3 successes (S) and $10 - 3 = 7$ failures. Successes occur with probability $\frac{1}{6}$, failures with probability $\frac{5}{6}$, and the trials are independent, so the probability of this sequence of successes and failures is $\left(\frac{1}{6}\right)^3 \left(\frac{5}{6}\right)^7$.

$$F \quad S \quad F \quad F \quad F \quad S \quad S \quad F \quad F \quad F$$

$$\frac{5}{6} \times \frac{1}{6} \times \frac{5}{6} \times \frac{5}{6} \times \frac{5}{6} \times \frac{1}{6} \times \frac{1}{6} \times \frac{5}{6} \times \frac{5}{6} \times \frac{5}{6}$$

But there are $\binom{10}{3}$ sequences like this containing 3 successes and 7 failures, so the probability of 3 successes in the 10 trials is

$$f(3) = P(X = 3) = \binom{10}{3} \left(\frac{1}{6}\right)^3 \left(\frac{5}{6}\right)^7 .$$

In general, if the random variable X is the number of successes in a sequence of n Bernoulli trials, the probability function of X is:

$$f(x) = P(X = x) = \begin{cases} \binom{n}{x} p^x q^{n-x}, & x = 0, 1, 2, \ldots, n, \\ \\ 0, \text{ otherwise} \end{cases}$$

where p is the probability of success, and $q = 1 - p$ is the probability of failure. Such a random variable is said to have a *binomial* distribution, and is identified as $b(n,p)$. The random variable of our example is $b(10, \frac{1}{6})$, and it has this probability function:

$$f(x) = \begin{cases} \binom{10}{x} \left(\frac{1}{6}\right)^x \left(\frac{5}{6}\right)^{10-x}, & x = 0, 1, 2, \ldots, 10 \\ \\ 0, \text{ elsewhere.} \end{cases}$$

Note that probabilities such as those shown in Figure 6.6 are given for a variety of binomial distributions in Table A-1.

The cumulative density function and its graph are shown in Figure 6.7.

FIGURE 6.6.　The probability function of the binomial distribution $b(10, \frac{1}{6})$.

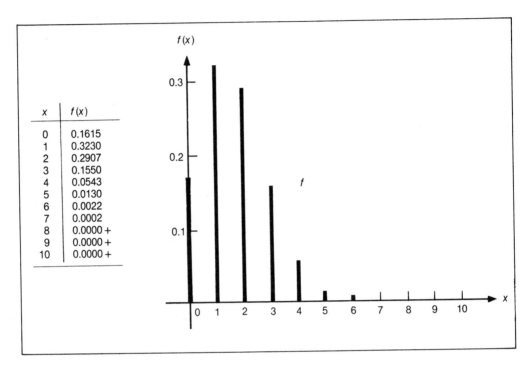

x	$f(x)$
0	0.1615
1	0.3230
2	0.2907
3	0.1550
4	0.0543
5	0.0130
6	0.0022
7	0.0002
8	0.0000 +
9	0.0000 +
10	0.0000 +

To find the mean and variance of a binomial distribution, we need not find $f(x)$ for all possible values of the random variable. Instead, we can develop some simple expressions which will give us the values of these parameters for *any* binomial distribution.

Let X be the number of successes in a sequence of n Bernoulli trials, with probability of success, p, and probability of failure, $q = 1 - p$. For $i = 1, 2, \ldots, n$, let X_i be the number of successes on the ith trial; then X_i has mean p and variance pq, as described on p. 98.

But $X = X_1 + X_2 + \ldots + X_n$, so

$$E(X) = E(X_2 + X_2 + \ldots + X_n)$$
$$= E(X_1) + E(X_2) + \ldots + E(X_n)$$
$$= p + p + \ldots + p$$
$$= np,$$

since the expected value of a sum of random variables is the sum of their expected values. Further, the Bernoulli experiments which comprise the sequence of trials are independent, so the variance of their sum is the sum of their individual variances:

$$\text{Var}(X) = \text{Var}(X_1 + X_2 + \ldots + X_n)$$
$$= \text{Var}(X_1) + \text{Var}(X_2) + \ldots + \text{Var}(X_n)$$

$$= \sigma_1^2 + \sigma_2^2 + \sigma_n^2$$
$$= npq. \text{ Also, } \sigma = \sqrt{npq}.$$

FIGURE 6.7. **Cumulative density function and its graph.**

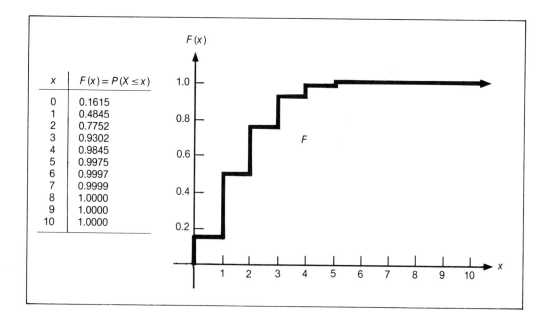

In our example, X is $b(10, \frac{1}{6})$, so:

$$\mu = E(X) = 10 \times \frac{1}{6} = \frac{10}{6} = \frac{5}{3} = 1.67,$$

$$\sigma^2 = \text{Var}(X) = 10 \times \frac{1}{6} \times \frac{5}{6} = \frac{50}{36} = \frac{25}{18} \doteq 1.389, \text{ and}$$

$$\sigma = \sqrt{1.389} \doteq 1.178.$$

The two parameters n and p determine which of the family of binomial distributions is being examined, and, therefore, the shape of the distribution. Figures 6.8 and 6.9 illustrate the distributions $b(10, 0.5)$ and $b(10, 0.8)$.

With $p = 0.5$, the distribution is symmetrical about its mean, $np = 10 \times 0.5 = 5$; when p is less than 0.5, the distribution is skewed to the right, and when p is greater than 0.5, the distribution is skewed to the left.

Example 1 Consider this example of a binomial distribution. A salesman will make 8 calls in a day. At each, he feels he has a 30% chance of making a sale, and whether he makes a sale on one particular call is independent of the others. If X is the number of sales the salesman will make, X has the binomial distribution $b(8, 0.3)$.

FIGURE 6.8.   Distribution of $b(10, 0.5)$.

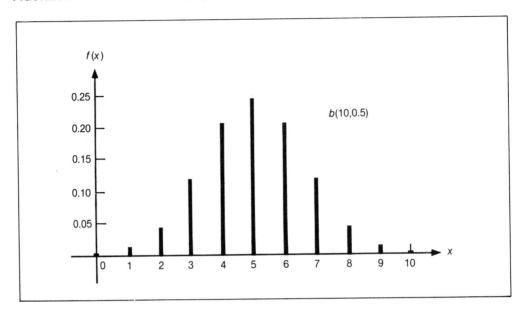

FIGURE 6.9.   Distribution of $b(10, 0.8)$.

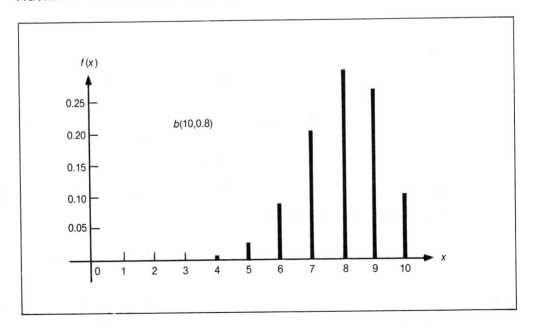

Solution

The probability function of X is $f(x) = \begin{pmatrix} 8 \\ x \end{pmatrix} 0.3^x \, 0.7^{8-x}$, and its values are given in Table 6.1 (as found in Table A-1).

TABLE 6.1. The binomial distribution $b(8, 0.3)$

x	$f(x) = P(X = x)$
0	0.058
1	0.198
2	0.296
3	0.254
4	0.136
5	0.047
6	0.001
7	0.000
8	0.000

The probability that the salesman will make 4 sales is $f(4) = 0.136$; the probability that he will make 3 or fewer sales is $f(0) + f(1) + f(2) + f(3) = 0.058 + 0.198 + 0.296 + 0.254 = 0.806$.

The expected number of sales is $8 \times 0.3 = 2.4$, with variance $8 \times 0.3 \times 0.7 = 1.68$ and standard deviation $\sqrt{1.68} \doteq 1.296$.

THE HYPERGEOMETRIC DISTRIBUTION

Situations which give rise to the distribution known as **hypergeometric** often resemble those of binomial distributions, but with an important difference. With the binomial distribution, the trials are *independent*, but in this new situation, the trials are *dependent* and in a specific way as shown in this example.

Example 2 An employer has 18 employees, 10 of whom are competent, with the remaining 8 incompetent. She will assign 7 employees, randomly chosen, to a certain task. What is the probability that 4 of those chosen will be competent, and the other 3 incompetent?

Solution

The outcomes in this experiment are groups of 7 employees, and there are $\begin{pmatrix} 18 \\ 7 \end{pmatrix}$ ways in which 7 employees can be chosen from the 18 (order of selection does not matter). We must count the number of outcomes—groups of 7 employees—containing 4 competent and 3 incompetent employees. The 4 competent employees may be chosen in $\begin{pmatrix} 10 \\ 4 \end{pmatrix}$ ways, the incompetents in $\binom{8}{3}$, so that there are $\binom{10}{4} \times \binom{8}{3}$ outcomes in this event, and its probability is

$$\frac{\binom{10}{4} \times \binom{8}{3}}{\binom{18}{7}} = \frac{210 \times 56}{31824} \doteq 0.3695.$$

In general, suppose that we are choosing a random group of n objects from a total of N objects, of which S are of a particular kind which we can call successes. The random variable X is the number of successes in the selected group. There are $\binom{N}{n}$ possible groups of N objects which can be chosen from N objects. To have exactly x successes in the chosen group requires that x objects be chosen from the S successes, and the remaining $(n - x)$ objects be chosen from the $(N - S)$ other objects; there are $\binom{S}{x}\binom{N-S}{n-x}$ ways for this to be done. We obtain this probability function for the random variable:

$$f(x) = P(X=x) = \begin{cases} \dfrac{\binom{S}{x}\binom{N-S}{n-x}}{\binom{N}{n}}, & x = \max(0, N-n+S), 1, 2, \ldots, \min(n,S), \\ 0, \text{ otherwise.} \end{cases}$$

DEFINITION

A random variable whose distribution can be described in this way is called **hypergeometric.**

The selection of a handful of n objects from N objects is equivalent to selecting one object n times, if the selected objects are *not* replaced. This is the most important application of the hypergeometric distribution: sampling *without* replacement.

Finding the mean and variance of a hypergeometric distribution by using expected values is tedious. Applying intuition will, however, lead us to a correct result.

If S of N objects are of the kind labeled success, and we take a random sample of size n, then we would expect the proportion of successes in the sample to be the same as that in the original collection of objects (the population). If X is a hypergeometric random variable, then

$$\mu = E(X) = n\,\frac{S}{N}.$$

It can also be shown, using techniques more sophisticated than those available to us here, that

$$\sigma^2 = \text{Var}(X) = n\frac{S}{N}\left(1 - \frac{S}{N}\right)\left(\frac{N-n}{N-1}\right).$$

In our example, the expected number of competent people to be assigned to the task is

$$E(X) = 7 \times \frac{10}{18} = 3.89, \text{ with}$$

$$\text{Var}(X) = 7 \ \frac{10}{18} \left(1 - \frac{10}{18}\right) \left(\frac{18-7}{18-1}\right) \doteq 1.118, \text{ and}$$

$$\sigma = \sqrt{1.118} = 1.058.$$

Figure 6.10 shows the probability functions of hypergeometric distributions with $N = 10$ and $n = 5$ for three values of S, the number of successes in the population.

FIGURE 6.10.   **Probability functions of hypergeometric distributions.**

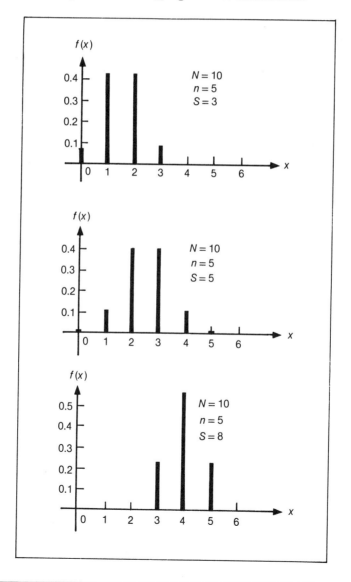

Consider selecting n times from a group of N objects of which S are successes, but *with replacement*, returning each selected object to the original group before the next selection. In this case, the successive drawings are independent, and the probability of drawing a success remains S/N throughout. This produces a sequence of n Bernoulli trials, and if X is the number of successes selected in the n drawings, X has the binomial distribution $b(n, S/N)$, with mean $n(S/N)$ and variance $n(S/N)(1-(S/N))$. This shows the relationship between sampling *without* replacement, which produces a hypergeometric distribution, and sampling *with* replacement, which gives rise to a binomial. The means of the corresponding distributions are the same, but the variance of the hypergeometric is less than the variance of the binomial by a factor of $(N-n)/(N-1)$.

Notice, too, that as the size N of the original group increases, $(N-n)/(N-1)$ nears 1, and the hypergeometric distribution approaches its related binomial.

Example 3 Hypergeometric distributions often occur in relation to quality control. Suppose that a manufacturer receives a certain component in lots of 10, and from each lot, he randomly selects 3 components to be tested. If one or more of the tested components are defective, the lot is rejected. If a particular lot contains 2 defective components, what is the probability that it will be rejected?

Solution

If X is the number of defective components in the sample of 3 which are tested, then X has a hypergeometric distribution with $N = 10$, $S = 2$, and $n = 3$. The lot will be rejected if the number of defectives is 1 or more, so

$$P(\text{Reject the lot}) = P(X \geq 1)$$
$$= 1 - P(X < 1)$$
$$= 1 - P(X = 0)$$
$$= 1 - \frac{\binom{2}{0}\binom{8}{3}}{\binom{10}{3}}$$
$$= 1 - \frac{1 \times 56}{120}$$
$$= 1 - \frac{56}{120} \doteq 1 - 0.467 = 0.533.$$

Note that the expected number of defective components in the sample is

$$E(X) = \mu = 3 \times \frac{2}{10} = 0.6, \text{ with variance}$$
$$\text{Var}(X) = \sigma^2 = 3 \times \frac{2}{10} \times \left(1 - \frac{2}{10}\right) \times \left(\frac{10-3}{10-1}\right)$$
$$= 0.3733, \text{ and standard deviation}$$
$$\sigma = \sqrt{0.3733} = 0.611.$$

THE POISSON DISTRIBUTION

Consider the flow of customers at a fast-food restaurant, arriving "at random" and being served at one of four counter locations. This is an example of a "waiting-line" problem, in which customers wait in one or more queues to be served by one or more servers.

Waiting-line systems like this occur in such situations as:

1. The assignment of operators to machines (one operator per machine). Here, the machines are customers, and the operators are servers.

2. The execution of programs on a time-sharing system. The programs are customers, while the computer is the single server.

3. The passage of motor vehicles through an intersection. Cars may form queues along the streets leading to the intersection.

In such situations, and many more like them, it is important to know how many customers might be expected to arrive, or how many service stations to set up so that waiting time is minimized without wasting resources. To answer these questions, and others related to waiting-line problems, we employ the *Poisson probability distribution*, which can be used to describe the arrivals of customers into queues.

The Poisson distribution applies only to situations called *Poisson processes*, and the occurrence of the event of interest—the submission of a job to a computer system, the arrival of a car at an intersection—in the process is called a *change*. We will be interested in the number of changes in an interval of time.

A Poisson process is defined by three conditions. When you are applying a Poisson distribution to a real-world situation, it is wise to check experimentally that these conditions hold. They are:

1. The numbers of changes in nonoverlapping intervals are independent.

2. The probability of exactly one change in an interval is proportional to the length of the interval.

3. In a small interval, the probability of more than one change is essentially 0.

Let α be the average number of changes in a one-unit interval of time, and let T be the length of an interval. Then the Poisson random variable X is the number of changes in the interval of length T, and its probability function is:

$$f(x) = P(X=x) = \begin{cases} \dfrac{(\alpha T)^x e^{-\alpha T}}{x!}, & x=0,1,2,3, \ldots, \\ 0, & \text{otherwise.} \end{cases}$$

The value of the parameter αT determines which Poisson distribution is being considered, and since this value is the average number of arrivals in time T, we would expect it to be the mean of X. Letting $\lambda = \alpha T$, we see that this is the case:

$$\mu = E(X) = \sum_x x \cdot f(x)$$

$$= \sum_{x=0}^{\infty} x \cdot \frac{\lambda^x e^{-\lambda}}{x!}$$

$$= \sum_{x=1}^{\infty} x \cdot \frac{\lambda^x e^{-\lambda}}{x!}$$

$$= \sum_{x=1}^{\infty} \frac{\lambda^x e^{-\lambda}}{(x-1)!}$$

$$= \lambda \sum_{x=1}^{\infty} \frac{\lambda^{x-1} e^{-\lambda}}{(x-1)!}$$

$$= \lambda \sum_{x-1=0}^{\infty} \frac{\lambda^{x-1} e^{-\lambda}}{(x-1)!} = \lambda \cdot 1 = \lambda.$$

(Since f is a probability function,

$$\sum_{x-1=0}^{\infty} f(x-1) = \sum_{x-1=0}^{\infty} \frac{\lambda^{x-1} e^{-\lambda}}{(x-1)!} = 1.)$$

A more surprising result (see Exercise 13) is that the variance of a Poisson distribution is also λ.

The graphs in Figure 6.11 show Poisson distributions with three different values for λ. When λ is small, the distribution is skewed to the right; as λ increases, it becomes more symmetrical.

Suppose that a motel receives an average of 1 call per minute in the course of a working day and that the arrivals of these calls constitute a Poisson process. (Note that this does not mean that the motel receives exactly 1 call every minute.) What is the probability that in a given 5-minute period, the motel receives 6 calls?

We first observe that $\alpha = 1$ and $T = 5$, so $\lambda = \alpha T = 5$. Then the probability of 6 calls in 5 minutes is:

$$f(6) = P(X=6) = \frac{5^6 e^{-5}}{6!} = \frac{5^6 \times 0.006738}{6!} = 0.1462.$$

(Note that Table A-3 provides values of $e^{-\lambda}$ for various λ, and Table A-2 gives probabilities from selected Poisson distributions.)

Likewise, the probability of 8, 9, or 10 calls in 10 minutes is:

$$P(8 \le X \le 10) = f(8) + f(9) + f(10)$$

$$= \frac{10^8 e^{-10}}{8!} + \frac{10^9 e^{-10}}{9!} + \frac{10^{10} e^{-10}}{10!}$$

$$= 0.1126 + 0.1251 + 0.1251 = 0.3628.$$

FIGURE 6.11(a)–(c). Poisson distributions with three different values for λ.

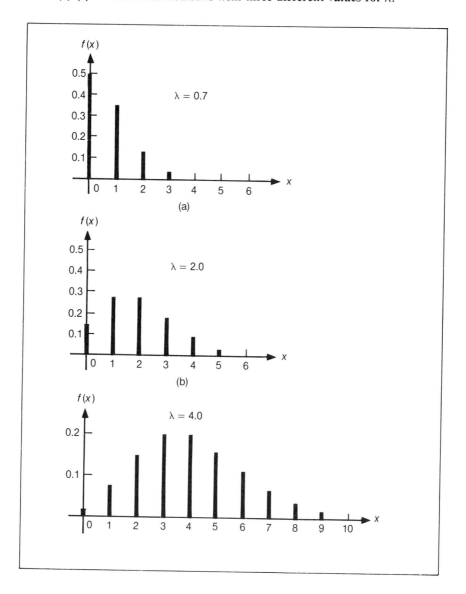

SUMMARY OF DISCRETE DISTRIBUTIONS

Table 6.2 summarizes the discrete distributions we have described.

TABLE 6.2. Summary of Discrete Distributions.

TYPE	PROBABILITY FUNCTION	$\mu = E(X)$	σ^2, σ	DISTRIBUTIONAL SHAPE
Uniform	$f(X) = 1/n$, where n is the number of values in the range of X.	Depends on values of X	Depends on values of X	Rectangular
Binomial	$f(X) = (n\ x)p_x q_{n-x}$, where n = number of trials, p = probability of success, and $q = 1 - p$.	np	$npq, \sqrt{npq}$	If $p = 1/2$, symmetrical, If $np > 5$, negatively skewed, If $np < 5$, positively skewed.
Hypergeo-metric	$f(x) = \dfrac{\left(\dfrac{S}{x}\right)\left(\dfrac{N-S}{n-x}\right)}{\left(\dfrac{N}{n}\right)}$, where n = size of sample, N = size of population, S = number of successes in population.	$n\dfrac{S}{N}$	$n\dfrac{S}{N}\left(1-\dfrac{S}{N}\right)$ $\left(\dfrac{N-n}{N-1}\right)$ $\sigma^2 = \sqrt{\sigma^2}$	Similar to binomial with $p = S/N$.
Poisson	$f(x) = \dfrac{\lambda^x e^{-\lambda}}{x!}$, where $\lambda = \alpha(T)$ average number of arrivals in time T.	λ	$\lambda, \sqrt{\lambda}$	Positively skewed for small λ, more symmetrical as λ increases.

Continuous Probability Distributions

While the range of a discrete random variable is finite or countably infinite, a set of distinct values each with positive probability, a *continuous* random variable—the height of a randomly chosen person or the time until a particular event occurs—can take on any value in an interval (or union of intervals). The range of a continuous random variable is *uncountably* infinite, and any assignment of probabilities which gives positive probability to each value in the range will result in the total probability associated with the experiment becoming infinite. This would violate the axioms on which we have built probability theory, so we must find a different way to assign probabilities to events which involve continuous random variables. In particular, the probability associated with any single value must be 0.

We saw when using Venn diagrams that the *area* of a portion of the Venn diagram representing an event could be used to represent the *probability* of that event. We now take the analogy of probability and area a step further by defining a function under which areas are *equal* to probabilities:

DEFINITION

Let X be a continuous random variable. The function f is the **probability function** of X if the probability that X takes on a value in the interval $[a,b]$ is

$$P(a \leq X \leq b) = \int_a^b f(x)dx.$$

The probability that the random variable X falls in the closed interval $[a,b]$ is equal to the area under the graph of the probability function and above $[a,b]$. (See Figure 6.12.)

Consider the simple experiment of spinning a toy spinner (Figure 6.13), with X the value at which it stops. X is continuous, and its range is the interval [0, 1]. The probability function of f, as shown in the graph, is

$$f(x) = \begin{cases} 1, 0 \leq x < 1, \\ 0, \text{ all other } x. \end{cases}$$

FIGURE 6.12 $P(a \leq x \leq b)$ **for a continuous random variable** X.

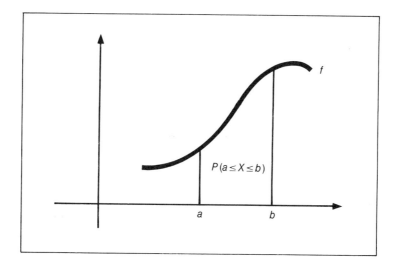

FIGURE 6.13 A toy spinner.

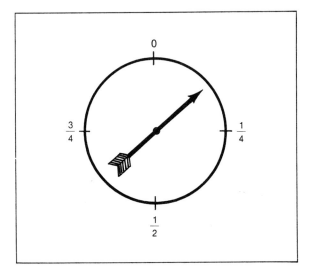

It is reasonable that $P(\frac{1}{4} \leq X \leq \frac{1}{2}) = \frac{1}{4}$, (see Figure 6.14) and this is the result we get when we integrate the probability function from $\frac{1}{4}$ to $\frac{1}{2}$:

$$P(\tfrac{1}{4} \leq X \leq \tfrac{1}{2}) = \int_{\frac{1}{4}}^{\frac{1}{2}} f(x)dx = \int_{\frac{1}{4}}^{\frac{1}{2}} 1\ dx = x \Big|_{\frac{1}{4}}^{\frac{1}{2}} = \tfrac{1}{2} - \tfrac{1}{4} = \tfrac{1}{4}.$$

Two observations should be made about a continuous probability function f:

FIGURE 6.14 P ($\frac{1}{4} \leq X \leq \frac{1}{2}$).

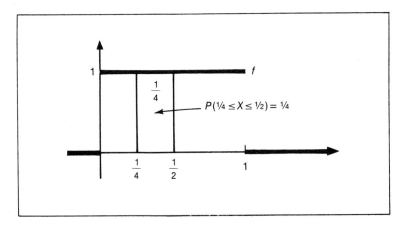

1. $f(x) \geq 0$ for all x.

If f were ever to dip below the x axis, then there would be a tiny interval in the dip over which the integral of f would be negative. This would say that the probability that X falls in that interval is negative, which cannot happen.

2. $\int_{-\infty}^{\infty} f(x)dx = 1.$

All the area under the graph of f, all the probability associated with the random variable, must total 1.

We can see, too, that probabilities assigned by integrating f satisfy the axioms of probability:

1. For every interval $[a,b]$,

$$P(a \leq X \leq b) = \int_{a}^{b} f(x)dx \geq 0, \text{ since } f(x) \geq 0.$$

2. $\int_{-\infty}^{\infty} f(x)dx = 1.$

$(P(S) = 1)$

3. For disjoint intervals $[a,b]$ and $[c,d]$,

$$P(a \leq X \leq b \text{ or } c \leq X \leq d) = \int_{a}^{b} f(x)dx + \int_{c}^{d} f(x)dx$$

$$= P(a \leq X \leq b) + (P(c \leq X \leq d).$$

$$[\text{If } A \cap B = 0, P(A \cup B) = P(A) + P(B).]$$

Finally, we can verify the earlier intuitive argument that for a continuous random variable, the probability associated with any single value must be 0:

$$P(X = c) = P(c \leq X \leq c) = \int_{c}^{c} f(x)dx = 0.$$

This means that including or excluding any single point (or countable set of points) from the possible values of X which describe an event will not change the probability of that event. In particular,

$$P(a \leq X \leq b) = P(a < X \leq b) = P(a \leq X < b) = P(a < X < b)$$

$$= \int_{a}^{b} f(x)dx.$$

We have seen that the cumulative density functions of discrete distributions are nondecreasing step functions because positive probabilities are associated with distinct individual values in the range of the random variable. In a continuous distribution, probability is not associated with particular values but is "spread out" over an interval; this changes the appearance of the cumulative density function.

DEFINITION

Let X be a continuous random variable with probability function f. The
cumulative density function F associated with X is defined to be:

$$F(x_0) = P(X \leq x_0) = \int_{-\infty}^{x_0} f(x)dx.$$

That is, the value of F at x_0 is the area under the graph of f and to the left of x_0.

Consider this example: X is a continuous random variable with the probability function
as shown in Figure 6.15. (It is left to the reader to verify that f has the properties of a continuous probability function.)

The value of the cumulative density function F at 1 is:

$$F(1) = \int_{-\infty}^{1} f(x)dx = \int_{0}^{1} \frac{x}{2}\,dx = \frac{x^2}{4}\Big|_0^1 = \frac{1}{4} - \frac{0}{4} = \frac{1}{4}.$$

In general,

$$F(x_0) = \int_{-\infty}^{x_0} f(x)dx = \int_{0}^{x_0} \frac{x}{2}\,dx = \frac{x^2}{4}\Big|_0^{x_0} = \frac{x_0^2}{4} - \frac{0}{4} = \frac{x_0^2}{4}, \text{ for } 0 \leq x_0 \leq 2,$$

and F has the graph as shown in Figure 6.16.

FIGURE 6.15. A Continuous probability function.

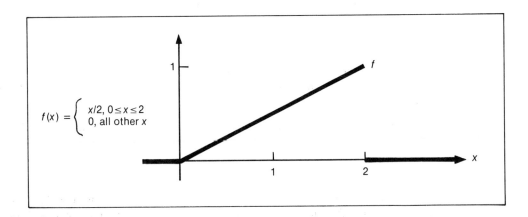

$$f(x) = \begin{cases} x/2, & 0 \leq x \leq 2 \\ 0, & \text{all other } x \end{cases}$$

FIGURE 6.16. Cumulative density function of a continuous random variable.

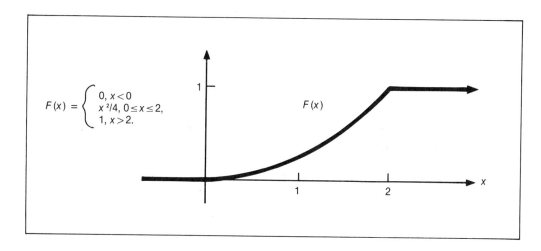

Note that the cumulative density function of a continuous random variable is always nondecreasing and continuous.

Expected Value and Variance of Continuous Distributions

From the development of the definite integral in calculus, we know that the integral is the limit of the sum of an increasing number of ever-smaller values. We have also seen the correspondence between the integrals involved in the current discussion and summation expressions related to discrete distributions. It is not surprising that the definitions of the mean and the variance of a continuous distribution are nearly identical to their discrete counterparts. We integrate over all possible values of the random variable rather than taking a sum:

DEFINITION

Let X be a continuous random variable with probability function f. The **expected value** or **mean** of X is defined to be

$$E(X) = \mu = \int_{-\infty}^{\infty} xf(x)dx.$$

<hr>

DEFINITION

The **variance** and **standard deviation** of X are

$$\text{Var}(X) = \sigma^2 = \int_{-\infty}^{\infty} (x-\mu)^2 f(x)dx, \text{ and}$$

$$\sigma = \sqrt{\sigma^2}.$$

<hr>

As with discrete distributions, there is a shortcut formula (see Exercise 19) which is generally used to find the variance:

$$\text{Var}(X) = \sigma^2 = E(X^2) - \mu^2$$

$$= \int_{-\infty}^{\infty} x^2 f(x)dx - \mu^2.$$

In our last example, where

$$f(x) = \begin{cases} \dfrac{x}{2}, & 0 \le x \le 2, \\ 0, & \text{elsewhere,} \end{cases}$$

$$E(X) = \mu = \int_{-\infty}^{\infty} xf(x)dx = \int_0^2 x \cdot \frac{x}{2} dx = \int_0^2 \frac{x^2}{2} dx$$

$$= \frac{x^3}{6} \Big|_0^2 = \frac{8}{6} - \frac{0}{6} = \frac{4}{3},$$

$$\text{Var}(X) = \sigma^2 = \int_{-\infty}^{\infty} x^2 f(x)dx - \mu^2 = \int_0^2 x^2 \cdot \frac{x}{2} dx - \mu^2$$

$$= \int_0^2 \frac{x^3}{2} dx - \mu^2 = \frac{x^4}{8} \Big|_0^2 - \mu^2$$

$$= \frac{16}{8} - \frac{0}{8} - \left(\frac{4}{3}\right)^2 = \frac{2}{9} \doteq 0.222, \text{ and}$$

$$\sigma = \sqrt{\sigma^2} = \sqrt{0.222} \doteq 0.471.$$

Particular Continuous Distributions

Certain continuous distributions occur in a wide variety of experimental situations. We examine three of them.

THE CONTINUOUS UNIFORM DISTRIBUTION

The range of this distribution is always an interval [*a,b*], and it is characterized by the fact that for any two intervals of equal width within [*a,b*], the probabilities that the random variable *X* will fall in each are equal. This is equivalent to saying that the areas under the graph of *f*, the probability function of *X*, above the two subintervals of equal width must always be equal. For this to be true, the graph of *f* must be a horizontal line above [*a,b*]. (See Figure 6.17.)

Let *X* be a continuous random variable with this probability function:

$$f(x) = \begin{cases} \dfrac{1}{b-a}, & a \leq X \leq b \\ \\ 0, & \text{all other } x. \end{cases}$$

Then *X* has a **continuous uniform distribution**. It is clear from the graph of *f* why such a distribution is also sometimes called *rectangular* (Figure 6.18).

FIGURE 6.17. Probability function of a uniform distribution.

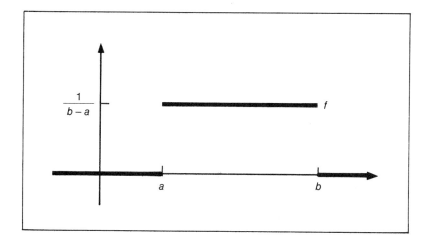

FIGURE 6.18.   Cumulative density function F of the continuous uniform distribution.

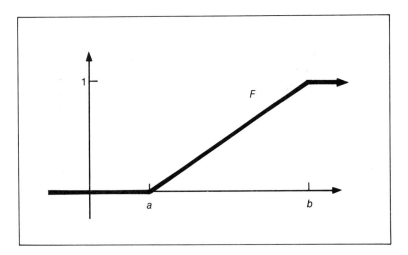

It is not difficult (See Exercise 23) to find the mean and variance of a uniform distribution:

$$\mu = \frac{b+a}{2} \text{ , and}$$

$$\sigma^2 = \frac{(b-a)^2}{12} \text{ .}$$

We have already seen a continuous uniform distribution in the toy spinner example of Figure 6.13. In that example, $a=0$, $b=1$, and the probability function is

$$f(x) = \begin{cases} \frac{1}{1-0}, 0 \leq X < 1 \\ \\ 0, \text{ all other } x. \end{cases}$$

For this distribution,

$$\mu = \frac{1+0}{2} = \frac{1}{2} \text{and } \sigma^2 = \frac{(1-0)^2}{12} = \frac{1}{12} \text{ .}$$

Random Numbers and Uniform Distributions

Random numbers, as listed in Table A-15, are generated by computer programs, and are intended to behave as though they come from uniform distributions. We can use these random numbers to generate numbers from other types of distributions, by taking advantage of the fact that the cumulative density function F of any continuous distribution is monotonically increasing and continuous, and therefore has a well-defined inverse F^{-1}. To find a nonuniform random number x, we generate a uniform random number r, and let $x = F^{-1}(r)$. (See Figure 6.19.)

FIGURE 6.19.

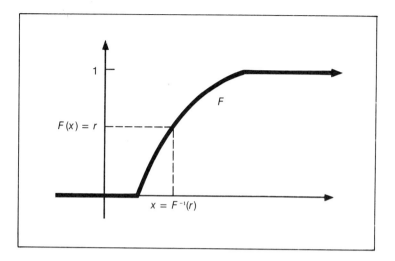

THE EXPONENTIAL DISTRIBUTION

Recall the Poisson process discussed earlier in this chapter (p. 107), the random arrivals of customers into a queue. Rather than count the number of arrivals in some interval of time T, let us now consider the length of time from some initial time t_o until the next customer arrives, that is, the time until a change occurs in a Poisson process.

Let the random variable X be the time until a change occurs in a Poisson process. Then X has an **exponential** distribution, and its probability function f is:

$$f(t) = \begin{cases} \alpha e^{-\alpha t}, t \geq 0 \\ 0, \text{ otherwise,} \end{cases}$$

where t = time until the next change, and α = average rate of changes per unit time. (Note that since time intervals of negative length are impossible, t is always nonnegative.)

We can find the mean of an exponential random variable by integration, according to the definition of the mean of a continuous distribution:

$$\mu = E(X) = \int_{-\infty}^{\infty} t \cdot f(t)dt = \int_{0}^{\infty} t(\alpha e^{-\alpha t})dt = \int_{0}^{\infty} \alpha t e^{-\alpha t}\, dt.$$

Integrating by parts, let

$$u = \alpha t, \text{ so that } du = \alpha dt, \text{ and let}$$

$$dv = e^{-\alpha t}\, dt, \text{ so that } v = \frac{-1}{\alpha}\, e^{-\alpha t}.$$

Then

$$\mu = \left[(\alpha t) \left(\frac{-1}{\bar{\alpha}} e^{-\alpha t} \right) \right]_0^\infty - \int_0^\infty \left(-\frac{1}{\alpha} e^{-\alpha t} \right) (\alpha dt)$$

$$= [-te^{-\alpha t}]_0^\infty - \left[\frac{1}{\alpha} e^{-\alpha t} \right]_0^\infty .$$

Using L'Hospital's rule,*

$$= \left[\lim_{t \to \infty} \frac{-t}{e^{\alpha t}} - 0 \right] - \left[\frac{1}{\alpha} \lim_{t \to \infty} \frac{1}{e^{\alpha t}} - \frac{1}{\alpha} \right]$$

$$= \left[\lim_{t \to \infty} \frac{-1}{\alpha e^{\alpha t}} \right] - \left[\frac{-1}{\alpha} \right] = \frac{1}{\alpha} .$$

This is reasonable; the average rate of change is α, and the expected time until a change is $1/\alpha$.

Two important results are left as exercises: that the variance of an exponential distribution is $1/\alpha^2$ (Exercise 28); and that the cumulative density function F is given by $F(t) = 1 - e^{-\alpha t}$, $t \geq 0$ (Exercise 29).

The graphs in Figure 6.20 show shapes of exponential distributions. Note that these curves always start at the point $(0, \alpha)$, and are steeper for smaller values of α.

This example illustrates the exponential model. Customers arrive at a bank at an average rate of 1.5 per minute, and their arrivals constitute a Poisson process. If one customer has just walked in, what is the probability that the next will follow in less than 1 minute?

Let X be the length of time until the next customer enters. Then X has an exponential distribution with $\alpha = 1.5$. We seek $P(X < 1)$:

$$P(X < 1) = F(1) = 1 - e^{-1.5 \times 1} = 1 - 0.2231 = 0.7769.$$

Similarly, the probability that the next customer will enter after a wait of 2 minutes or more is:

$$P(X \geq 2) = 1 - P(X < 2) = 1 - F(2)$$

$$= 1 - (1 - e^{-1.5 \times 2}) = e^{-1.5 \times 2} = 0.0498.$$

The expected length of time between customers is $1/\alpha = 1/1.5 = 0.67$ minutes, with variance $1/\alpha^2 = 1/1.5^2 = 0.44$ and standard deviation $1/\alpha = 0.67$.

THE NORMAL DISTRIBUTION

Consider a person who is throwing darts, and aiming at the bull's-eye. If the thrower is acquainted with the game, he or she will "on the average" hit the bull's-eye, and large deviations from it will occur less often than will small ones. Also, deviations above and below the bull's-eye will occur with approximately equal frequency.

*As applied here, if $\lim_{x \to \infty} f(x) = \infty$ and $\lim_{x \to \infty} g(x) = \infty$, then $\lim_{x \to \infty} \frac{f(x)}{g(x)} = \lim_{x \to \infty} \frac{f'(x)}{g'(x)}$.

FIGURE 6.20(a)–(c). Shapes of exponential distribution.

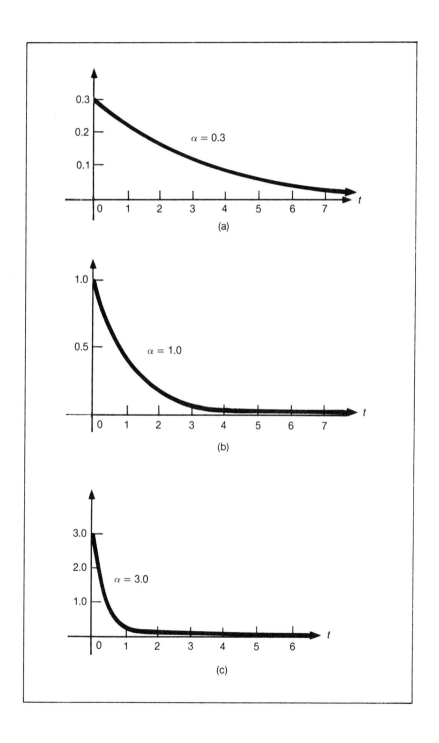

These characteristics—values clustered symmetrically around a single central value, small deviations more common than large ones—can be observed in a great variety of phenomena. Measurements of men's heights are an example; very few men are more than a foot shorter or taller than the average height of 5'10", but many are within two inches of this value. Other quantities which behave in this way are scores on standardized tests, men's and women's weights, and dimensions of manufactured objects.

In the eighteenth century, Carl Friedrich Gauss derived a probability function which provided a model for many variables with the characteristics described above. This function, whose graph is called the *normal curve*, combines with the Central Limit Theorem, a result which will be discussed in Chapter 7, to form the foundation of statistical inference, and it is given in this definition:

DEFINITION

Let X be a continuous random variable whose probability function f is

$$f(x) = \frac{1}{\sqrt{2\pi}\sigma}[\exp - (x-\mu)^2/2\sigma^2], \quad -\infty < x < \infty.$$

Then X has a *normal* distribution with mean μ and variance σ^2, denoted $N(\mu,\sigma^2)$.

The probability function in Figure 6.21 describes what is traditionally known as the "bell-shaped curve."

For a given value d, $f(\mu - d) = f(\mu + d)$; that is, the probability function is symmetrical

FIGURE 6.21.

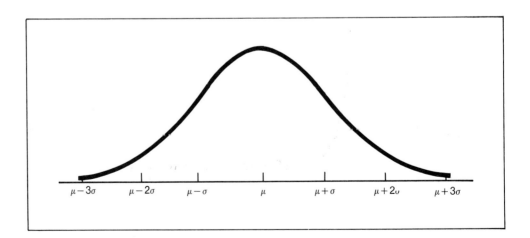

around the mean μ. Also, the quantity $\exp_e\left[-(x-\mu)^2/2\sigma^2\right]$ attains its maximum value of 1 when $x = \mu$, so the maximum value of f is $f(\mu) = 1/\sqrt{2\pi}\sigma$. Finally,

$$\lim_{x\to\infty} f(x) = \frac{1}{\sqrt{2\pi}\sigma} \lim_{x\to\infty} e^{\frac{-(x-\mu)^2}{2\sigma^2}} = 0, \text{ and}$$

$$\lim_{x\to-\infty} f(x) = \frac{1}{2\sigma} \lim_{x\to\infty} e^{\frac{-(x-\mu)^2}{2\sigma^2}} = 0;$$

the graph of f approaches the x axis as a limit away from μ.

The shape of a normal distribution is determined by its standard deviation, while its location is determined by its mean (Figure 6.22).

FIGURE 6.22.

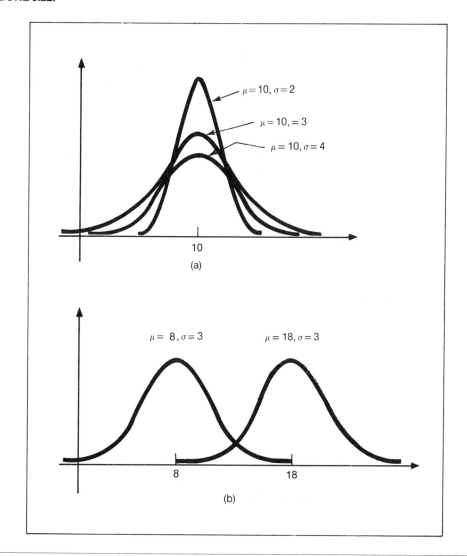

$\mu = 10, \sigma = 2$

$\mu = 10, = 3$

$\mu = 10, \sigma = 4$

10

(a)

$\mu = 8, \sigma = 3$ $\mu = 18, \sigma = 3$

8   18

(b)

The cumulative density function for a normal distribution is defined as:

$$F(x) = \int_{-\infty}^{x} f(y)dy = \int_{-\infty}^{x} \frac{1}{\sqrt{2\pi\sigma}} \exp - [(y-\mu)^2/2\sigma^2] \, dy,$$

and its graph, called an *S curve*, is shown in Figure 6.23.

There are an infinite number of normal distributions, each one characterized by its mean and standard deviation. This fact has great impact on finding probabilities involving normal random variables.

In our previous work with continuous random variables, we have calculated probabilities by evaluating definite integrals, a process that involves finding the antiderivative of the probability function. We cannot do this with normal distributions, however, because there is no function whose derivative is $\exp(-x^2)$. The normal probability function has no antiderivative.

However, there are numerical methods for approximating areas under a curve (the trapezoidal approximation, Simpson's rule, and others), and these methods can be used to build tables of probabilities involving normal distributions. Since there are an infinite number of normal distributions, we build such a table for only one, then relate all other normal distributions to it. The following definition describes this fundamental distribution.

DEFINITION

The **standard normal** distribution, whose random variable is generally indicated by Z, has mean 0, variance and standard deviation 1, and probability function:

$$f(x) = \frac{1}{\sqrt{2\pi}} \exp(-x^2/2), \quad -\infty < x < \infty.$$

FIGURE 6.23.   *S*-curve graph of the cumulative density function for a normal distribution.

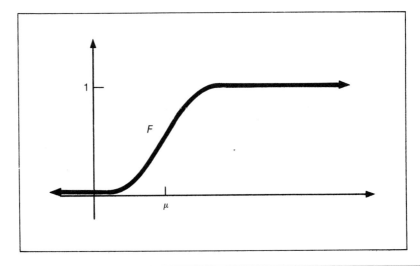

That is, Z is $N(0, 1)$.

The standard normal random variable can be transformed into another normal random variable with mean μ and standard deviation σ by multiplying by σ and adding μ. That is, if X is $N(\mu, \sigma^2)$,

$$X = \sigma Z + \mu.$$

Likewise, X can be transformed into Z by subtracting μ and dividing by σ:

$$(X - \mu)/\sigma = Z.$$

The pair of graphs in Figure 6.24 show the relationship between a normal distribution, with mean 10 and variance 9, and the standard normal distribution. Note that $X = 3Z + 10$ and $(X - 10)/3 = Z$.

A normal random variable X with mean μ and variance σ^2 is related to the standard normal random variable Z through the equation $(X - \mu)/\sigma = Z$. We will use this relationship to find probabilities involving normal random variables, but first we must become familiar with the distribution of Z. We do this through a series of examples.

FIGURE 6.24.   **Relationship between a normal distribution, with mean 10 and variance 9, and standard normal distribution.**

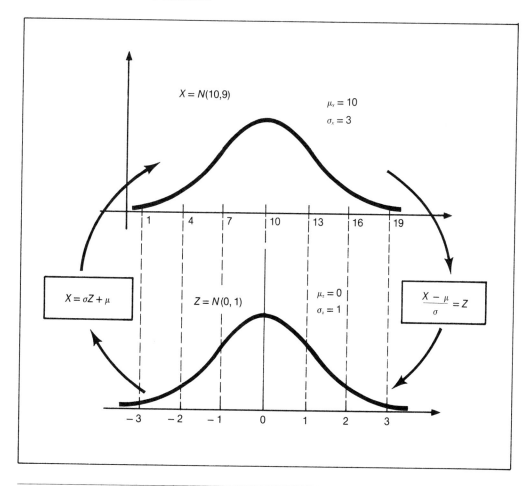

Example 4 Let Z be $N(0, 1)$, the standard normal random variable. Find $P(0 \leq Z < 1)$.

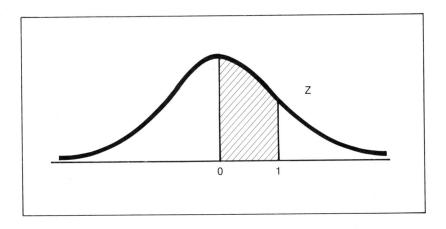

Solution

Table A-4 gives values of $P(0 \leq Z \leq z)$ for values of z between 0 and 3. From the table, $P(0 \leq Z < 1) = .3413$. Note that, since there is zero probability associated with any single value in a continuous distribution, $P(0 \leq Z < 1) = P(0 < Z < 1) = P(0 < Z \leq 1) = P(0 \leq Z \leq 1)$.

Example 5 Find $P(Z \geq 0)$.

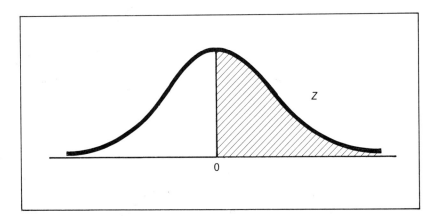

Solution

We have seen that normal distributions are symmetric around their means. Therefore, Z is symmetric around 0. The total area under the curve is 1, so $P(Z \geq 0) = 0.5000$. Likewise, $P(Z \leq 0) = 0.5000$.

Example 6 Find $P(Z > 1.5)$.

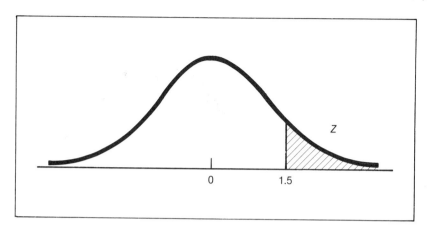

Solution

$P(Z > 1.5)$ is not given in the table. However, $P(Z > 1.5) = P(Z > 0) - P(0 < Z < 1.5)$, and this latter value is 0.4332. Then

$$P(Z > 1.5) = 0.5000 - P(0 < Z < 1.5)$$
$$= 0.5000 - 0.4332 = 0.0688.$$

Example 7 Find $P(1.25 \leq Z \leq 2.10)$.

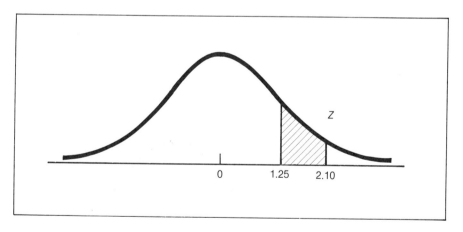

Solution

To find this probability, this area, we look up two values in Table A-4, $P(0 \leq Z < 1.25)$ and $P(0 \leq Z < 2.10)$. Then,

$$P(1.25 \leq Z \leq 2.10) = P(0 \leq Z \leq 2.10) - P(0 \leq Z \leq 1.25)$$
$$= 0.4821 - 0.3944 = 0.0877.$$

Example 8 Find $P(-1 \leq z \leq 0)$.

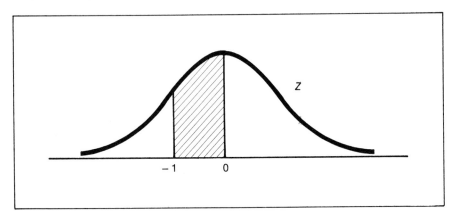

Solution

Because of the symmetry of Z around 0,

$$P(-1 \leq Z < 0) = P(0 < Z \leq 1) = 0.3413.$$

Likewise:

$$P(Z < -1.5) = P(Z \leq 0) - P(-1.5 < Z < 0)$$
$$= 0.5000 - 0.4332 = 0.0688$$

and

$$P(-2.10 \leq Z \leq -1.25) = P(-2.10 \leq Z \leq 0) - P(-1.25 \leq Z \leq 0)$$
$$= 0.4821 - 0.3944 = 0.0877.$$

Example 9 Find $P(Z \geq -1.2)$.

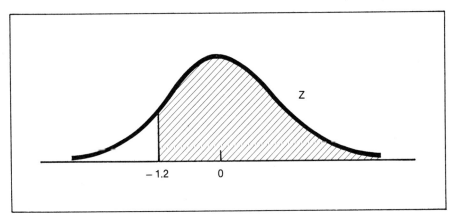

Solution

Here, we must add two probabilities:

$$P(Z \geq -1.2) = P(-1.2 \leq Z \leq 0) + P(Z > 0)$$
$$= 0.3849 + 0.5000 = 0.8849.$$

Example 10 Find $P(Z < 1.66)$.

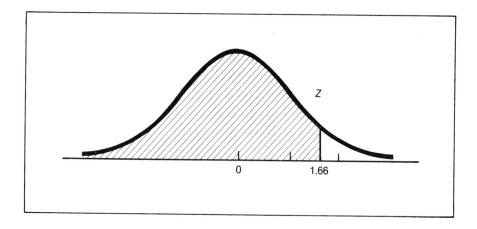

Solution

This example is essentially the same as Example 9:

$$P(Z < 1.66) = P(Z < 0) + P(0 \leq Z \leq 1.66)$$
$$= 0.5000 + 0.4575 = 0.9515.$$

Example 11 Find $P(-0.57 < Z < 1.05)$.

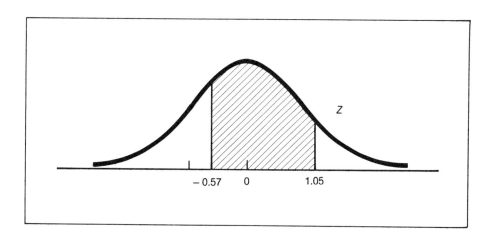

Solution

Again we add two probabilities:

$$P(-0.57 < Z < 1.05) = P(-0.57 < Z \leq 0) + P(0 \leq Z < 1.05)$$
$$= 0.2157 + 0.3531 = 0.5688.$$

It should be clear from these examples that the graph of Z is a useful aid in finding probabilities involving the standard normal distribution.

Given a probability, we can also describe a corresponding event, as in this next example.

Example 12 Find c so that $P(-c \leq Z \leq c) = 0.8132$.

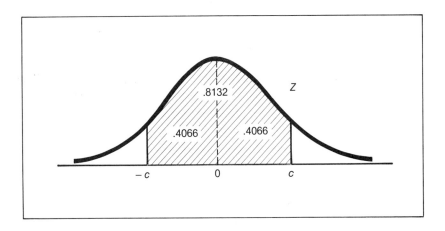

Solution

We are looking for c to fit into this graph; if it does, then $P(0 \leq Z \leq c) = 0.4066$. From Table A-4, $c = 1.32$.

With this knowledge of the distribution of Z, we can use the relationship $(X - \mu)/\sigma = Z$ to find probabilities for any normal random variable $X = N(\mu, \sigma^2)$.

Consider this next example.

Example 13 A production process uses bearings whose lives in service are normally distributed with mean 740 hours and standard deviation 30 hours. What proportion of the bearings last more than 700 hours?

Solution

This is the same as the probability that a randomly selected bearing will last more than 700 hours. That is, if X is the lifetime of a randomly selected bearing, X is $N(740, 900)$, and we seek $P(X > 700)$.

If $X > 700$, then $(X - 740) > (700 - 740)$, and

$$\frac{X - 740}{30} > \frac{700 - 740}{30} \ .$$

That is,

$$P(X > 700) = P\left(\frac{X - 740}{30} > \frac{700 - 740}{30}\right).$$

But 740 is the mean of X, and 30 its standard deviation. We know that $(X - \mu)/\sigma = Z$ for any normal random variable, so:

$$P(X>700)=P\left(\frac{X-740}{30}>\frac{700-740}{30}\right)=P(Z>-1.33),$$

and

$$P(Z>-1.33)=P(-1.33<Z\le0)+P(Z>0)$$

$$=0.4082+0.5000=0.9082.$$

That is $P(X > 700) = 0.9082$; approximately 91% of the bearings last 700 hours or more.

In general, if X is $N(\mu,\sigma^2)$, a normal random variable with mean μ and variance σ^2,

$$P(a\le X\le b)=P\left(\frac{a-\mu}{\sigma}<\frac{X-\mu}{\sigma}<\frac{b-\mu}{\sigma}\right)=P\left(\frac{a-\mu}{\sigma}<Z<\frac{b-\mu}{\sigma}\right).$$

Using the fact that $(X-\mu)/\sigma = Z$, we can reduce any problem involving a normal random variable to one involving the familiar Z.

Example 14 Consider again the bearing example. If management wants to replace the bearings after no more than 2% of them have worn out, how long should they remain in service?

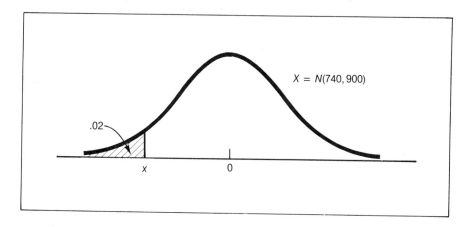

Solution

Again let X be the lifetime of a randomly chosen bearing. X is $N(740, 900)$, and we seek the value x for which $P(X < x) = 0.02$.

This kind of problem is also solved by changing it to one involving Z.

$$P(X<x)=.02 \quad\longrightarrow\quad P\left(\frac{X-740}{30}<\frac{x-740}{30}\right)=0.02.$$

$$\longrightarrow P\left(Z<\frac{x-740}{30}\right)=0.02.$$

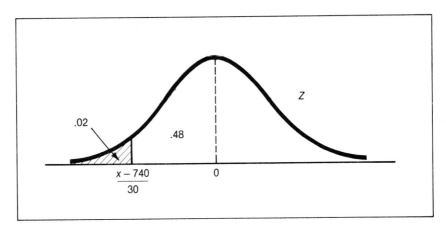

Then

$$P\left(\frac{x-740}{30} < Z < 0\right) = 0.5 - 0.02 = 0.48,$$

so from Table A-4, $(x-740)/30 = -2.06$. Therefore, $x = (30 \cdot -2.06) + 740 = 678.2$. If the bearings are replaced after 678.2 hours, only 2% of them will wear out in service.

SUMMARY OF CONTINUOUS DISTRIBUTIONS

TABLE 6.3. Summary of continuous distributions.

TYPE	PROBAILITY FUNCTION	μ	σ^2, σ	DISTRIBUTION SHAPE
Uniform	$f(x) = \dfrac{1}{b-a}$, $a \le X \le b$	$\dfrac{a+b}{2}$	$\dfrac{(b-a)^2}{12}$, $\dfrac{(b-a)\sqrt{3}}{6}$	
Exponential	$f(t) = \alpha e^{-\alpha t}, t \ge 0$, where α is the average number of arrivals in a unit of time	$1/\alpha$	$1/\alpha^2, 1/\alpha$	
Normal	$f(x) = \dfrac{1}{\sqrt{2\pi}\,\sigma} \exp[-(x-\mu)^2/2\sigma^2]$ $-\infty < x < \infty$	μ	σ^2, σ	

Distributional Approximations

Though the underlying concepts of quantities such as time and length are continuous, in practice we measure these with discrete approximations; tenths of a second, hundredths of an inch. In this section, we investigate using one distribution to approximate another.

DISCRETE APPROXIMATIONS OF DISCRETE DISTRIBUTIONS

We have already commented on similarities between the binomial and hypergeometric distributions. If we select a random sample of n items from a population of N items, of which S are of a particular type, the binomial distribution corresponds to sampling *with* replacement, and the hypergeometric, *without*. Their means are identical—$\mu = np = n(S/N)$—while the variance of the hypergeometric is less than that of the binomial:

$$\sigma^2_{\text{binomial}} = np(1-p)$$

$$\sigma^2_{\text{hypergeometric}} = n\,\frac{S}{N}\left(1 - \frac{S}{N}\right)\left(\frac{N-n}{N-1}\right).$$

The difference is the factor $(N-n)/(N-1)$, which is near 1 when N is large compared to n. In this situation, the hypergeometric distribution can be approximated with the binomial, as in this next example.

Example 15 In a production lot of 200 integrated circuits, 30 are defective. If 10 circuits are randomly chosen to be tested, what is the probability that no more than 2 of those tested are defective?

Solution

Note than $N = 200$, $S = 30$, and $n = 10$. Then

$$p = \frac{30}{200} = 0.15.$$

Using the binomial distribution $b(10, 0.15)$, we approximate the desired probability:

$$f(0) + f(1) + f(2) = \binom{10}{0}0.15^0\,0.85^{10} + \binom{10}{1}0.15^1\,0.85^9 + \binom{10}{2}0.15^2\,0.85^8$$

$$= 0.1969 + 0.3474 + 0.2759 = 0.8202.$$

In general, if $p = S/N$ is near $\frac{1}{2}$, the hypergeometric and binomial distributions have the relationship shown in Figure 6.25.

FIGURE 6.25.   Comparison of hypergeometric and binomial distributions.

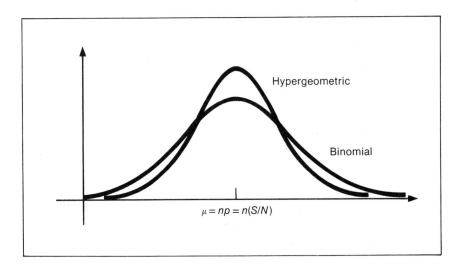

The Poisson distribution is also related to the binomial. When the probability of success p is small (that is, when $np \leq 5$), the binomial may be approximated by the Poisson distribution whose mean is the same as that of the binomial. The Poisson distribution has $\lambda = np$, as shown in this next example.

Example 16 The probability that a clock radio will require return to the factory for repairs is 0.04. If a department store has sold 50 of these clock radios, approximate the probability that more than one will be returned to the factory.

Solution

First, P (more than one will be returned) $= 1 - P$ (none or one will be returned). We approximate this latter probability, using the Poisson distribution with $\lambda = np = 50 \times 0.04 = 2$. (Since $np \leq 5$, we may do this.)

$$f(0) + f(1) \doteq \frac{2^0 e^{-2}}{0!} + \frac{2^1 e^{-2}}{1!} \doteq 0.1353 + 0.2707 = 0.4060.$$

The probability we seek is $1 - 0.4060 = 0.5940$.

In general, the graphs of a binomial distribution and its corresponding Poisson have the relationship shown in Figure 6.26.

FIGURE 6.26. **Comparison of binomial and Poisson distributions.**

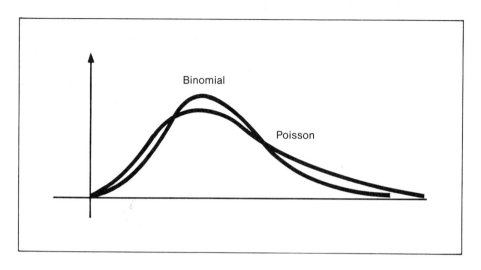

CONTINUOUS APPROXIMATIONS OF DISCRETE DISTRIBUTIONS

We have seen that as λ increases, the Poisson distribution becomes less skewed and more bell-shaped (Figure 6.27).

FIGURE 6.27. **The Poisson distribution for several values of λ.**

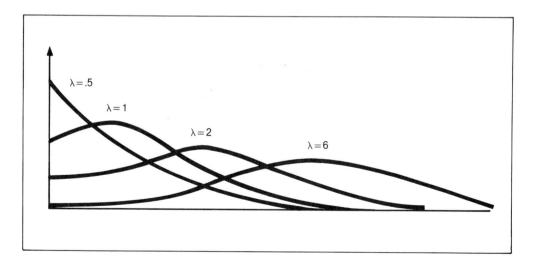

This suggests that if λ is large enough, a normal distribution may be used to approximate the Poisson, and this is the case. If $\lambda \geq 25$, then the Poisson may be approximated by the normal distribution with the same mean and variance, $N(\lambda,\lambda)$.

In a discrete distribution, there are positive probabilities associated with individual values in the range of the random variable, while in a continuous distribution there are not. When using a continuous distribution to approximate a discrete one, we include an interval of width 1 around each value of the random variable in the event whose probability we seek. That is, if continuous Y is used to approximate discrete X, $P(X = 10) \doteq P(9.5 \leq Y \leq 10.5)$. This adjustment is called the **continuity correction** (Figure 6.28).

Now, let X be a Poisson random variable with $\lambda = 50$. We use the normal distribution $Y = N(50, 50)$ to approximate $P(52 \leq X \leq 60)$.

$$P(52 \leq X \leq 60) \doteq P(51.5 < Y < 60.5)$$

$$= P\left(\frac{51.5 - 50}{\sqrt{50}} < \frac{Y - 50}{\sqrt{50}} < \frac{60.5 - 50}{\sqrt{50}}\right)$$

$$= P(0.21 < Z < 1.48)$$

$$= 0.4306 - 0.0832 = 0.3474.$$

The binomial distribution is also bell-shaped when p is near $\frac{1}{2}$ or when n is large. When these conditions occur (when np and $n(1 - p) \geq 5$), the binomial distribution $b(n, p)$ can be approximated with the normal distribution with mean (np) and variance (npq), $N(np, np(1-p))$. Again, the continuity correction is used, as shown in this example.

FIGURE 6.28.

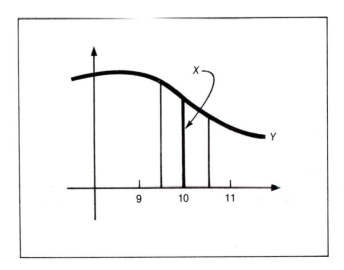

Example 17 In a large city, 35% of the households have two incomes. If we randomly select 100 households, what is the probability that 40 or more will have two incomes?

Solution

Let X be the number of sampled households with two incomes. Then X is $b(100, 0.35)$, with mean 35 and variance 22.75. We seek $P(X \geq 40)$, and we will approximate X with $Y = N(35, 22.75)$.

$$P(X \geq 40) = P(Y > 39.5) = P\left(\frac{Y-35}{\sqrt{22.75}} > \frac{39.5-35}{\sqrt{22.75}}\right)$$

$$= P(Z > 0.94) = 0.5000 - 0.3264 = 0.1736.$$

EXERCISES

1. An acquaintance offers to bet you $5 against your $3 that a certain event A will not occur. You estimate the probability of A to be 0.25, and must decide whether to accept the bet. Describe the steps you take in creating a model of this situation, solving the model, and applying that solution to the original problem.

2. We draw a single card from a well-shuffled deck, and X is its rank.
   a. Give the probability function of X, and its graph.
   b. Draw the graph of the cumulative density function of X.
   c. Find $E(X)$, Var (X), and σ_x.
   d. Find $P(3 \leq X < 7)$.

3. We flip a coin 10 times, and Y is the number of tails that appear.
   a. Describe the distribution of Y.
   b. Find the mean and variance of Y.
   c. What is the probability that exactly 6 tails appear?
   d. Find $P(Y \leq 4)$.
   e. Find $P(6 < Y < 9)$.
   f. Find $P(3 \leq Y \leq 5)$.

4. At an automobile dealership, the probability that a salesperson will sell a car is 0.20 for each customer that walks in, and the customers behave independently of each other. If 20 customers enter the dealership on a given morning,
   a. What is the probability that 5 will buy cars?
   b. What is the probability that fewer than 4 will buy cars?
   c. What is the expected number of sales? The variance?

5. Use the definition of $E(X)$ and the probability function of the binomial distibution to show that if X is $b(n, p)$, $E(X) = np$.

6. Use the definition of the mean of a random variable to verify that the mean of the binomial random variable in the example on p. 99 is 5/3. Verify also that Var$(X) = 25/18$.

7. Repeat Exercise 6 for the example on p. 103, verifying that $E(X) = 2.4$ and $\text{Var}(X) = 1.68$.

8. Shipments of raw materials arrive at a factory on time with probability 0.80. Assume that the shipments arrive independently.

   a. What is the probability that of 8 shipments, fewer than 6 arrive on time?

   b. Is the assumption in the statement of the problem likely to be valid? Why or why not?

9. A group of 20 job applicants contains 12 college graduates. If 6 are chosen at random to be interviewed,

   a. What is the probability that 3 are college graduates?

   b. What is the probability that fewer than 3 are college graduates?

   c. What is the expected number of college graduates in the group? Its variance and standard deviation?

10. Use the definition of $E(X)$ to verify that the mean of the distribution in the example on p. 103 is 3.89. Verify also that $\sigma^2 = 1.118$.

11. Repeat Exercise 10 for the example on p. 104, verifying that $\mu = 0.6$ and $\sigma^2 = 0.3733$.

12. In a batch of 25 calculators, 3 are defective. If 5 are selected at random and tested,

    a. What is the probability that none of the defectives is found?

    b. What is the probability that all the defectives are found?

    c. What is the probability that at least one is found?

    d. What is the expected number of defectives found, and the variance?

13. Let X be a Poisson random variable with mean λ. Show that $\text{Var}(X) = \lambda$.

14. Customers arrive at a shoe store at an average rate of 0.3 per minute. In a 10-minute period,

    a. What is the probability that exactly 4 customers arrive?

    b. What is the probability that 2 or less customers arrive?

    c. What is the probability that more than 2 customers arrive?

    d. What is the expected number of customers? The variance and standard deviation?

15. In Exercise 14, what is the expected number of customers in 5 minutes? 1 hour?

16. Poisson distributions often occur in association with intervals of some quantity other than time:

    In a factory which makes coaxial cable, flaws occur with a Poisson distribution at an average rate of 1 per 1000 feet of cable. In a 5000-foot roll, then

    a. What is the expected number of flaws?

    b. What is the probability of 3 or fewer flaws?

17. Accidents occur at a nuclear plant with a Poisson distribution with an average rate of one every 100 years. If 5 plants are operating independently, what is the probability that there will be no accidents at any of them in a 20-year period?

18. Let $f(x) = \begin{cases} 2x, 0 < x < 1 \\ 0, \text{ all other } x. \end{cases}$

    a. Verify that f is acceptable as the probability function of a continuous random variable, and draw its graph.

    b. Find $P(0 \le X \le \frac{1}{2})$.

    c. Find $P(\frac{1}{2} < X < 2)$.

    d. State the cumulative density function of X and draw its graph.

    e. Find $E(X)$.

    f. Find σ^2 and σ.

19. Use calculus to prove that if f is continuous, and $f(x_0) < 0$ for some x_0, then there must be an interval $[a, b]$ for which $\int_a^b f(x)dx < 0$. (*Hint:* Use the definition of continuity.)

20. If X is a continuous random variable, show that $\text{Var}(x) = \int_{-\infty}^{\infty} x^2 f(x)dx - \mu^2$. (*Hint:* Use properties of the definite integral. This derivation is very much like the one for discrete distributions.)

21. Let X be a continuous random variable, with mean μ and variance σ^2. Prove the following results:
   a. If $Y = ax + b$, $E(Y) = a\mu + b$.
   b. If $Y = ax + b$, $\text{Var}(Y) = a^2\sigma^2$.

22. Let X and Y be continuous random variables. Prove that $E(X + Y) = E(X) + E(Y)$. (*Hint:* Look at the proof for discrete random variables.)

23. Let X be a continuous uniform random variable on the interval $[a, b]$. Show that $E(X) = (a+b)/2$ and $\text{Var}(X) = (b-a)^2/12$. (*Hint:* To find $\text{Var}(X)$, use the results of Exercise 20.)

24. Let the random variable X be uniformly distributed on the interval $[0, 3]$.
   a. Give the probability function of X and draw its graph.
   b. Find $P(1 \le X \le 2)$.
   c. Find $P(2.5 \le X)$.
   d. Find $E(X)$ and $\text{Var}(X)$, using the formulas developed in Exercise 23 and using integrals.

25. Oil deliveries at a refinery are uniformly distributed between 50,000 and 100,000 barrels per day.
   a. What is the probability that between 50,000 and 60,000 barrels will be delivered on a given day?
   b. If X is the amount of oil delivered on a given day, what is the probability function of X?
   c. What is the probability that more than 80,000 barrels will be delivered?
   d. What is the expected amount of a daily delivery? The variance and standard deviation?

26. Consider again Exercise 14. If the store opens precisely at 9:00 A.M.,
   a. What is the probability that the first customer enters within 1 minute?
   b. What is the expected length of time until the first customer enters? The variance?

27. Cars arrive at an intersection from the west at an average rate of 4 per minute.
   a. What is the expected length of time between two arrivals from the west?
   b. One car has just driven through the intersection from the west. What is the probability that another car will pass within 15 seconds? After at least a minute?
   c. Two cars have passed with a gap between them of only 5 seconds. What is the probability that another will pass within 15 seconds?

28. Let X have an exponential distribution with rate α. Use integration to show that $\text{Var}(X) = 1/\alpha^2$.

29. Let X be an exponential random variable with probability function f. Show that its cumulative density function is $F(t) = 1 - e^{-\alpha t}$ for $0 \le t < \infty$.

30. Z is the standard normal random variable, N(0, 1). Find the following probabilities:
   a. $P(0 \le Z < 1.3)$
   b. $P(Z \ge 2.15)$
   c. $P(1.1 \le Z < 2.27)$
   d. $P(-2.89 < Z \le 0)$
   e. $P(Z < -1.77)$
   f. $P(-2.14 < Z < -1.65)$
   g. $P(Z \ge -1.00)$
   h. $P(Z < 0.24)$
   i. $P(-1.03 < Z \le 0.79)$
   j. $P(Z > 3.0)$

31. Z is the standard normal random variable. In each of the following, find the value of z for which the statement is true:

   a. $P(0 \leq Z \leq z) = 0.3925$ e. $P(Z < z) = 0.1093$

   b. $P(Z > z) = 0.0281$ f. $P(Z \leq z) = 0.5239$

   c. $P(Z \leq z) = 0.8888$ g. $P(-z \leq Z \leq z) = 0.6826$

   d. $P(z \leq Z < 0) = 0.1064$ h. $P(-z < Z < z) = 0.7372$

32. The probability function of Z is $f(x) = 1/\sqrt{2\pi} \exp(-x^2/2)$. Use calculus to show that this function attains its maximum value at $x = 0$.

33. Use calculus to show that the inflection points of the probability function of Z occur at $x = -1$ and $x = 1$.

34. X has a normal distribution with mean 50 and standard deviation 5; that is, X is $N(50, 25)$. Find the following probabilities:

   a. $P(50 < X < 57)$ f. $P(X \leq 51.7)$

   b. $P(X > 57.5)$ g. $P(X \geq 48.6)$

   c. $P(57.2 \leq X \leq 62)$ h. $P(41.7 < X < 49.3)$

   d. $P(42 < X \leq 50)$ i. $P(45 < X \leq 52)$

   e. $P(X < 40)$ j. $P(46.4 \leq X \leq 57.6)$

35. X is a normal random variable with mean 70.5 and standard deviation 7.6. Find X so that:

   a. $P(70.5 < X \leq x) = 0.3686$ e. $P(X \leq x) = 0.2776$

   b. $P(X > x) = 0.3594$ f. $P(X < x) = 0.7224$

   c. $P(X \geq x) = 0.9082$ g. $P(70.5 - x < X < 70.5 + x) = 0.6922$

   d. $P(x \leq X \leq 70.5) = 0.1628$

36. Bricks made at the Stonehenge Brickyard have weights which are normally distributed around a mean of 8.0 pounds, with standard deviation 0.25 pounds.

   a. We select one brick at random. What is the probability that its weight is greater than 7.9 pounds?

   b. What proportion of all the bricks have weights between 7.6 and 8.2 pounds?

   c. 90% of the bricks have weights greater than what value?

37. Times required to take an aptitude test are normally distributed, with mean 110 minutes and standard deviation 18 minutes.

    a. A person takes the test. What is the probability she finishes in less than 2 hours?

    b. What proportion of those who take the test require more than 90 minutes but less than 125?

    c. It is desired that 80% of those taking the test finish it. How long should be allowed?

38. Verify the *"empirical rule"* given in Chapter 4, p. 45.

39. A pile of 60 tests contains five with scores of 100. If 7 tests are randomly selected, what is the probability that none or one of those selected have scores of 100? Solve this problem in two ways, using the hypergeometric distribution and the appropriate binomial approximation.

40. From a group of 54 men and 46 women, a committee of 10 people is randomly assigned. Use both the hypergeometric distribution and the binomial approximation to find the probability that 5 members of the committee are women.

41. Two percent of the lightbulbs produced by the Acme Light Company are defective. Use a Poisson distribution to approximate the probability that in a box of 100 bulbs, less than 4 are defective.

42. Having completed a special training program, the probability that a salesman will stay with the Ace Home Products Company is 96%. Use a Poisson distribution to approximate the probability that in a training group of 70 salesmen, less than 68 will stay with the company. (*Hint:* The probability that a salesman will leave is 4%.)

43. Cars arrive at a drive-up bank in a Poisson process with an average rate of 50 per hour. Use a normal distribution to approximate the probability that in an hour between 45 and 60 cars arrive.

44. Jobs are submitted at the input window of a computer center in a Poisson process with an average rate of 15 per hour. Use a normal distribution to approximate the probability that in a 4-hour period, more than 70 jobs are submitted.

45. We flip a coin 50 times, and X is the number of heads. Use a normal distribution to approximate these probabilities:

    a. $P(20 \leq X \leq 30)$.

    b. $P(20 < X < 30)$.

    c. $P(X > 34)$.

    d. $P(X < 27)$.

46. The probability that a person will pass a particular standardized test is 0.65. If 90 people take this test, use a normal distribution to approximate the probability that between 50 and 60 (inclusive) pass.

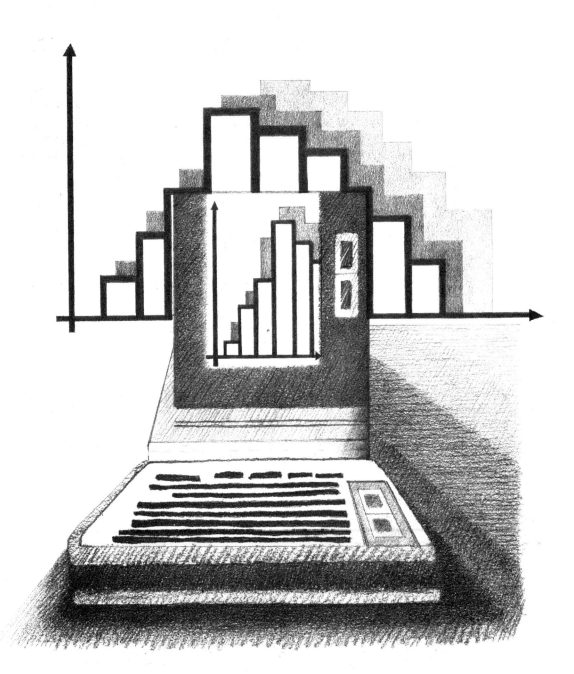

7

Sampling Distributions

INTRODUCTION

In Chapter 3 we considered descriptive statistics, methods by which large amounts of data are condensed and made more intelligible. This is a deductive *process, reasoning from the whole to some part or characteristic of the whole. We now begin our examination of* inductive *reasoning, from a part to the whole, from a sample to the population from which it came; we begin to explore* statistical inference.

The need for such processes is clear. Populations of values in which we are interested might be too difficult, expensive, or time-consuming to obtain, or simply too large to be efficiently analyzed. Instead of investigating the entire population of values, we select a *sample* from it, analyze the sample, and from it *infer* characteristics of the population. By examining the incomes of 100 factory workers in Connecticut, for example, we can estimate the incomes of all factory workers there.

In general, our process is this:

1. From a population of values, select a sample.

2. Compute one or more statistics of the sample.

3. Use these statistics to estimate or draw conclusions about parameters of the population.

In order to perform the third step, the sample must be related to the population in a way in which we can reason from the sample to the population. This can be done with samples chosen so that no element of the population is more likely than any other to be selected:

DEFINITION

A **simple random sample** is chosen from a population when elements of the population have the same probability of being included in the sample.

That is, the elements of the sample are randomly selected.

The selection of a random sample is a random experiment, upon which we may define random variables. In particular, we can consider statistics of the sample—its standard deviation, or mean—as random variables, and examine their distributions:

DEFINITION

The distribution of a random variable which is a statistic of a sample is called a **sampling distribution**.

In this chapter we describe some sampling distributions.

An Example of a Sampling Distribution

The most widely applied statistic is the sample mean, $\bar{X}$. We examine the sampling distribution of $\bar{X}$ by generating many samples from a known population and by comparing the observed distribution of sample means with the population distribution.

For illustration a population composed of real numbers from 60 to 110 was generally possessing the distribution shown in Figure 7.1. The mean of the population is 87.60, and the standard deviation is 14.63.

The authors used the matrix manipulation language FOSOL, mentioned in Chapter 2, to select 200 random samples from this population of values and to compute their sample

FIGURE 7.1.   Population from which samples of size 50 were drawn.

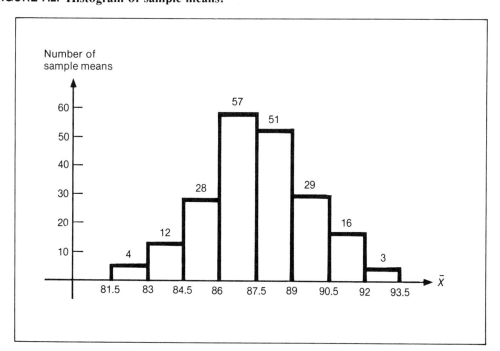

means. A histogram of the sample means, which represents 200 sample means based on samples of size 50 from this given population, is shown in Figure 7.2.

FIGURE 7.2. Histogram of sample means.

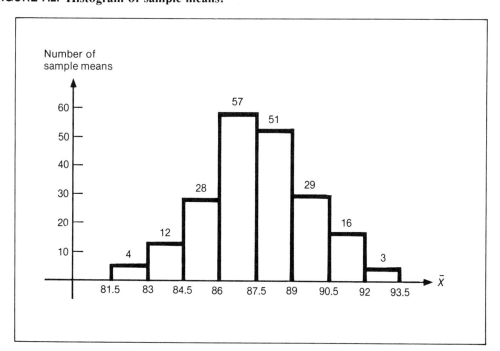

Three important observations can be made about the distribution of sample means:

1. It appears normal, since it is symmetrical and bell-shaped.

2. The mean of the distribution of sample means was found to be 87.5, very near the population mean of 87.6. This suggests that the expected value of $\bar{X}$, as a random variable, is near the population mean μ.

3. The standard deviation of the sample means was found to be 2.17, much less than the population standard deviation 14.63. That is, the distribution of sample means shows less variability than does the population.

We now consider the sampling distribution of $\bar{X}$ in general, and the theoretical foundations of the above observations.

The Sampling Distribution of $\overline{X}$

Suppose we have a population of values with mean μ, variance σ^2, and standard deviation σ. From this population, we select *one* element at random, and its value is the random variable X_1. It should be clear that the distribution of X_1 is identical to that of the population: X_1 has mean μ, variance σ^2, and standard deviation σ.

Now select a random sample of n elements from the population. Let the random variables $X_1, X_2, \ldots, X_n$ be their values; the distributions of these random variables are identical to the distribution of X_1. All have mean μ, variance σ^2, and standard deviation σ.

The sample mean, $\bar{X}$, is the mean of the elements of the sample, so

$$\bar{X} = \frac{X_1 + X_2 + \ldots + X_n}{n} = \frac{1}{n}\left(\sum_{i=1}^{n} X_i\right);$$

the mean of the distribution of $\bar{X}$ is:

$$E(\bar{X}) = E\left(\frac{1}{n}\sum_{i=1}^{n} X_i\right) = \frac{1}{n}E\left(\sum_{i=1}^{n} X_i\right) = \frac{1}{n}\sum_{i=1}^{n}E(X_i) = \frac{1}{n}\sum_{i=1}^{n}\mu = \frac{1}{n} \times n\mu = \mu.$$

As predicted by our experience with the 200 samples of values, the expected value of the sample mean is equal to the population mean; the sample mean is "aimed at" the population mean.

To consider $\text{Var}(\bar{X})$, first assume that the population size is large relative to the sample size. Then the elements of the sample will be independent, and we can apply a result from Chapter 5, that the variance of a sum of independent random variables is the sum of their variances:

$$\text{Var}(\bar{X}) = \text{Var}\left(\frac{1}{n}\sum_{i=1}^{n} X_i\right) = \frac{1}{n^2}\text{Var}\left(\sum_{i=1}^{n} X_i\right) = \frac{1}{n^2}\sum_{i=1}^{n}\text{Var}(X_i)$$

$$= \frac{1}{n^2}\sum_{i=1}^{n}\sigma^2 = \frac{1}{n^2} \times n\sigma^2 = \frac{\sigma^2}{n}.$$

Also, $\sigma_{\bar{X}} = \sigma/\sqrt{n}$. ($\sigma_{\bar{X}}$ is often called the *standard error of the mean*, and is the value labeled STD ERR in SPSS printouts. SPSS assumes that the values on which it is operating are a sample from some population.)

If the population is small relative to the sample size, then the elements of the sample are *not* independent, and we must include a correction factor in the calculations of $\sigma_{\bar{X}}^2$ and $\sigma_{\bar{X}}$ for when $n \geq 5\%$ of N,

$$\sigma_{\bar{X}}^2 = \frac{\sigma^2}{n} \times \frac{N-n}{N-1} \text{ and } \sigma_{\bar{X}} = \frac{\sigma}{\sqrt{n}} \, [(N-n)/(N-1)]^{\frac{1}{2}}.$$

(Note that the relationship between the values of $\sigma_{\bar{X}}^2$ for independent and dependent sample elements is precisely the relationship between the variance of a binomial distribution and its corresponding hypergeometric.)

In our example, the standard error of the mean is

$$\sigma_{\bar{X}} = \frac{14.63}{\sqrt{50}} \doteq 2.07$$

This is very close to the observed standard deviation of the 200 sample means.

An important result in probability theory, though one whose proof is beyond the scope of this text, is that any linear combination of normal random variables is itself normally distributed. That is, if Y_i is $N(\mu_i, \sigma_i^2)$ for $i = 1, 2, \ldots, n$, and the a_i are constants, then $X = \sum_{i=1}^{n} a_i Y_i$ will have a normal distribution. (The mean of X will be $\sum_{i=1}^{n} a_i \mu_i$, but the variance of X will not be $\sum_{i=1}^{n} a_i^2 \sigma_i^2$ unless the Y_i are independent.)

Since the mean of a sample is a linear combination of random variables (each of the coefficients is n^{-1}, a consequence of the above result is this: When sampling from a normally distributed population, the sampling distribution of $\bar{X}$ is also normal (with mean μ and variance σ^2/n).

For example, suppose that we take a random sample of 15 elements from a normally distributed population with mean 140 and variance 36. Then the sampling distribution of $\bar{X}$ is $N(140, 36/15)$, and calculations like these may be performed:

$$P(\bar{X} > 141) = P\left(\frac{\bar{X} - 140}{6/\sqrt{15}} > \frac{141 - 140}{6/\sqrt{15}} \right) = P(Z > 0.65)$$

$$= 0.5000 - 0.2422 = 0.2578.$$

The Central Limit Theorem

Populations need not be normally distributed, of course, and samples are taken from those which are not, so it is pleasant and surprising to find that $\bar{X}$ is approximately normal in a much wider range of situations. This remarkable result is the foundation of statistical inference.

THEOREM 7.1

The Central Limit Theorem: If we take a sample of size $n \geq 30$ from a population with mean μ and standard deviation σ, then the sampling distribution of $\bar{X}$ is approximately normal with mean μ and standard deviation $\sigma/\sqrt{n}$.

That is, for a large enough sample, the sampling distribution of $\bar{X}$ is always approximately normal, regardless of the shape of the population distribution. With samples of size less than 30 (a value dictated by experience, not theory), the formulas for $E(\bar{X})$ and $\text{Var}(\bar{X})$ hold, but the distribution of $\bar{X}$ is not near enough to normal to be useful.

The graphs in Figure 7.3 illustrate the central limit theorem. Note that as the sample sizes increase, the distributions of the sample means become closer to normal distributions.

FIGURE 7.3(a)-(c). Graphs illustrating the central limit theorem.

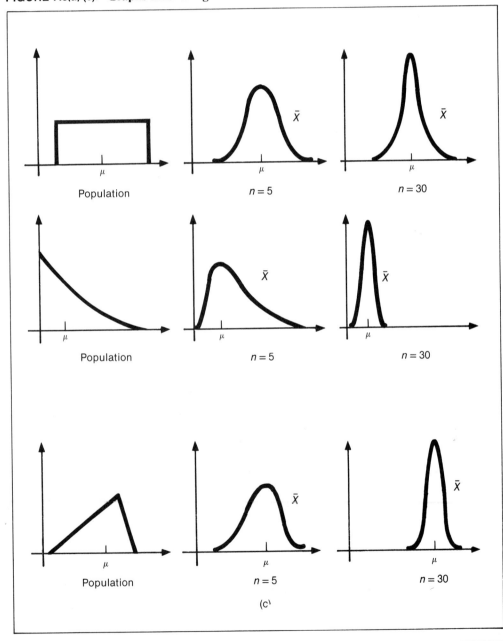

These graphs also illustrate that as the size of the sample increases, $\sigma_{\bar{X}} = \sigma/\sqrt{n}$ decreases. The larger the sample size, the smaller the standard error, and the more closely packed around $E(\bar{X}) = \mu$ is the distribution of $\bar{X}$. For a larger sample, $\bar{X}$ is more likely to be near the population mean μ.

Given the parameters of a population, we can use the central limit theorem to find probabilities involving $\bar{X}$ as in this example.

Example 1 Fluorescent bulbs manufactured by the All-Night Light Company have mean lifetime 1700 hours, with standard deviation 300 hours. If 100 bulbs are tested, what is the probability that their mean lifetime falls between 1650 and 1725 hours?

Solution

We are seeking $P(1650 < \bar{X} < 1725)$, and we know that $\bar{X}$ is approximately normal with mean 1700 and standard deviation $300/\sqrt{100}$. Therefore,

$$P(1650 < \bar{X} < 1725) \doteq P\left(\frac{1650 - 1700}{300/\sqrt{100}} < \frac{\bar{X} - 1700}{300/\sqrt{100}} < \frac{1725 - 1700}{300/\sqrt{100}} \right)$$

$$= P(-1.67 < Z < 0.83) = 0.4525 + 0.2967 = 0.7492.$$

Similarly, the probability that the mean lifetime of the sample is more than 1720 hours is:

$$P(\bar{X} > 1720) = P\left(\frac{\bar{X} - 1700}{300/\sqrt{100}} > \frac{1720 - 1700}{300/\sqrt{100}} \right)$$

$$= P(Z > 0.67) = 0.5000 - 0.2486 = 0.2514.$$

In general, if $\bar{X}$ is the mean of a sample of size $n \geq 30$ from a population with mean μ and standard deviation σ,

$$P(a < \bar{X} < b) \sim \left(\frac{a - \mu}{\sigma/\sqrt{n}} < Z < \frac{b - \mu}{\sigma/\sqrt{n}} \right).$$

Had the sample of 100 bulbs been taken from a production lot of 500 bulbs, we would include the correction factor $[(N - n)/(N - 1)]^{1/2}$ in finding $\sigma_{\bar{X}}$:

$$\sigma_{\bar{X}} = \frac{300}{\sqrt{100}} \times [(500 - 100)/(500 - 1)]^{1/2} \doteq 26.86.$$

Then, for example,

$$P(\bar{X} > 1720) = P\left(\frac{\bar{X} - 1700}{26.86} > \frac{1720 - 1700}{26.86} \right)$$

$$= P(Z > 0.74) = 0.5000 - 0.2704 = 0.2296.$$

Sampling in SPSS

When you are analyzing a very large data file, it is sometimes convenient to work with a randomly selected subset of the file. In SPSS, cases are randomly chosen with the SAMPLE command, which operates in this way:

For each case, a random number between 0 and 1 is chosen from a uniform distribution. If the number chosen for a particular case is less than a value specified in the SAMPLE command, that case is included in the sample. With the command

SAMPLE   0.10

every case has probability 0.10 of being included in the sample. Note that the number of cases in the sample is a binomial random variable, and the number of cases included may vary from one execution to another of an SPSS program containing such a statement. This is not the most efficient way to generate samples of a previously determined size; other packages generate samples by other methods.

This program selects a sample of approximately one-fifth of the cases in the USNEWS file and calculates the mean and median of X3 and X5 for the chosen cases:

```
?10     GET FILE;USNEWS
?20     SAMPLE;0.20
?30     CONDESCRIPTIVE;X3,X5
?40     STATISTICS; 1,3
```

The Distribution of the Sample Median

We have seen that for a unimodal, symmetrical distribution, the mean and median are equal. When sampling from such a population, we might approximate the population mean with the sample median.

When a population is symmetrically distributed, the expected value of the sample median is the population mean. While the standard error of $\bar{X}$ is $\sigma/\sqrt{n}$, where σ is the population standard deviation, that of the median is larger, $1.253\sigma/\sqrt{n}$, so that the sample mean can be expected to be closer than the sample median to the population mean. The distribution of the sample median tends toward normality as the sample size increases, and when sampling from a normal population, the distribution of the sample median is itself normal, with mean equal to the population mean and standard deviation $1.253\sigma/\sqrt{n}$. Figure 7.4 compares the sampling distributions of the mean and the median.

Sampling Distributions of Measures of Dispersion

Often the researcher will wish to investigate the variance or standard deviation of a population. In bottling soft drinks, for example, variations in the amounts in the bottles are of great importance. We consider some parameters of the distributions of the sample variance and sample standard deviation, though detailed examination of these distributions is postponed.

THE EXPECTED VALUE OF THE SAMPLE VARIANCE

While the definitions of the sample and population means are identical except for the symbols used, there is a significant difference in the calculations of variances. The population variance

$$\frac{1}{N}\sum_{i=1}^{N}(X_i-\mu)^2$$

FIGURE 7.4.   **Sampling distributions of the mean and the median.**

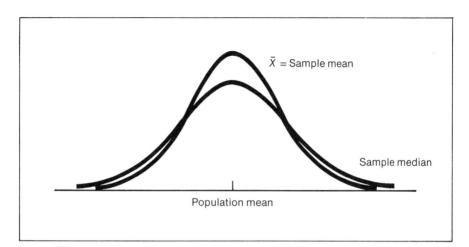

is the mean of the squared deviations from the mean, but the sample variance

$$\frac{1}{n-1} \sum_{i=1}^{n} (X_i - \bar{X})^2$$

is computed with a division by $n-1$, rather than by n. When these definitions were made, the difference was justified by claiming that the sample variance was thereby made a better estimator of its corresponding population variance. We now have the mathematical tools to prove that claim. Because of the division by $n-1$ in the definition of the sample variance S^2, its expected value is the population variance σ^2:

$$E(S^2) = E\left(\frac{1}{n-1} \sum_{i=1}^{n} (X_i - \bar{X})^2\right)$$

$$= E\left(\frac{1}{n-1} \sum_{i=1}^{n} X_i^2 - \frac{n}{n-1} \bar{X}^2\right)$$

$$= \frac{1}{n-1} E\left(\sum_{i=1}^{n} X_i^2\right) - \frac{n}{n-1} E(\bar{X}^2).$$

But $\mathrm{Var}(X_i) = E(X_i^2) - (E\,X_i^2)$, so that

$$E(X_i^2) = \mathrm{Var}(X_i) + (E\,X_i^2)$$

$$= \sigma^2 + \mu^2.$$

Then

$$E(S^2) = \frac{1}{n-1} \sum_{i=1}^{n} (\sigma^2 + \mu^2) - \frac{n}{n-1} E(\bar{X}^2)$$

$$= \frac{n}{n-1} (\sigma^2 + \mu^2) - \frac{n}{n-1} E(\bar{X}^2).$$

But $\mathrm{Var}(\bar{X}) = E(\bar{X}^2) - E(\bar{X})^2$, so that

$$E(\bar{X}^2) = E(\bar{X})^2 + \mathrm{Var}(\bar{X})$$

$$= \mu^2 + \sigma^2/n.$$

Then

$$E(S^2) = \frac{n}{n-1} (\sigma^2 + \mu^2) - \frac{n}{n-1} \left(\mu^2 - \frac{\sigma^2}{n} \right)$$

$$= \frac{n}{n-1} \sigma^2 - \frac{n}{n-1} \times \frac{\sigma^2}{n}$$

$$= \frac{n}{n-1} \sigma^2 - \frac{1}{n-1} \sigma^2 = \frac{n-1}{n-1} \sigma^2 = \sigma^2.$$

That is, $E(S^2) = \sigma^2$.

Later we will consider sampling and S^2 in more detail, relating the distribution of S^2 to a positively skewed continuous distribution with the name "chi-square."

Intuitively, it might seem that the shape and parameters of the distribution of the sample variance would dictate the shape and parameters of the distribution of the sample standard deviation, but this is not entirely so. In particular, the expected value of the square root of a random variable is not necessarily the square root of the expected value. Though we have shown that $E(S^2) = \sigma^2$, we cannot necessarily say that $E(S) = \sigma$.

The Sample Range

The sample range provides an example of a sample statistic whose expected value is not, in general, equal to its corresponding population parameter, since the sample range will equal the population range only if the sample contains the two extreme values of the population. This is an unlikely situation, so the expected value of the sample range is always less than the population range.

The Distribution of the Sample Proportion

Frequently, sampling is used to estimate the proportion of cases P in a population which have a particular characteristic. If we associate the value 1 with those cases which have the characteristic, and the value 0 with those that do not, the mean of this population of 0's and 1's is the proportion P of the original population which has the characteristic of interest. If we take a sample of n cases from the population, the proportion $\hat{p}$ of the sample having the characteristic, the mean of the corresponding sample of 0's and 1's, is an estimate of the population proportion P.

The number of elements R of the sample having the characteristic has the binomial distribution $b(n, P)$, which has mean nP and variance $nP(1 - P)$. The *proportion p* of the sample having the characteristic is R/n, so:

$$E(\hat{p}) = E\left(\frac{R}{n}\right) = \frac{1}{n}\ E(R) = \frac{1}{n}\ nP = P,$$

and

$$\mathrm{Var}(\hat{p}) = \mathrm{Var}\left(\frac{R}{n}\right) = \frac{1}{n^2}\ \mathrm{Var}(R) = \frac{1}{n^2}\ nP(1 - P) = \frac{P(1 - P)}{n}\ .$$

The distribution of the sample proportion has expected value P, the population proportion, and standard deviation $[P(1 - P)/n]^{1/2} = \sigma_{\hat{p}}$.

Also, we know by the central limit theorem that as n increases the distribution of $\hat{p}$ becomes approximately normal. Therefore, for large samples, the distribution of the sample proportion is approximately normal with mean P and standard deviation $[P(1 - P)/n]^{1/2}$.

Example 2 Thirty-eight percent of all registered voters are Democrats. If we interview 300 randomly selected registered voters, what is the probability that between 36% and 40% of them are Democrats?

Solution

We seek $P(0.36 < \hat{p} < 0.40)$, where p is approximately $N(0.38, (0.38 \times 0.62)/300)$:

$$P(0.36 < p < 0.40) = P\left(\frac{0.36 - 0.38}{\left(\frac{0.38 \times 0.62}{300}\right)^{1/2}} < \frac{p - 0.38}{\left(\frac{0.38 \times 0.62}{300}\right)^{1/2}} < \frac{0.40 - 0.38}{\left(\frac{0.38 \times 0.62}{300}\right)^{1/2}}\right)$$

$$= P(-0.71 < Z < 0.71)$$

$$= 0.2611 + 0.2611 = 0.5222.$$

Finally, if the sample size is large relative to the size of the population, we must adjust the standard error $\sigma_{\hat{p}}$. For example, if a population of 500 elements has proportion 0.71, and we take a sample of 100 elements from it, the distribution of the sample proportion $\hat{p}$ is approximately normal with mean 0.71 and standard deviation

$$\sigma_{\hat{p}} = \left(\frac{0.71 \times 0.29}{100}\right)^{1/2} \times \left(\frac{500 - 100}{500 - 1}\right)^{1/2}$$

EXERCISES

1. Appendix B contains statistics of 100 samples of size 7 of values of X5 from the USNEWS file. The samples were generated with the program given on p. 150.

   a. Use a RECODE command and the FREQUENCIES subprogram to produce a histogram of all the values of X5, and their mean and standard deviation.

   b. Build a histogram of the means of the 100 samples of values of X5, and find the mean and standard deviation of the sample means.

   c. Compare the histograms and statistics of the population and the distribution of sample means. What observations can you make?

2. Select a variable from the USNEWS file or some other data file. Treating the values of this variable as a population, find its parameters. Then generate many random samples of size 10, and find their statistics. Use these statistics to examine the sampling distributions of:

   a. the mean.

   b. the median.

   c. the variance.

   d. the standard deviation.

   Compare the observed values of parameters of these distributions with the values predicted by theory, where we have discussed them, and observe the shapes of the sampling distributions. (*Note:* The repetition involved in generating many random samples suggests having each member of the class generate, say, five samples, then using all the samples to construct sampling distributions.)

3. Consider generating random samples from a population of size 74 with the command SAMPLE 0.10.

   a. If X is the number of elements selected for the sample on one run of the program, what is the distribution of X, its expected value, and its variance?

   b. What is the probability that a particular run will produce a sample of size 7?

4. A normally distributed population of values has mean 56 and standard deviation 12. If X is the mean of a random sample of size 9 from this population, find

   a. $P(54 < \bar{X} < 59)$.

   b. $P(\bar{X} \geq 62.4)$.

   c. $P(47 < \bar{X} < 53)$.

   d. x for which $P(\bar{X} > x) = 0.0228$.

5. The widths of metal parts produced by Northern Metal Products are normally distributed with mean 10.6 cm and standard deviation 0.5 cm. A sample of 16 parts is measured, and $\bar{X}$ is the mean of their widths. Find:

   a. $P(10.5 < \bar{X} < 10.7)$.

   b. $P(\bar{X} \leq 10.4)$.

   c. $P(\bar{X} \geq 10.3)$.

   d. x for which $P(\bar{X} > x) = 0.0228$.

6. We take a sample of size n from a population which has mean 100 and standard deviation 20. Find $P(98 < \bar{X} < 102)$ if

   a. $n = 50$.

   b. $n = 100$.

   c. $n = 200$.

7. We take a sample of size n from a population which has mean 5.75 and standard deviation 0.32. Find $P(5.70 < \bar{X} < 5.80)$ if:

   a. $n = 40$.

   b. $n = 80$.

   c. $n = 150$.

8. Solar cells manufactured by the Solar Products Company have mean power output 1.75 watts in direct sunlight, with standard deviation 0.25 watts. We take a random sample of 75 cells.

   a. Describe the distribution of the mean power output of the sample of cells, and sketch its graph.

   b. What is the probability that the mean power output of the sample is between 1.70 and 1.80 watts? Greater than 1.72 watts?

   c. A sample is taken, and $\bar{X}$ is 1.82 watts. What would the probability have been of getting a value of $\bar{X}$ at least this large?

   d. How large a sample should be taken so that the probability that the mean power output of the sample is between 1.73 and 1.77 watts is at least 0.95?

9. Salaries of white-collar workers in a large city have standard deviation $14,300. We take a random sample of 200 such workers.

   a. Describe the distribution of the mean salary of the sample, and sketch its graph.

   b. What is the probability that the mean salary of the sample is within $500 of the mean salary of all white-collar workers in the city?

   c. How large a sample should be taken so that the probability that the sample mean is within $250 of the population mean is at least 0.90?

10. Canisters of fertilizer have mean weight 80 pounds, with standard deviation 2.5 pounds.

    a. If a sample of 50 canisters is weighed, what is the probability that the mean weight of the sample is between 79.2 and 80.2 pounds?

    b. How large a sample should be taken so that the probability that the mean weight of the sample is greater than 80.3 pounds is less than 0.05?

    c. A sample of 50 canisters has mean weight 79.1 pounds. What would the probability have been that $\bar{X}$ would be no more than this value?

    d. If a sample of size 50 is taken from a production lot of 250 canisters, what is $P(79.2 \leq \bar{X} \leq 80.2)$?

11. A large company has 140 accounts receivable, which have a mean value of $210, with standard deviation $75. If 35 accounts are randomly selected, what is the probability that their mean amount is:

    a. between $190 and $230?

    b. less than $202?

12. In Exercise 8, assuming that the population of solar cell power outputs is normally distributed, we take a random sample of 75 cells.

    a. Describe the sampling distribution of the median, and sketch its graph.

    b. What is the probability that the median of the sample will fall between 1.70 and 1.80 watts?

    c. How large a sample should be taken so that the probability that the median power output of the sample is between 1.73 and 1.77 watts is at least 0.95?

13. In Exercise 9, assuming that the population of white-collar incomes is normally distributed,

    a. Describe the sampling distribution of the median, and sketch its graph.

    b. What is the probability that the median of a sample of size 200 will fall within $500 of the mean salary of all white-collar workers?

    c. How large a sample should be taken so that the probability that the sample median is within $250 of the population mean is at least 0.90?

    d. Is the assumption of a normally distributed population likely to be valid here? Why or why not?

14. Appendix B contains sample medians of 100 samples of size 7 from the 73 values of X5 in the USNEWS file. Find the mean and standard deviation of the sample medians, construct a histogram of the values, and compare this distribution to that of the values of X5. (*Hint:* It might be easiest to use SPSS or to approximate the statistics of the sampling distribution by calculating them from the grouped values corresponding to the histogram.)

15. You take a sample of size 50 from a large symmetrically distributed population, and compute the mean and median of the sample. Which is likely to be closer to the population mean? Why?

16. Repeat Exercise 14 for the standard deviations of the 100 samples, as given in Appendix B. Compare the results with the standard deviation of X5.

17. Approximately 10% of all Americans are left-handed. If we take a sample of 500 Americans, then,

   a. Describe the distribution of the proportion of the sample which is left-handed, and sketch its graph.

   b. What is the probability that between 8% and 12% of the sample is left-handed?

   c. What is the probability that more than 11.5% of the sample is left-handed?

   d. The sample proportion is 11.7%. What is the probability of getting a value at least this large for p?

18. At a university computer center, 52% of all jobs submitted are in the BASIC programming language. If a random sample of 300 jobs is taken:

    a. Describe the distribution of the proportion of the sample which is in BASIC, and sketch its graph.

    b. What is the probability that the proportion of the sample in BASIC is between 45% and 55%?

    c. What is the probability that the proportion of the sample in BASIC is less than 50%?

    d. How large a sample should be taken so that the probability that the proportion of the sample in BASIC is between 49% and 55% is at least 0.90?

19. At a political rally of 450 people, 75% support the candidate who speaks. If 100 people are randomly selected from the crowd and interviewed:

    a. What is the probability that between 70% and 80% of the sample support the candidate?

    b. What is the probability that less than 68% support the candidate?

20. Of a production run of 1000 gold-plated widgets, 190 are imperfect. If 200 widgets are randomly selected and tested:

    a. What is the probability that more than 23% of the sample is imperfect?

    b. What is the probability that between 15% and 20% of the sample is imperfect?

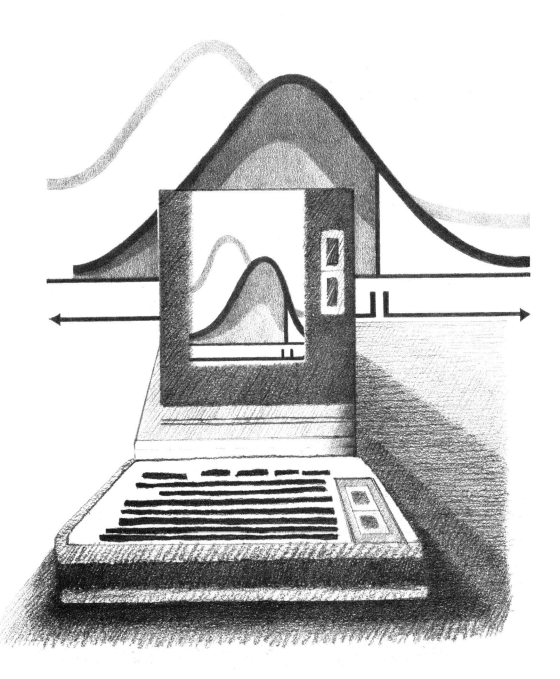

Estimation Procedures

INTRODUCTION

Having examined the sampling distributions of several important statistics in Chapter 7, we now apply this concept to estimation, *the process of predicting population parameters by analyzing a random sample from the population. The three techniques of this analysis are* point estimation, interval estimation, *and* hypothesis testing.

A point estimate of a population parameter is a single sample statistic used to represent the true value of the parameter, and we endeavor to find the "best" point estimate for a given parameter. In one point estimation method, a population parameter is estimated simply with its corresponding sample statistic; the sample mean estimates the population mean, and so on. We will discuss several criteria for determining a "best" point estimator.

A point estimate of a population parameter—"We estimate the mean of this population to be 24.6"—might be precise or inaccurate; we have no way of knowing. We would like to have an indication of the accuracy of the estimate. This gives rise to confidence intervals *or* interval estimates *of parameters. The researcher defines sample statistics* L_L *and* L_U, *for which*

$$P(L_L < \text{true value of parameter} < L_U) = C,$$

where C is a value between 0 and 1, usually 90% or 95%. When the sample is taken and L_L *and* L_U *computed, we can be C confident that the interval between them contains the parameter we seek;* L_L *and* L_U *are the limits of a confidence interval for the parameter with confidence level C.*

In hypothesis testing, we pose two mutually exclusive statements—hypotheses—about a population parameter, then develop a decision rule, *a procedure based on one or more statistics of a random sample that will determine which of the two hypotheses we should accept as true.*

Point Estimation

Consider the following example, in which we estimate a population proportion P.

Example 1 The manager of a hardware store has bought 200 lightbulbs, and wishes to know what proportion of them will burn out prematurely. Testing them all is futile, so she decides to test a random sample of 20 of them. She will use the proportion $\hat{p}$ of the sample which is defective as an estimate of the defective proportion P in the population of 200 bulbs. If 4 of the sampled bulbs are defective, $p = 4/20 = 0.20$, and she will estimate that 20% of the entire shipment is defective.

$\hat{p}$ is a point estimator of P, but is it a "good" point estimator? Might some other statistic of the sample, say $\hat{p} - .01$, be "better"? In evaluating a point estimator, or comparing point estimators, these five factors are often examined: *bias, efficiency, mean-square error, consistency,* and *sufficiency.* We consider them in order:

DEFINITION

The sample statistic T is an **unbiased** estimator of the population parameter Θ if for all possible values of Θ, $E(T) = \Theta$. If $E(T) \neq \Theta$, then the **bias** of T as an estimator of Θ is $B = E(T) - \Theta$.

Example 2 Consider a dart thrower. If, on the average, his throws cluster around the specified target, we would say he is an unbiased thrower. An unbiased estimator is "aimed at" its corresponding parameter.

In Chapter 7 we demonstrated that $E(\bar{X}) = \mu$; the sample mean is an unbiased estimator of the population mean. We also showed that $E(S^2) = \sigma^2$, that

$$S^2 = \frac{1}{n-1} \sum_{i=1}^{n} (X_i - \bar{X})^2$$

is an unbiased estimator of the population variance. Note that division simply by n in the sample variance produces a biased estimator of the population variance σ^2:

$$\text{Let } S_1^2 = \frac{1}{n} \sum_{i=1}^{n} (X_i - \bar{X})^2.$$

$$\text{Then } S_1^2 = \frac{n-1}{n} \times \frac{1}{n-1} \sum_{i=1}^{n} (X_i - \bar{X})^2$$

$$= \frac{n-1}{n} S^2.$$

Therefore $E(S_1^2) = E\left(\dfrac{n-1}{n} S^2\right)$

$$= \dfrac{n-1}{n} E(S^2) = \dfrac{n-1}{n} \sigma^2.$$

S_1^2 is a biased estimator of σ^2, with a tendency to underestimate the population variance, while S^2 is an unbiased estimator of σ^2. This justifies the definition of sample variance made in Chapter 4.

But is our unbiased dart thrower necessarily the best contestant? Not when competing against someone who throws a tighter pattern. We would also prefer estimators with the property of **minimum variance** or **efficiency**.

DEFINITION

Let T_1 and T_2 be point estimators of the population parameter Θ. We say that T_1 is more **efficient** than T_2 if $\mathrm{Var}(T_1) < \mathrm{Var}(T_2)$.

Given two unbiased estimators, the one with smaller variance is more likely to be close to the estimated parameter.

Consider drawing a random sample of size 2 from a population which has mean μ and variance 1. These three quantities are possible point estimators of μ:

(1) $\quad \bar{X} = \dfrac{X_1 + X_2}{2}$,

(2) $\quad X_1$,

(3) $\quad \dfrac{X_1 + 2X_2}{3}$,

where X_1 and X_2 are the values in the sample.

You can show that each of these quantities is an unbiased estimator of μ, but consider their variances:

(1) $\quad \mathrm{Var}(\bar{X}) = \dfrac{\sigma^2}{n} = \dfrac{1}{2}$

(2) $\quad \mathrm{Var}(X_1) = \sigma^2 = 1$

(3) $\quad \mathrm{Var}\left(\dfrac{X_1 + 2X_2}{3}\right) = \dfrac{1}{9} \mathrm{Var}(X_1) + \dfrac{4}{9}(X_2)$

$$= \dfrac{1}{9} \times 1 + \dfrac{4}{9} \times 1 = \dfrac{5}{9}.$$

We prefer the estimator with minimum variance, so we use $\bar{X}$; of these three estimators of μ, it is the most efficient.

There are situations in which a more efficient estimator is biased, as when estimating the population proportion P. We have shown that $E(\hat{p}) = P$; the sample proportion $\hat{p} = X/n$ is an unbiased estimator of P. Consider, however, the estimator $\hat{p}_1 = (X+1)/(n+2)$, where X again is the number of elements in the sample of size n with the characteristic of interest.

$$E(\hat{p}_1) = E\left(\frac{X+1}{n+2}\right) = \frac{E(X+1)}{n+2} = \frac{E(X)+1}{n+2}$$

$$= \frac{nP+1}{n+2}.$$

This quantity is equal to P only if $P = \frac{1}{2}$, so in general $\hat{p}_1$ is a biased estimator of P. However,

$$\text{Var}(\hat{p}) = \text{Var}\left(\frac{X}{n}\right) = \frac{1}{n^2}\text{Var}(X), \text{ while}$$

$$\text{Var}(\hat{p}_1) = \text{Var}\left(\frac{X+1}{n+2}\right) = \frac{1}{(n+2)^2}\text{Var}(X+1)$$

$$= \frac{1}{(n+2)^2}\text{Var}(X) < \frac{1}{n^2}\text{Var}(X) = \text{Var}(\hat{p}),$$

so $\hat{p}_1$ is a more efficient estimator of P than is $\hat{p}$.

In any estimation process, we would like to minimize the error between the estimating statistic and the parameter. That error is measured in this way:

DEFINITION

T is an estimator of the parameter Θ. Then T has **mean-square error** (MSE)

$$\text{MSE}(T) = E[(T-\Theta)^2].$$

The mean-square error measures the expected deviation of the estimator not from the center of its sampling distribution, which may not be equal to the parameter being estimated, but from the parameter itself. In fact, if the statistic T is an estimator of the parameter Θ, it is always true that $\text{MSE}(T) = \text{Var}(T) + B^2$, where $B = E(T) - \Theta$ is the bias of T as an estimator of Θ. [The proof of this result is left as an exercise. Note that if T is an unbiased estimator of Θ, $B=0$ and $\text{MSE}(T) = \text{Var}(T)$.] Both the variance and bias of the estimator contribute to its mean-square error.

In the previous example, in which $\hat{p} = X/n$ and $\hat{p}_1 = (X+1)/(n+2)$ are considered as estimators of the population proportion P, it can be shown that though $\hat{p}$ is unbiased while $\hat{p}_1$ is not, $\text{MSE}(\hat{p}_1) < \text{MSE}(\hat{p})$.

Not only should our dart enthusiast maintain a tight pattern and bunch his darts around the target, but we should require that he improve with practice. The corresponding characteristic in an estimator is called *consistency*.

A sample statistic T is a consistent estimator of the parameter θ if, as the sample is increased, T becomes more likely to be near θ.

Consider $\bar{X}$ as an estimator of μ. $\bar{X}$ is unbiased, so its sampling distribution is centered on μ: $\sigma_{\bar{x}} = \sigma/\sqrt{n}$, so as n increases, the standard deviation of the distribution of $\bar{X}$ decreases, and the distribution becomes more tightly packed around μ. As n increases, $\bar{X}$ becomes more likely to be near μ, so $\bar{X}$ is a consistent estimator of μ.

If we wanted to estimate the variability of a population, we could take a random sample of n values from the population and compute the range of the sample, the difference between its two extreme values. But this statistic uses only two of the sample values; it ignores the information represented by the other values. To take advantage of all the information in the sample which we have probably gone to some trouble to collect, we should estimate the variability of the population with a statistic which is computed from *all* the elements of the sample, such as the sample variance. No other point estimator contains more information from the sample, and the sample variance is then called *sufficient*:

If a sample statistic contains all the information the sample provides about a parameter σ, then T is called a **sufficient** statistic for σ.

The usefulness of sufficiency derives from a theorem in statistics which states that any function of a sufficient statistic is itself sufficient, and this suggests the following strategy for finding a "best" estimator for a population parameter:

1. Consider the set of all possible estimators of the parameter (for example, μ, the population mean).

2. In this set, find a sufficient statistic (for example, $\sum_{i=1}^{n} X_i$).

3. Find a function of the sufficient statistic which is an unbiased estimator of the parameter ($\bar{X} = 1/n \sum_{i=1}^{n} X_i$).

4. Under certain conditions, this statistic will also be consistent and of minimum variance.

Note that since estimators for the most prominent population parameters are established ($\bar{X}$ for μ, S^2 for σ^2, etc.), we will not in fact employ this strategy.

Interval Estimation

Though the sample mean $\bar{X}$ is a "good" estimator of the population mean μ, $\bar{X}$ is not likely to be equal to μ, and we know nothing about the accuracy of such a point estimate. However, we can use our knowledge of the sampling distribution of $\bar{X}$ to construct an *interval* around the point estimate and to state our degree of certainty that the population mean μ is in that interval. This is the process of *interval estimation*.

Consider the construction of an interval estimate for the population mean. Suppose that we take a random sample of size $n \geq 30$ from a large population with mean μ and standard deviation σ. From the central limit theorem we know that the sampling distribution of $\bar{X}$ is approximately normal with mean μ and standard deviation $\sigma/\sqrt{n}$.

An immediate consequence of this, which we used in the last chapter, is that

$$\frac{\bar{X} - \mu}{\sigma/\sqrt{n}} \doteq Z,$$

where Z is the standard normal random variable whose distribution is $N(0, 1)$.

Given a value C (like 0.90 or 0.95), we can find, using Table A-3, a value z for which

$$P(-z < Z < z) = C.$$

That is (see Figure 8.1),

$$P\left(-z < \frac{\bar{X} - \mu}{\sigma/\sqrt{n}} < z\right) = C.$$

FIGURE 8.1.

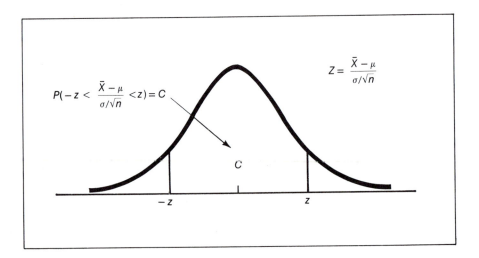

For example, if $C = 0.90$, then:

$$P(-z < Z < z) = 0.90$$

$$\rightarrow \quad P(0 < Z < z) = 0.4500$$

$$\rightarrow \quad z = 1.645.$$

Algebraic manipulations produce the following sequence of equivalent statements:

$$P\left(-z < \frac{\bar{X} - \mu}{\sigma/\sqrt{n}} < z\right) = C$$

$$\rightarrow \quad P\left(-z \frac{\sigma}{\sqrt{n}} < \bar{X} - \mu < z \frac{\sigma}{\sqrt{n}}\right) = C$$

$$\rightarrow \quad P\left(-\bar{X} - z \frac{\sigma}{\sqrt{n}} < -\mu < -\bar{X} + z \frac{\sigma}{\sqrt{n}}\right) = C$$

$$\rightarrow \quad P\left(\bar{X} - z \frac{\sigma}{\sqrt{n}} < \mu < \bar{X} + z \frac{\sigma}{\sqrt{n}}\right) = C.$$

From this development we construct the estimate. Since $\bar{X}$ is a random variable, so are the two quantities $\bar{X} \pm z(\sigma/\sqrt{n})$. The last equation above says that the probability that we will catch the population mean in the interval between these quantities is C. If we were to take a great number of random samples, each would produce a value of $\bar{X}$ at the center of one of these intervals, and the proportion of them which contain μ would be approximately C. If we have one sample with its associated interval, we can think of that interval as having been randomly selected from all possible intervals associated with samples of size n, and we can be C confident that our particular interval does contain μ (see Figure 8.2).

This development gives us the following definitions:

DEFINITION

The **confidence level** or **degree of confidence** C associated with an interval estimate of μ is the proportion of all such interval estimates which contain μ. The interval $[\bar{X} - z(\sigma/\sqrt{n}), \bar{X} + z(\sigma/\sqrt{n})]$ is a $C \times 100\%$ **confidence interval** for the population mean; the endpoints of this interval, $\bar{X} \pm z(\sigma/\sqrt{n})$, are called **confidence limits**.

Example 3 Suppose we take a random sample of size 100 from a population which has standard deviation 25, and we compute $\bar{X} = 78$. We construct a 90% confidence interval for the population mean.

Solution

Corresponding to $C = 0.90$, we have seen that $z = 1.645$. The confidence limits are

$$78 \pm 1.645 \frac{25}{\sqrt{100}} \doteq 78 \pm 4.11 = 73.9, 82.1,$$

FIGURE 8.2.   **90% confidence intervals generated by repeated sampling.**

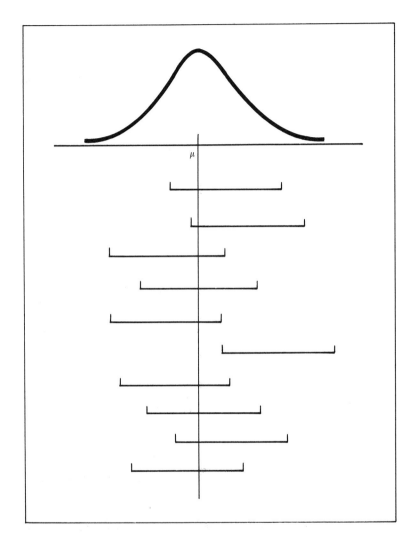

and [73.9, 82.1] is a 90% confidence interval for the population mean.

This means that we are 90% confident or 90% certain that μ is between 73.9 and 82.1. It does *not* mean that there is a 90% probability that the interval contains μ. The value of μ, though unknown to us, is fixed, and either is in the interval or is not, so the concept of probability does not apply.

When estimating the value of μ with a large sample ($n \geq 30$), we are not likely to know the value of the population standard deviation σ. When this happens, as it usually does, we can estimate σ with the sample standard deviation s; no other changes need be made.

Example 4 Using the sample of 74 American cities in the USNEWS file, we can estimate the mean income of factory workers in all American cities.

The variable of interest is once again X5, whose 73 nonmissing values comprise our sample. The SPSS subprogram CONDESCRIPTIVE provides these statistics,

VARIABLE X5   AVERAGE INCOME OF FACTORY WORKERS

MEAN	15065.918	STD ERR	270.903	STD DEV		2314.596
VARIANCE	5357352.382	KURTOSIS	− .550	SKEWNESS		− .126
MINIMUM	10337.000	MAXIMUM	20378.000	SUM		1099812.000
C.V. PCT	15.363	.95 C.I.	14525.883		TO	15605.953
VALID CASES	73	MISSING CASES	1			

from which we see that $\bar{X} = 15065.918$ and $s = 2314.596$. Using s to estimate σ, we find a 95% confidence interval for μ, the mean income of factory workers in all American cities.

Solution

Corresponding to $C = 0.95$, $z = 1.96$, and the confidence limits are:

$$\bar{X} \pm z \frac{s}{\sqrt{n}} = 15065.918 \pm 1.96 \frac{2314.596}{\sqrt{73}}$$

$$= 15065.918 \pm 1.96 \times 270.903$$

$$= 15065.918 \pm 530.970$$

$$\doteq 14535, 15597.$$

Note that the value $2314.596/\sqrt{73} = 270.903$ is given in the printout as STD ERR, an estimate of the standard deviation of the sampling distribution of $\bar{X}$, $\sigma_{\bar{X}} = \sigma/\sqrt{n}$, and that the values we have just found are approximately equal to those given as .95 C.I.* (The slight difference occurs because the SPSS subprogram uses, rather than z, values from what is called a t distribution, to be discussed later. For large sample sizes, that distribution is approximately equal to Z.)

To find a 99% confidence interval, we need only change the value of z to 2.58 in the above calculations:

$$15065.918 \pm 2.58 \frac{2314.596}{\sqrt{73}} = 14367, 21038.$$

We can be 99% certain that the average income of factory workers in United States cities is between \$14,367 and \$21,038.

In moving from 95% to a 99% confidence interval, we have increased the certainty that the interval contains μ, but at the expense of precision; the last interval is larger than the earlier one. This is the first occurrence of a trade-off we will encounter repeatedly: for a given sample size, we can increase confidence at the cost of precision, or improve accuracy by sacrificing certainty. The only way to improve one without harming the other, or to improve both, is to increase the sample size, or to collect more information about the population.

---

*Some implementations of SPSS do not calculate the confidence interval.

Given a desired confidence level and precision, how large a sample should be taken? Consider this example:

Example 5 We plan to estimate the mean of a population of values whose standard deviation we know to be 32.6. How large a sample should we take to obtain a 95% confidence interval no more than 3 units wide?

Solution

First, $C = 95\% \rightarrow z = 1.96$.
Then, for a confidence interval 3 units wide,

$$z\,\frac{\sigma}{\sqrt{n}} = 1.5 \rightarrow 1.96\,\frac{32.6}{\sqrt{n}} = 1.5.$$

Solving for n, we obtain

$$n = \frac{1.96 \times 32.6}{1.5} = 42.60 \rightarrow n = 1814.5.$$

To meet the specified conditions, we should take a sample of at least 1815 elements.

Our ability to construct confidence intervals for the population mean is based on knowledge of the distribution of the sample mean. Analogous processes allow us to generate confidence intervals for other parameters as long as the sampling distributions of their estimators are known, as in this example:

Example 6 Of a random sample of 300 Chicago voters, 84 support George Leroy Tyrebiter for mayor. We seek a 90% confidence interval for the true proportion of Chicago voters who support the illustrious Mr. Tyrebiter.

Solution

We will find a confidence interval for the population proportion P, based on a sample proportion $p = X/n$. We know that for a large sample, the distribution of $\hat{p}$ is approximately normal with mean P and standard deviation $p = P/[P(1-P)/n]^{1/2}$. Therefore,

$$\frac{\hat{p}-P}{[P(1-P)/n]^{1/2}} \doteq Z,$$

and corresponding to the confidence level C we can find z so that

$$P\left(-z < \frac{\hat{p}-P}{[P(1-P)/n]^{1/2}} < z\right) = C.$$

Algebraically manipulating the expression within parentheses, we obtain:

$$P(\hat{p} - z[P(1-P)/n]^{1/2} < P < \hat{p} + z[P(1-P)/n]^{1/2}) = C.$$

As in the development of confidence intervals for μ, the confidence limits are $\hat{p} \pm z[P(1-P)/n]^{1/2}$. We are estimating P with $\hat{p}$, however, so the confidence limits become

$$\hat{p} \pm z[\hat{p}(1 - \hat{p})/n]^{1/2},$$

where z corresponds to the confidence level C.

In the example, $C = 90\%$, so $z = 1.645$, $n = 300$, and $\hat{p} = 84/300 = 0.28$. 90% confidence limits for P are then

$$0.28 \pm 1.645 \left(\frac{0.28 \times 0.72}{300} \right)^{1/2} = 0.28 \pm .043$$

$$= 0.237, 0.323.$$

We can be 90% certain that between 23.7% and 32.3% of Chicago's voters support George Leroy Tyrebiter.

Hypothesis Testing

Example 7 Suppose that drug A has been used to treat a particular disease. A group of medical researchers develops drug B, which they hope will be more useful in combating the disease, but before it can be commonly used, the researchers must demonstrate that their new medicine is more effective than drug A. They will conduct an experiment to evaluate the two treatments, and will perform a *hypothesis test*.

Solution

In hypothesis testing, two mutually exclusive hypotheses are advanced. The **null hypothesis** (designated H_o) states the *status quo*, the condition the researcher usually wishes to disprove, while the **alternative hypothesis** (H_a) is generally what the researcher wishes to show, that is, what will be true if conditions have changed. In the drug example, the hypotheses are these:

H_o: Drug A is at least as effective as drug B.

H_a: Drug B is more effective than drug A.

Hypothesis testing is a conservative process, based on the presumption that H_o is true. The researcher hopes to reject H_o and accept H_a on the basis of experimental (sample) evidence, but if H_a is not decisively shown to be true, then the researcher fails to reject H_o, and no definite conclusion is reached.

What is needed is a procedure for deciding between the null and alternative hypotheses, a well-defined rule by which the decision, based on sample information, can be made. Consider another example:

Example 8 A manufacturer of TV picture tubes has been producing a 19-inch tube which has expected life 3100 hours, with standard deviation 450 hours. An engineer suggests a modification which she believes will increase the lifetimes of such tubes. To test her contention, she tests a sample of 100 tubes built with the modification, and finds that the sample has mean lifetime 3225 hours. Is this convincing evidence that the modification should be incorporated in all the company's picture tubes? Does it demonstrably extend expected tube life?

Solution

The hypotheses being tested are:

H_o: The modification causes no improvement.

H_a: The modification extends tube life.

Rephrasing them in terms of a parameter, the population mean,

H_o: $\mu \leq 3100$

H_a: $\mu > 3100$,

where μ is the expected lifetime of picture tubes incorporating the modification. (Recall that the expected life of tubes without the modification was 3100 hours.)

The engineer must decide whether to reject H_o and accept H_a (conclude that the modification is effective) or not reject H_o (conclude that lifetimes of tubes incorporating the modification are no longer than those of the original design). The four possible situations are shown in Figure 8.3.

Failing to reject H_o when H_o is true and rejecting H_o when it is false are correct decisions. The other two possibilities represent errors. Rejecting the null hypothesis when it is, in fact, true is a *Type I error*, while failing to reject a false null hypothesis is a *Type II error*. In our

FIGURE 8.3.   Outcomes of a hypothesis test.

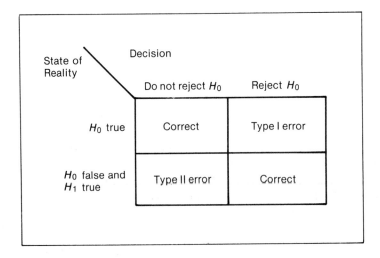

example, a Type I error is committed if the engineer concludes that the modification is effective in prolonging tube life when it is not, while failing to conclude that the modification is effective when, in fact, it constitutes a Type II error.

The Type I error is generally more serious—in the example it might result in spending time and money to incorporate an ineffective modification—so hypothesis tests are constructed to control the probability of a Type I error. This probability is indicated by α, while the probability of a Type II error, to be discussed in Chapter 9, is β.

We are constructing a rule by which we can decide if μ is greater than 3100. We have seen the utility of estimating μ with the sample mean $\bar{X}$, so we will be convinced to reject the null hypothesis and accept the alternative if $\bar{X}$ is big enough, but how big is big enough? Since we also want to control $\alpha = P(\text{Type I error})$, the probability that H_o is rejected when it is true, consider the sample distribution of $\bar{X}$ if H_o is true. (See Figure 8.4.)

If H_o is true, then $\mu = 3100$, and we know from the central limit theorem that the sampling distribution of X is approximately normal with mean 3100 and standard deviation $450/\sqrt{100}$ (recall that $n = 100$). Choose a value for α, say 5%, and cut off an area of size α in the upper tail of the graph of this distribution. Consider the value $\bar{X}^*$ which marks off its lower edge.

If H_0 is true, the probability that $\bar{X}$ will be greater then $\bar{X}^*$ is α, which we have chosen to be only 5%, so if the observed value of $\bar{X}$ is greater than $\bar{X}^*$, we can reject H_0 and accept H_a, concluding that the modification extends tube life, with only a 5% probability of making a Type I error. We need only find the value of $\bar{X}^*$:

$$\alpha = P(\text{Type I error}) = P(\text{Reject } H_0 | H_0)$$
$$= P(\bar{X} > \bar{X}^* \mid H_0) = 5\% = 0.05$$

FIGURE 8.4.

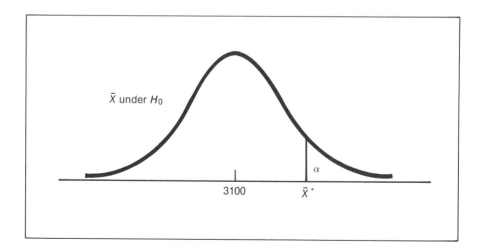

$$\rightarrow \quad P\left(\frac{\bar{X} - 3100}{450/\sqrt{100}} > \frac{\bar{X}^* - 3100}{450/\sqrt{100}} \ \Big| \ \mu = 3100\right) = 0.05$$

$$\rightarrow \quad P\left(Z > \frac{\bar{X} - 3100}{450/\sqrt{100}}\right) = 0.05$$

$$\rightarrow \quad P\left(0 < Z < \frac{\bar{X}^* - 3100}{450/\sqrt{100}}\right) = 0.4500$$

$$\rightarrow \quad \frac{\bar{X}^* - 3100}{450/\sqrt{100}} = 1.645$$

$$\rightarrow \quad \bar{X}^* = 3100 + 1.645 \ \frac{450}{\sqrt{100}} = 3174.025.$$

This value completes construction of the *decision rule:*

If the sample mean $\bar{X}$ is greater than $\bar{X}^* \doteq 3174$, then we reject H_o and accept H_a. If $\bar{X} < (\bar{X}^* = 3174)$, we cannot reject H_o.

The probability α is called the *significance level* of the test, and $\bar{X}^*$ is the *critical value.* Because our conclusion will be determined by the value of $\bar{X}$, the sample mean is called the *test statistic.* The interval above the critical value in this *upper-tail test* is called the *region of rejection*, since we reject H_o when $\bar{X}$ falls in this region, while the interval below $\bar{X}^*$ is called the *region of acceptance.*

In her examination of the 100 tubes made with the modification, the engineer found that $\bar{X} = 3225$, which is greater than the critical value of 3174. According to the decision rule, she can reject H_o at the 5% level of significance, and conclude that the modification does extend the expected life of picture tubes.

In general, suppose we take a sample of size $n > 30$ to perform an upper-tail test of the form

$$H_o: \mu \leq \mu_o$$

$$H_a: \mu > \mu_o$$

We choose the significance level α, and must find the critical value $\bar{X}^*$ around which to build the decision rule. If H_o is true, the distribution of the sample mean $\bar{X}$ is approximately normal with mean μ_o and standard deviation $\sigma/\sqrt{n}$, where σ is the population standard deviation and n is the sample size. Then

$$\alpha = P(\text{Type I error}) = P(\text{Reject } H_o | H_o) = P(\bar{X} > \bar{X}^* | H_o)$$

$$P\left(\frac{\bar{X} - \mu_0}{\sigma/\sqrt{n}} > \frac{\bar{X}^* - \mu_0}{\sigma/\sqrt{n}} \ | H_0\right) = \alpha \rightarrow P\left(Z > \frac{\bar{X}^* - \mu_0}{\sigma/\sqrt{n}}\right) = \alpha$$

$$\rightarrow \quad \frac{\bar{X}^* - \mu_0}{\sigma/\sqrt{n}} = z_\alpha \ ,$$

where z_a, called the *critical normal deviate*, is chosen so that $P(Z > z_a) = \alpha$. Then $\bar{X}^* = \mu_o + z_\alpha(\sigma/\sqrt{n})$.

If $\bar{X} > \bar{X}^*$, we reject H_o and accept H_a; if $\bar{X} < \bar{X}^*$, we fail to reject H_o, and no conclusion is reached. (See Figure 8.5.)

If the population standard deviation is unknown, as it often is in sampling situations, the sample standard deviation s is used instead of σ. Then $\bar{X}^* = \mu_o + z_\alpha(s/\sqrt{n})$.

The decision rule for this kind of hypothesis test can be phrased in two alternative but equivalent ways. First, notice that if $\bar{X} > \bar{X}^* = \mu_o + z_\alpha(\sigma/\sqrt{n})$, then $(\bar{X} - \mu_o)/\sigma/\sqrt{n} > z_\alpha$. That is, we can compare the value of the normal deviate $(\bar{X} - \mu_o)/\sigma/\sqrt{n}$ to the critical normal deviate (see Figure 8.6):

$$\text{If } \frac{\bar{X} - \mu_0}{\sigma/\sqrt{n}} > z_\alpha, \text{ reject } H_0 \text{ and accept } H_a;$$

$$\text{if } \frac{\bar{X} - \mu_0}{\sigma/\sqrt{n}} < z_\alpha, \text{ do not reject } H_0.$$

In our example, $\alpha = 5\%$ so $z_a = 1.645$.

$$\frac{\bar{X} - \mu_0}{\sigma/\sqrt{n}} = \frac{3225 - 3100}{450/\sqrt{100}} = 2.78 > 1.645.$$

As before, we conclude that the engineer should reject the null hypothesis and conclude that the modification does increase expected tube life.

Statistical package programs, such as some we will examine in Chapters 9 and 10, often calculate the normal deviate or its estimate $(\bar{X} - \mu_o)/(s/\sqrt{n})$, in which the population standard deviation σ is approximated by the sample standard deviation s.

FIGURE 8.5.   An upper-tail hypothesis test for μ.

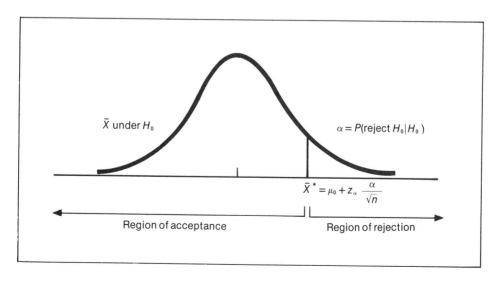

FIGURE 8.6.

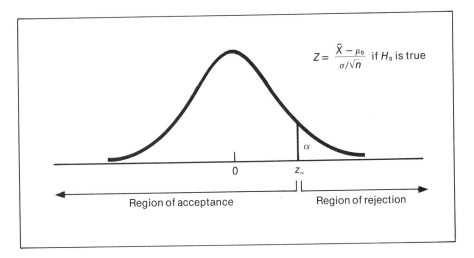

The second alternative rephrasing of the decision rule compares the area under the graph of the distribution of $\bar{X}$ when H_o is true above the observed value of $\bar{X}$ to the significance level α. If this area is less than α, then $\bar{X}$ itself must be above the critical value $\bar{X}^*$, and we reject H_o; if the area above the observed value of $\bar{X}$ is greater than α, $\bar{X}$ must be below $\bar{X}^*$, and we cannot reject H_o. (See Figure 8.7.) The area above $\bar{X}_o$, the observed value of $\bar{X}$, is $P(\bar{X} > \bar{X}_o | H_o)$.

Again looking at the example, $\alpha = 5\% = 0.05$, and the observed value of $\bar{X}$ was $\bar{X}_o = 3225$. Then:

$$P(\bar{X} > 3225 | H_0) = P\left(\frac{\bar{X} - 3100}{450/\sqrt{100}} > \frac{3225 - 3100}{450/\sqrt{100}} | H_0 \right)$$

$$= P(Z > 2.78) = 0.5000 - 0.4973 = 0.0027.$$

$$0.0027 < 0.05 = \alpha, \text{ so we reject } H_0.$$

The value $P(\bar{X} > \bar{X}_o | H_o)$, sometimes called a *one-tail probability*, is also commonly provided by packaged programs dealing with hypothesis tests. Note that this is the probability that a value of $\bar{X}$ at least as extreme as the observed value will occur if H_o is true.

Other forms of hypothesis tests are possible. We can perform *lower-tail* or *two-tail* tests with the population mean μ, as well as tests involving other population parameters. Any statistical test of hypotheses, however, will contain these elements:

1. A formal statement of the null and alternative hypotheses, H_o and H_a.

FIGURE 8.7.

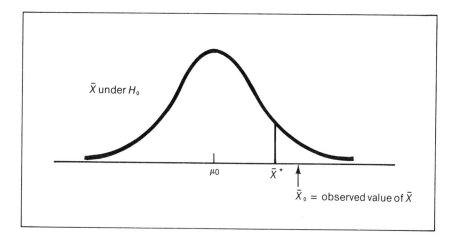

2. A test statistic and its sampling distribu..on.

3. A chosen level of significance, α.

4. A decision rule which defines the critical value(s) of the test statistic and the regions of acceptance and rejection.

5. A random sample from which to obtain the observed value of the test statistic.

EXERCISES

1. What is meant by statistical bias?

2. Under what conditions is the sample median an unbiased estimator of the population mean?

3. You wish to estimate the mean educational level of the adults in a large city from statistics of a random sample of 500 adults. You may estimate the population mean with either the mean or the median of the sample. Which do you choose, and why?

4. Of these sample statistics, which are sufficient: mean, standard deviation, range, maximum, skewness?

5. From any of the data bases in Appendix C, choose a variable and find the best estimates for the mean, variance, and standard deviation of the population from which the values came.

6. Show that $E(S_1^2) = (n - 1/n)\sigma^2$, without using the fact that $E(S^2) = \sigma^2$. [*Hint:* See the proof that $E(S^2) = \sigma^2$ in Chapter 7.]

7. Suppose that 5 numbers are randomly drawn *with* replacement from the integers $1, 2, 3, \ldots, N$.
   a. Select an estimator of the population maximum N.
   b. Find the mean of the distribution of the estimator chosen in part (a).

8. If T is an estimator of the parameter Θ, show that the $\text{MSE}(T) = \text{Var}(T) + B^2$, where $B = E(T) - \Theta$ is the bias of T as an estimator of Θ.

9. Consider the sample proportion $\hat{p}$ as an estimator of the population proportion P. Of the desirable characteristics a point estimator might have, which are associated with $\hat{p}$?

10. In general and for practical purposes, what are the most desirable estimators of the population parameters μ, σ^2, σ, and P?

11. What is a confidence interval?

12. What is the importance of the central limit theorem in constructing confidence intervals for the population mean?

13. For $C = 80\%, 90\%, 95\%, 98\%$, and 99%, find z so that $P(-z < Z < z) = C$.

14. A random sample of size n from a large population has mean 58 and standard deviation 27. Find a 90% confidence interval for μ if:
   a. $n = 50$.
   b. $n = 100$.
   c. $n = 200$.
   d. What general observation do these results suggest?

15. A random sample of 75 elements from a large population has mean $\bar{X} = 426$ and standard deviation $s = 91$. Find a confidence interval for the population mean μ if:
   a. $C = 80\%$.
   b. $C = 90\%$.
   c. $C = 99\%$.
   d. What general observation do these results suggest?

16. Consider the cities in the USNEWS data file as a sample of all U.S. cities. Use these statistics for the values of X3, unemployment, to find a 95% confidence interval for unemployment in all U.S. cities.

VARIABLE X3   UNEMPLOYMENT

MEAN	5.691	STD ERR	.197	STD DEV		1.647
VARIANCE	2.712	KURTOSIS	2.700	SKEWNESS		1.215
MINIMUM	3.000	MAXIMUM	12.000	SUM		398.400
C.V. PCT	28.936	.95 C.I.	5.299		TO	6.084
VALID CASES	70	MISSING CASES	4			

Several features of this exercise should be noted. First, because four values are missing, the sample size is 70. Second, note that STD ERR = STD DEV/$\sqrt{n}$. Third, compare your computed results to the .95 C.I. given above.

17. The SPSS subprogram CONDESCRIPTIVE produced these statistics of the 74 values of X7 in the USNEWS file. From these statistics, find confidence intervals for the mean change in construction activity for all U.S. cities for confidence levels 90%, 95%, and 98%.

VARIABLE X7   CHANGE IN CONSTRUCTION ACTIVITY

MEAN	5.523	STD ERR	1.081	STD DEV	9.301
VARIANCE	86.517	KURTOSIS	1.974	SKEWNESS	.924
MINIMUM	− 15.400	MAXIMUM	40.600	SUM	408.700
C.V. PCT	168.414	.95 C.I.	3.368	TO	7.678
VALID CASES	74	MISSING CASES	0		

Note the values which occur both in your computations and in the printout.

18. Use the SPSS subprograms CONDESCRIPTIVE or FREQUENCIES (or their equivalent on your system) to find 95% confidence intervals for means of other populations of values using the USNEWS file or one of the other data files given in Appendix C.

19. When sampling from a normally distributed population with variance σ^2, the distribution of s has mean σ and standard deviation $\sigma/\sqrt{2n}$, and becomes normal as the sample size increases. Use this result to find a 95% confidence interval for the standard deviation of the incomes of workers in American cities. (Use the statistics of X5 in the USNEWS file.)

20. Based on the data in the USNEWS file, find a 95% confidence interval for the proportion of U.S. cities in which construction activity had risen 10% or more from the previous year. The following program uses a RECODE to group the sample values of X7, then plot a histogram and calculate statistics.

```
? 10 GET FILE;USNEWS
? 20 RECODE;X7(LOW THRU 9.999 = 0)(10 THRU HI = 1)
? 30 FREQUENCIES;GENERAL = X7
? 35 STATISTICS;1,2,5
? 40 OPTIONS;7,8
```

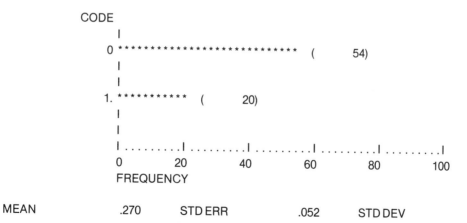

MEAN	.270	STD ERR	.052	STD DEV	.447
VALID CASES	74	MISSING CASES	0		

21. Use the printout below to find 90%, 95%, and 99% confidence intervals for the proportion of all U.S. cities in which unemployment is less than 4.0%, based on the sample of cities in the USNEWS file. (Note the values in your computations which appear in the printout.) Also, write an SPSS program which would produce this output.

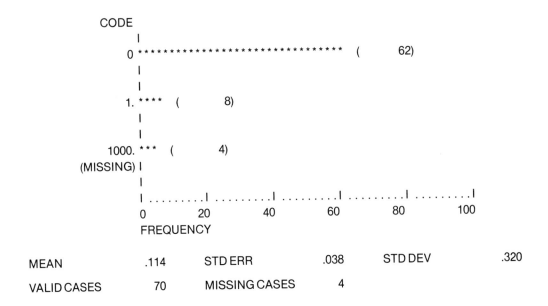

X3   UNEMPLOYMENT

MEAN	.114	STD ERR
VALID CASES	70	MISSING CASES

MEAN .114   STD ERR .038   STD DEV .320

VALID CASES 70   MISSING CASES 4

22. Find a 95% confidence interval for the proportion of all U.S. cities in which department store sales increased by 10% or more, based on the USNEWS file.

23. Find a 98% confidence interval for the proportion of all U.S. cities in which nonfarm employment had increased by at least 5%, based on the USNEWS file.

24. For $\alpha = 1\%, 2\%, 5\%$, and 10%, find z so that $P(Z > z) = \alpha$.

25. In the drug example on p. 169, what is done if
   a. the null hypothesis is rejected?
   b. the null hypothesis is not rejected?
   c. a Type I error is committed?
   d. a Type II error is committed?

26. Using data from the USNEWS file, test the null hypothesis that department store sales grew by no more than 8% against the alternative hypothesis that they grew by more than 8%. Test at the 5% significance level. These are the statistics of variable X2.

VARIABLE X2		CHANGE IN DEPARTMENT STORE SALES			
MEAN	10.180	STD ERR	.799	STD DEV	6.684
VARIANCE	44.682	KURTOSIS	.042	SKEWNESS	− .089
MINIMUM	− 7.300	MAXIMUM	24.300	SUM	712.600
C.V. PCT	65.663	.95 C.I.	8.586	TO	11.774
VALID CASES	70	MISSING CASES	4		

27. Using these statistics of the variable X4 from the USNEWS file given here, test the null hypothesis that nonfarm employment in U.S. cities rose no more than 2.2% against the alternative that the rise was greater than 2.2%.

    a. Use $\alpha = 5\%$.

    b. Use $\alpha = 10\%$.

    c. Why is there a difference in the results of parts (a) and (b)? What generalization can be made?

VARIABLE X4		CHANGE IN NONFARM EMPLOYMENT			
MEAN	2.699	STD ERR	.328	STD DEV	2.820
VARIANCE	7.950	KURTOSIS	2.984	SKEWNESS	.937
MINIMUM	− 4.400	MAXIMUM	13.900	SUM	199.700
C.V. PCT	104.483	.95 C.I.	2.045	TO	3.352
VALID CASES	74	MISSING CASES	0		

28. Using a sample of size 150, these hypotheses will be tested at the 5% level of significance:

$$H_o: \mu \leq 86$$

$$H_a: \mu > 86$$

    a. If $\bar{X} = 89.3$, and $s = 19.6$, what conclusion is reached?

    b. Consider the statistic $(\bar{X} - \mu_o)/(s/\sqrt{n})$. Compare this value to the critical normal deviate x_α corresponding to $\alpha = 5\%$. How can the decision rule be rephrased?

29. Using the data given in Exercise 21, test the null hypothesis that the *proportion* of U.S. cities in which unemployment is less than 4% is no more than 10% against the alternative that the proportion is no more than 10%. Use $\alpha = 5\%$. (*Hint:* $H_o: P \leq 10\%$, $H_a: P > 10\%$. Find a critical value p^* for which $P(\hat{p} > \hat{p}^* | H_o) = \alpha$.)

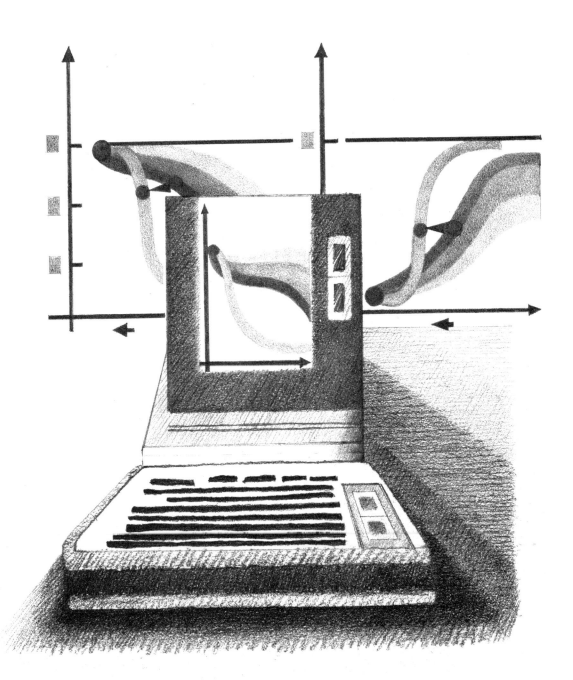

One-Sample
Hypothesis
Testing

INTRODUCTION

In Chapter 8 we introduced hypothesis testing as a form of estimation in which we derive a rule that enables us to choose between two mutually exclusive statements about a population parameter, the null hypothesis H_0, and the alternative hypothesis H_a. We take a random sample from the population and compute a test statistic. The decision rule, chosen to control the probability of a Type I error, describes the values of the test statistic for which we reject H_0 and accept H_a, and the values for which we are unable to reject H_0. In this chapter, we consider the probability of a Type II error, of failing to reject a false null hypothesis, and its influence on selecting or evaluating a test, and we extend our repertoire of techniques for performing hypothesis tests.

β, The Probability of a Type II Error

We have seen that the decision rule of a hypothesis test is developed to correspond to our choice of the significance level, the probability of a Type I error. We determine, and keep small, the probability of rejecting the null hypothesis when it is true. Suppose, however, that we fail to reject H_o. How much confidence can we have that we have not committed a Type II error? We must investigate β, the probability of failing to reject the null hypothesis when it is false.

Reconsider the major example of Chapter 8 (Example 8), in which the engineer hopes to demonstate, using a sample of size 100, that a modification to her company's 19-inch picture tubes will increase their expected life beyond 3100 hours. The population standard deviation is 450 hours and the hypotheses, to be tested at the 5% significance level, are these:

$$H_o: \mu \leq 3100$$

$$H_a: \mu > 3100$$

The critical value of this test is

$$\bar{X}^* = 3100 + 1.645 \; \frac{450}{\sqrt{100}} \; \doteq 3174,$$

and the decision rule is:

If $\bar{X} > 3174$, reject H_o and accept H_a;

if $\bar{X} < 3174$, do not reject H_o.

Suppose that the expected lifetime of tubes incorporating the modification is 3200 hours. Then $\mu = 3200$, H_o is false, and H_a is true. In this situation, the sampling distribution of $\bar{X}$ is approximately normal with mean 3200 and standard deviation $450/\sqrt{100}$, and the probability of a Type II error, of failing to conclude that H_o is false even though, since $\mu = 3200$, H_a is true, is:

$$\beta = P(\text{Type II error}) = P(\bar{X} < 3174 | H_a \text{ with } \mu = 3200)$$

$$= P \left(\frac{\bar{X} - 3200}{450/\sqrt{100}} < \frac{3174 - 3200}{450/\sqrt{100}} \; | H_a \text{ with } \mu = 3200 \right)$$

$$= P(Z < -0.58) = 0.5000 - 0.2190 = 0.2810.$$

That is, if the true mean lifetime of tubes incorporating the new process is 3200 hours, the probability that our test will nonetheless fail to reject H_o is 28.10%. This probability is represented graphically in Figure 9.1 as the area under the curve of the distribution of $\bar{X}$ if $\mu = 3200$ to the left of $\bar{X}^* = 3174$.

In general, for an upper-tail test of this kind,

$$\beta = P(\bar{X} < \bar{X}^* | H_a \text{ with } \mu = \mu_a) = P\left(Z < \frac{\bar{X}^* - \mu_a}{\sigma/\sqrt{n}}\right),$$

where μ_a is a possible value of μ for which H_o is false and H_a true.

It is fundamental to observe that the value of β depends on the true value of the population mean μ, which is also the center of the distribution of $\bar{X}$. If μ is, in fact, 3125,

FIGURE 9.1.

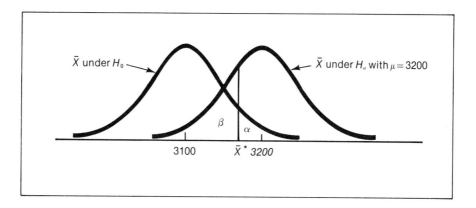

$$\beta = P(\bar{X} < 3174|H_a \text{ with } \mu = 3125)$$

$$= P\left(\frac{\bar{X} - 3125}{450/\sqrt{100}} < \frac{3174 - 3125}{450/\sqrt{100}} \, |H_a \text{ with } \mu = 3125\right)$$

$$= P(Z < 1.09) = 0.5000 + 0.3621 = 0.8621.$$

If the true mean is 3125, the probability that we will mistakenly fail to reject H_o is 86.21%, as shown in Figure 9.2.

FIGURE 9.2.

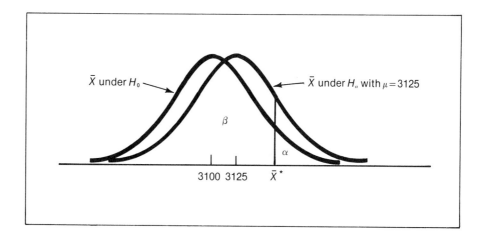

There is a value of β corresponding to every possible value of μ for which H_o is false. Since there are an infinite number of such values, it would be futile to attempt to calculate them all. We can, however, represent graphically the relationship between P(Accept H_o) and all possible values of μ.

If $\mu = 3100$, then H_o is true, and P(Accept $H_o|H_o$) $= 1 - P$(Reject $H_o|H_o$) $= 1 - \alpha = 0.95$. Calculations like those above fill in the rest of Table 9.1 which shows representative values.

TABLE 9.1

μ	$P(ACCEPT H_0)$
3100	0.9500
3125	0.8621
3150	0.7019
3174 = $\bar{X}^*$	0.5000
3200	0.2810
3225	0.1292
3250	0.0455

Note that if $\mu = \bar{X}^*$, if the population mean is equal to the critical value, the value of β is 0.5000.

From these values, we draw Figure 9.3 relating P(Accept H_o) to possible values of μ. It is called the *operating characteristic curve* or *OC curve* for the test, and at each possible value of μ except 3100 (where H_o is true), the height of the graph is the probability of a Type II error, β. The *OC* curve always begins at the point (μ_o, $1 - \alpha$), where μ_o is the null hypothesis value of μ, and passes through the point ($\bar{X}^*$, 0.5).

Used as often as the operating characteristic curve is the *power curve*, which plots $1 - \beta$

FIGURE 9.3. Operating characteristic curve.

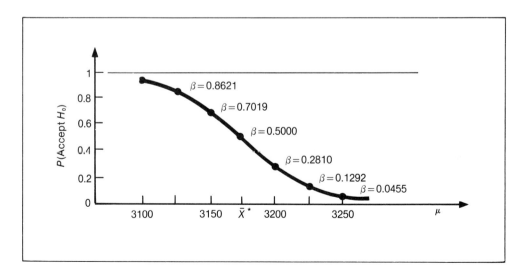

FIGURE 9.4.   Power curve.

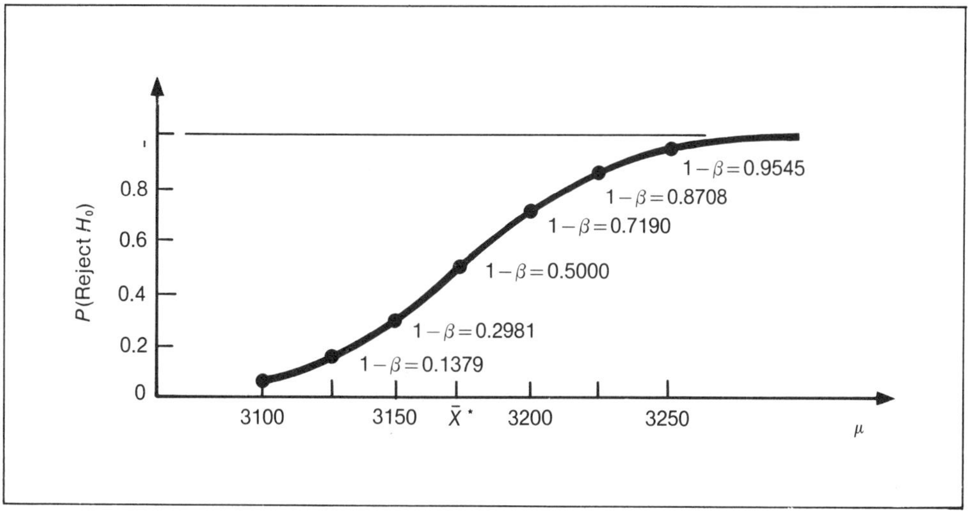

= P(Reject H_o) against possible values of μ (Figure 9.4). Note that it begins at the point (μ_o, α), also passes through ($\bar{X}^*$, 0.5), and contains the same information as the operating characteristic curve.

The graphs in Figures 9.3 and 9.4 tell us the probabilities of accepting or rejecting the null hypothesis for all possible values of μ; they tell us how likely it is that our test will distinguish between μ_o(3100 in the example) and other possible values of μ.

These curves can also be used to illustrate how a test is affected by changes in significance level or sample size. For example, in an upper-tail test for μ of the type we have been considering, an increase in the significance level α will shift the starting point of the OC curve down, that of the power curve up, and will decrease the critical normal deviate z_α so that $\bar{X}^* = \mu_o + z_\alpha \, \sigma/\sqrt{n}$ also decreases.

In Figure 9.5 the OC curve is moved down, the power curve up. That is, if the significance level is increased, the probability of a Type II error, β, is decreased. Conversely, decreasing α increases β. Again we encounter a trade-off, like that between certainty and precision in confidence intervals.

With confidence intervals, we found that we could improve certainty without degrading precision, or *vice versa*, by collecting more information, by increasing the sample size n. In our upper-tail hypothesis test for μ, increasing n will decrease the standard deviation of $\bar{X}$, $\sigma/\sqrt{n}$, and will reduce the critical value $\bar{X} = \mu_o + z_\alpha(\sigma/\sqrt{n})$. The starting points of the two curves, (μ_o, 1 − α) and (μ_o, α), will not be affected, so the curves will become steeper (Figure 9.6). Except in a small region near μ_o, β will be reduced without degrading α; for most values of μ above μ_o, the test with larger n is more likely to be accurate.

In general, if the OC and power curves of one test are steeper than those of another, we say that the first test is more *powerful*, better able to distinguish between μ_o and other possible values of μ. We can increase the power of a test—its ability to distinguish between H_o and H_a—by enlarging the sample.

FIGURE 9.5.   **Effects on the OC and power curves of increasing α.**

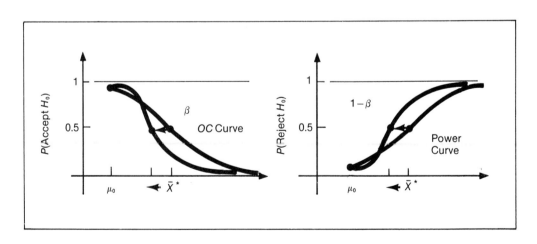

Lower-Tail Tests for the Population Mean

A *lower-tail* test for the population mean has the form

$$H_o: \mu \geq \mu_o$$

$$H_a: \mu < \mu_o$$

FIGURE 9.6.   **Effects on the OC and power curves of increasing n.**

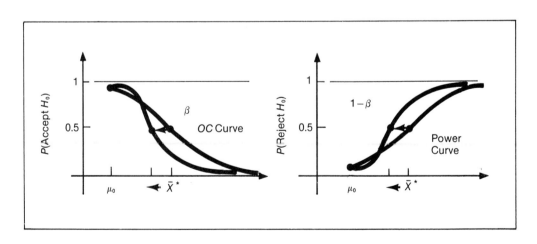

In such a test we are attempting to determine if the population mean is *less than* some value μ_o; our test statistic is the sample mean, and we will be convinced that $\mu < \mu_o$ if $\bar{X}$ is *small* enough. The critical value $\bar{X}^*$ is *below* μ_o, and we will reject the null hypothesis and accept the alternative if $\bar{X} < \bar{X}^*$. The decision rule is derived as in the upper-tail test, but with all the inequalities reversed, assuming a sample of size $n \geq 30$ and invoking the central limit theorem.

We select the significance level α, and must find the critical value $\bar{X}^*$ so that P(Type I error) $= \alpha$:

$$P(\text{Type I error}) = P(\text{Reject } H_o | H_o)$$

$$= P(\bar{X} < \bar{X}^* | H_o) = P\left(\frac{\bar{X} - \mu_o}{\sigma/\sqrt{n}} < \frac{\bar{X}^* - \mu_o}{\sigma/\sqrt{n}} \mid H_o \right)$$

$$= P(Z < \frac{\bar{X}^* - \mu_o}{\sigma/\sqrt{n}}) = \alpha.$$

Therefore,

$$\frac{\bar{X}^* - \mu_o}{\sigma/\sqrt{n}} = -z_\alpha, \text{ so } \bar{X}^* = \mu_o - z_\alpha \frac{\sigma}{\alpha/\sqrt{n}},$$

where z_α corresponds to α as before. (See Figure 9.7.) Note that α and $\bar{X}^*$ are in the *lower* tail of the distribution of $\bar{X}$ under H_o. The decision rule is this:

If $\bar{X} < \bar{X}^*$, reject H_o and accept H_a;

if $\bar{X} > \bar{X}^*$, do not reject H_o.

Again we can compare the normal deviate $(\bar{X} - \mu_o)/(\sigma/\sqrt{n})$ to the critical normal deviate $-z_\alpha$, or the one-tail probability $P(\bar{X} < \bar{X}_o | H_o)$ to the significance level α, and if σ is unknown, we estimate it with s.

FIGURE 9.7. Lower-tail test for the population mean.

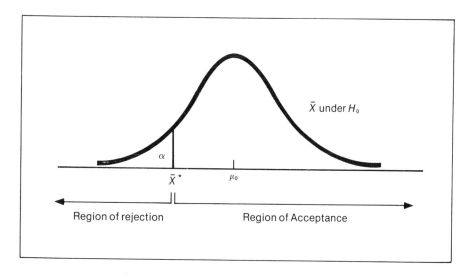

Example 1   Using the data in the USNEWS file, we wish to determine if average unemployment is below 5.8% in American cities, and we wish to limit the risk of making such a claim falsely to 2%.

Solution

If μ is the mean unemployment in American cities, then we are testing these hypotheses at the 2% significance level:

$$H_o: \mu \geq 5.8$$

$$H_a: \mu < 5.8$$

The statistics of the sample of values of X3, unemployment, are these:

VARIABLE X3   UNEMPLOYMENT

MEAN	5.691	STD ERR	.197	STD DEV	1.647
VARIANCE	2.712	KURTOSIS	2.700	SKEWNESS	1.215
MINIMUM	3.000	MAXIMUM	12.000	SUM	398.400
C.V. PCT	28.936	.95 C.I.	5.299	TO	6.084
VALID CASES	70	MISSING CASES	4		

Using the sample standard deviation to estimate σ, $z_\alpha = 2.05$ corresponding to $\alpha = 2\%$, and noting that the sample size is 70,

$$\bar{X}^* = 5.8 - 2.05 \frac{1.647}{\sqrt{70}} = 5.396.$$

We will reject H_o if $\bar{X}$ is less than $\bar{X}^* = 5.396$. The observed value of $\bar{X}$ is $5.691 > 5.396$, so we are unable to reject H_o; we cannot conclude that average unemployment in American cities is below 5.8%.

[Equivalently, the z statistic $(\bar{X} - \mu_o)/(s/\sqrt{n}) = (5.691 - 5.8)/(1.647/\sqrt{70}) = -0.55$ is not less than the critical normal deviate -2.05, and the one-tail probability $P(\bar{X} < 5.691|H_o) = P(Z < -0.55) = 0.2912$, is not less than $\alpha = 0.02$. We cannot reject H_o.]

β, the probability of a Type II error, is found as in an upper-tail test; inequalities are again reversed, as shown in this calculation of β for the above example if $\mu = 5.65$:

$$\beta = P(\text{Type II error}|H_a \text{ with } \mu = 5.65)$$

$$= P(\bar{X} > \bar{X}^*|H_a \text{ with } \mu = 5.65)$$

$$= P\left(\frac{\bar{X} - 5.65}{1.647/\sqrt{70}} > \frac{5.396 - 5.65}{1.647/\sqrt{70}} \Big| H_a \text{ with } \mu = 5.65\right)$$

$$= P(Z > -1.29) = 0.4015 + 0.5000 = 0.9015.$$

The operating characteristic and power curves of a lower-tail test are the mirror images of those for an upper-tail test. The OC curve begins at $(\mu_o, 1 - \alpha)$, and the power curve at (μ_o, α), and both pass through the point $(\bar{X}^*, 0.5)$. (See Figure 9.8.)

FIGURE 9.8.   *OC* and power curves of a lower-tail test for μ.

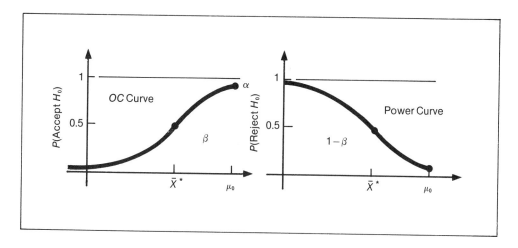

Two-Tail Tests for the Population Mean

Often we are concerned only that the population mean is *different* from some particular value, and not with the direction of that difference. Thus we have *two-tail* tests:

$$H_o: \mu = \mu_o$$

$$H_a: \mu \neq \mu_o$$

In such cases, we will be convinced that H_o is false if the test statistic $\bar{X}$ is far enough away from μ_o. There are two critical values, $\bar{X}_1^*$ and $\bar{X}_2^*$, again chosen to control α, and generally equidistant from μ_o. Because of this symmetry,

$$P(\bar{X} < \bar{X}_1^* | H_o) = \frac{\alpha}{2} \text{ and } P(\bar{X} > \bar{X}_2^* = | H_o) = \frac{\alpha}{2} \ .$$

Derivations identical to those for the two one-tail tests give us formulas for the critical values:

$$\bar{X}_1^* = \mu_o - z_{\alpha/2} \frac{\sigma}{\sqrt{n}} \text{ and } \bar{X}_2^* = \mu_o + z_{\alpha/2} \frac{\sigma}{\sqrt{n}} \ .$$

We reject H_o if $\bar{X}$ is less than the lower critical value or greater than the upper. If $\bar{X}$ falls between the critical values, we cannot reject H_o.

Again, we may compare $(\bar{X} - \mu_o)/(\sigma/\sqrt{n})$ with $\pm z_{\alpha/2}$, or α with the *two-tail probability* that a value of $\bar{X}$ at least as far away from μ_o as that observed would occur (Figure 9.9). Calculations of β and the forms of the *OC* and power curves are left as exercises.

FIGURE 9.9.

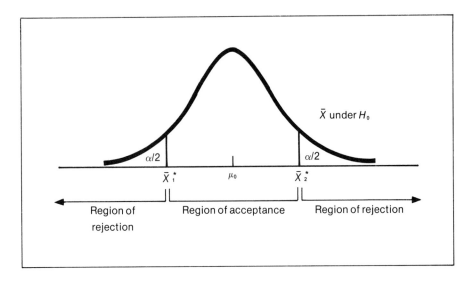

Example 2 A process makes machined parts to a mean diameter of 25.70 cm, with standard deviation 0.01 cm. Periodically a sample of 50 parts is measured to see if the process requires adjustment, a task management wishes to perform unnecessarily only 5% of the time. If one such sample has mean diameter 25.704 cm, should the process be stopped for adjustment?

Solution

The hypotheses are these:

$$H_o: \mu = 25.70$$

$$H_a: \mu \neq 25.70$$

We are told that management wishes to limit the probability of a Type I error to 5%, so $z_{\alpha/2}$ = 1.96, and the critical values are:

$$\mu_o \pm z_{\alpha/2} \frac{\sigma}{\sqrt{n}} = 25.70 \pm 1.96 \frac{0.01}{\sqrt{50}} = 25.6972, 25.7028.$$

The observed value of $\bar{X}$, 25.704, does not fall between the critical values so management rejects H_o, and adjusts the manufacturing process.

Hypothesis Tests and Program Packages

When looking over statistical software packages, the reader will find that most contain no procedure specifically designed to perform one-sample hypothesis tests. SPSS is one of these;

we can extract the necessary statistics from outputs of the CONDESCRIPTIVE and FRE-QUENCIES subprograms, but must perform the hypothesis tests ourselves. The recent proliferation of mini- and microcomputers, however, has provided great stimulus to the software industry, and many packages have been written for these smaller systems. Here we have an opportunity to examine one of them.

The Radio Shack TRS-80 Statistical Analysis package consists of eight programs stored on cassette tapes and the accompanying manual. The package includes five programs which describe data sets and perform statistical analysis, two which prepare, update, and list data files, and one which generates random samples. All are written in BASIC to be run on the TRS-80 Model I computer; the package costs about $20, and the computer, $600.

Consider again the USNEWS data file. Suppose we wish to test the null hypothesis that the mean income of factory workers in U.S. cities is $15,500 against the alternative that it is not at the 5% significance level. Formally, the hypotheses are:

$$H_o: \mu = 15,500$$

$$H_a: \mu \neq 15,500$$

and this is the procedure to carry out this test using the TRS-80 statistical package:

1. Turn on the video monitor.

2. Turn on the CPU console and press the BREAK key simultaneously. This loads the BASIC compiler into memory and prepares the system for use.

3. Place the T-TEST* cassette in the recorder and depress the PLAY level.

4. Load the T-TEST program by typing CLOAD followed by the ENTER key. When the program has been successfully loaded, '*' will appear on the screen.

5. To execute the program, type RUN and hit the ENTER key. The program will then guide you through the session shown below:

We see that the probability of a sample mean at least as far from 15500 as 15065.9 is greater than 5%, therefore we do not reject H_o. At the 5% significance level, we cannot conclude that the average income of factory workers in American cities is not $15,500.

Note that this program is limited to two-tailed tests for the population mean at the 5% significance level. Statistics on which to base other tests can be found using the Descriptive Statistics program in the TRS-80 package.

T-TEST

HOW WILL DATA BE ENTERED (1 = KEYBOARD, 2 = DATA FILE)?*1*
ENTER THE POPULATION MEAN (MU) OR YOUR QUALITY CONTROL VALUE? *15500*

---

* Test of population means are often called *t*-tests because they involve the *t* distribution, to be discussed shortly.

```
BEGIN ENTERING THE DATA FOR YOUR SAMPLE.
SIGNAL END OF DATA WITH 99999.
?   13432
?   16178
        .
        .
        .
?   17639
?   13459
?   99999
```

When you have signaled the end of your data with 99999, these results will appear:

```
            T-TEST RESULTS

MU OR CONTROL VALUE  =  15500

SAMPLE MEAN                =  15065.9
STANDARD DEVIATION         =  2314.60
SAMPLE SIZE (N)            =     73

T-RATIO                    =  − 1.60235
DIFFERENCE (MEAN-MU)       =  − 434.082
DEGREES OF FREEDOM         =     72

WITH A SAMPLE OF 74 AT RATIO OF (+ OR −) 1.60235

COULD HAVE OCCURRED BY CHANCE MORE THAN 5 TIMES IN 100

WANT TO RUN ANOTHER SET OF DATA (1 = YES, 2 = NO)?
```

Small-Sample Tests for the Population Mean

In our work so far we have taken large samples and have found in the central limit theorem (CLT) the description of the sampling distribution of $\bar{X}$ and the equivalent statement that $(\bar{X} - \mu)/(\sigma/\sqrt{n}) = Z$. Samples are not always of a size to allow us to invoke the CLT, however, though we often want to estimate μ with $\bar{X}$ when n is less than 30.

When we are dealing with small samples we apply a result published in 1908 by the chemist William S. Gossett, who had studied the distribution of the quantity $t = (\bar{X} - \mu)/$

$(s/\sqrt{n})$ for many values of n. Gossett's employers, a brewery in Dublin, Ireland, did not wish him to publish, so this result first appeared under the pen name "Student":

> If we take a sample of size n from a normally distributed population with mean μ, the quantity $t = (\bar{X} - \mu)/(s/\sqrt{n})$ has a **t distribution** [also called **student's-t distribution**] with $n - 1$ degrees of freedom.

The mean of every t distribution is 0, around which it is symmetrical and bell-shaped, though flatter than the standard normal distribution. The "number of degrees of freedom" is the parameter which determines which t distribution is being considered, and refers to the fact that for a given value of $\bar{X}$, $n - 1$ of the sample elements are arbitrary; the last is determined by the others and $\bar{X}$. Mathematically, the t random variable is the standard normal random variable Z divided by the square root of a chi-square random variable (the chi-square distribution will be discussed shortly) over its own number of degrees of freedom.

As the number of degrees of freedom increases (as the sample size increases), the t distribution approaches that of Z. For 30 or more degrees of freedom, they are approximately equal, so we generally apply the t distribution to situations involving small samples, of size less than 30 (Figure 9.11).

Though Gossett's result assumed that the population from which the sample was taken was normally distributed, further research has shown that this condition may be dropped. The t statistic is *robust*: it produces accurate results even without the assumption of normality, and we may use the t statistic without too much concern for the distributional shape of the underlying population.

Suppose that we wish to test these hypotheses at the 10% significance level, and we plan to take a sample of size 15:

$$H_o: \mu = 52$$

$$H_a: \mu \neq 52$$

FIGURE 9.11. *t* **distributions and the standard normal distribution.**

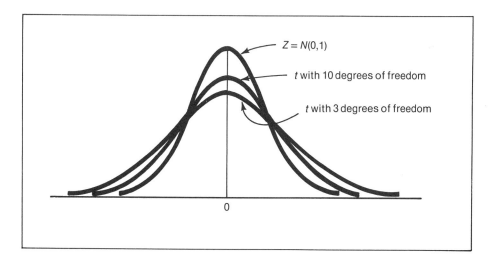

If H_o is true, the sampling distribution of the t statistic $(\bar{X} - 52)/(s/\sqrt{n})$ is a t distribution with $15 - 1 = 14$ degrees of freedom. We will be convinced to reject H_o if t is less than $-t_{\alpha/2}$ or greater than $t_{\alpha/2}$. From Table A-5, corresponding to $\alpha = 10\%$ and 14 degrees of freedom, $t_{\alpha/2} = 1.761$ (Figure 9.12).

We take the sample and find that $\bar{X} = 47$ and $s = 12$. Then $t = (47 - 52)/(12/\sqrt{15}) = 1.614$ is between the critical deviates, and we are not able to reject H_o.

Note that upper- and lower-tail tests may be performed in the obvious ways using the t statistic, and that it is tests of this type which are performed by the TRS-80 program T-TEST.

Hypothesis Tests with the Population Variance

Not all inferences concern the population mean; we are also interested in the value of the population variance. To investigate σ^2, we apply this result:

> If we take a sample of size n from a normally distributed population with variance σ^2, then the quantity $\chi^2 = (n-1)S^2/\sigma^2$ has a **chi-square distribution** with $n - 1$ degrees of freedom.

The value $X^2 = (n-1)S^2/\sigma^2$ is called the *chi-square statistic*, and its distribution, like that of the t statistic, depends on the sample size; the number of degrees of freedom determines which chi-square distribution is being considered. (A chi-square distribution with r degrees of freedom has mean r and variance $2r$.) A chi-square random variable cannot take on negative values, and its distribution is skewed to the right, becoming more symmetrical and bell-shaped as the number of degrees of freedom increases (Figure 9.13).

Using the chi-square distribution, we can construct confidence intervals and perform hypothesis tests for the population variance, and because the chi-square statistic, like the t statistic, is robust, we are not restricted to normally distributed populations. Consider finding a $C \times 100\%$ confidence interval for σ^2 based on S^2, the variance of a sample of size n.

From Table A-6, we can find values $\chi^2_{1-\alpha/2}$ and $\chi^2_{\alpha/2}$ corresponding to $n - 1$ degrees of freedom so that

$$P(\chi^2_{1-\alpha/2} < \frac{(n-1)S^2}{\sigma^2} < \chi^2_{\alpha/2}) = C,$$

as illustrated in Figure 9.14.

Algebraic manipulations transform the above statement into this:

$$P\left(\frac{(n-1)S^2}{\chi^2_{\alpha/2}} < \sigma^2 < \frac{(n-1)S^2}{\chi^2_{1-\alpha/2}} \right) = C.$$

Of all such intervals computed from observed values of s^2, the proportion C contains σ^2. If we have found one such interval, we can be C certain that it contains σ^2, so a $C \times 100\%$

FIGURE 9.12.   Critical values of the t statistic.

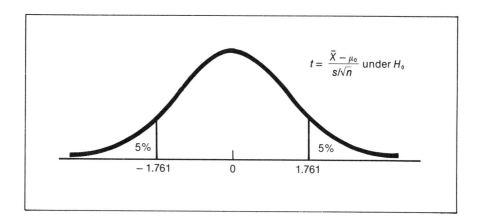

confidence interval for the population variance is

$$\left[\frac{(n-1)s^2}{x^2_{\alpha/2}} \; , \; \frac{(n-1)s^2}{x^2_{1-\alpha/2}} \right].$$

FIGURE 9.13.   Chi-square distributions associated with various sample sizes.

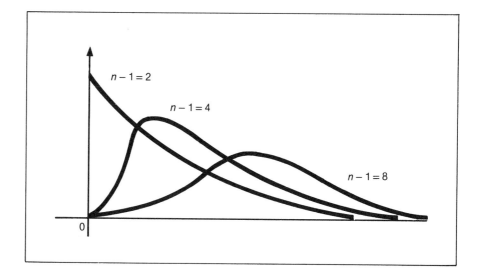

FIGURE 9.14.

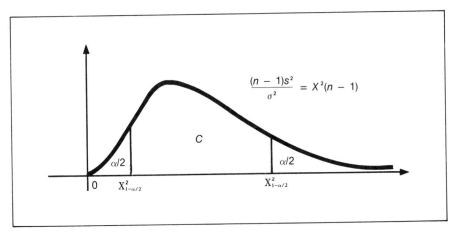

Consider finding a 90% confidence interval for the variance of unemployment in eastern U.S. cities, based on the sample of 25 nonmissing values of X3 eastern cities in the USNEWS data file. Corresponding to $C = 0.90$ and $25 - 1 = 24$ degrees of freedom, the appropriate values from Table A-6 are $\chi^2_{0.95} = 14.611$ and $\chi^2_{0.05} = 37.652$. The variance s^2 of the sample of 25 values of X3 is 1.567^2, so the 90% confidence interval is:

$$\left[\frac{(25-1)\ 1.567^2}{36.415} \ , \ \frac{(25-1)\ 1.567^2}{13.848} \right] = [1.618,\ 4.256].$$

Note that the corresponding confidence interval for σ is:

$$\left[\sqrt{1.618}\ , \sqrt{4.256}\ \right] = \left[1.272,\ 2063\right].$$

FIGURE 9.15.   Two-tail test of the population variance.

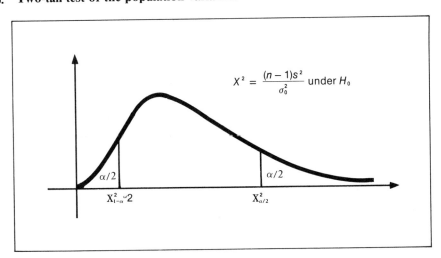

To test hypotheses concerning σ^2, we use the above result to select critical values. Consider a two-tailed test of the form

$$H_o: \sigma^2 = \sigma_o^2$$

$$H_a: \sigma^2 \neq \sigma_o^2,$$

with significance level α. We will reject H_o if s^2 is far enough from σ_o^2, which will cause the value of $(n-1)s^2/\sigma_o = X^2$ to be extreme. We select critical values for the χ^2 statistic by chopping off an area of $\alpha/2$ in each tail. (See Figure 9.15.)

The critical values are $\chi_{1-\alpha/2}^2$ and $\chi_{\alpha/2}^2$; we reject H_o if the observed value of $\chi^2 = (n-1)s^2/\sigma_o^2$ is not between them.

For example, we test at the 2% significance level the null hypothesis that the variance in unemployments in eastern cities is 2.0 against the alternative that it is not, using the 27 eastern cities of the USNEWS file.

$$H_o: \sigma^2 = 2.0$$

$$H_a: \sigma^2 \neq 2.0$$

Corresponding to $\alpha = 2\%$ and $25 - 1 = 24$ degrees of freedom, $\chi_{0.99}^2 = 10.856$ and $\chi_{0.01}^2 = 42.980$. The observed value of the chi-square statistic is

$$\chi^2 = \frac{(25-1)\,1.567^2}{2.0} = 29.47.$$

This value is between the critical values, so we cannot reject H_o; we cannot conclude that this variance is not 2.0.

For upper-tail tests, the critical value is simply χ_α^2, and for lower-tail tests, $\chi_{1-\alpha}^2$, with the appropriate changes in the decision rule.

EXERCISES

1. We will take a sample of size 80 from a population whose standard deviation we know to be 56, to test these hypotheses:

$$H_o: \mu \leq 300$$

$$H_a: \mu > 300$$

   a. At the 5% significance level, find the critical value and state the decision rule.
   b. For these possible values of μ, find the probability of a Type II error: 305, $\bar{X}^*$, 315, 320.
   c. Use the values found in part(b) to sketch the operating characteristic and power curves of the test.
   d. Find the critical value corresponding to $\alpha = 1\%$, restate the decision rule, and find β if $\mu = 315$. What generalization does this suggest about α and β?
   e. In the original test, $\bar{X} = 312$. What conclusion is reached? What type of error might have been made?
   f. Perform the test in (e) again by comparing the z statistic to the critical normal deviate, then by comparing the probability that $\bar{X}$ would be at least as large as its observed value with α.

2. We will take a sample of size 150 from a population whose standard deviation is 410 to test at the 5% significance level these hypotheses:

$$H_o: \mu \leq 1650$$

$$H_a: \mu > 1650$$

   a. Find the critical value and state the decision rule.
   b. For these possible values of μ, find β: 1675, $\bar{X}^*$, 1725, 1750.
   c. Use the values found in part (b) to sketch the OC and power curves of the test.
   d. Suppose that the sample is of size 300 rather than 150. Find the new critical values, restate the decision rule and find β if $\mu = 1675$ and if $\mu = 1725$. What generalization does this suggest about β and n?
   e. In the original test, $\bar{X} = 1661$. What conclusion is reached, and what type of error might have been made?

3. Show that for any upper-tail test of the population mean, the value of β if $\mu = \bar{X}^*$ is 0.5000.

4. What factors influence the power of a test of hypotheses?

5. Use the following data from the USNEWS file to test the claim that construction in U.S. cities was up by more than 5% at the 5% significance level.

VARIABLE X7   CHANGE IN CONSTRUCTION ACTIVITY

MEAN	5.523	STD ERR	1.081	STD DEV	9.301
VARIANCE	86.517	KURTOSIS	1.947	SKEWNESS	.924
MINIMUM	− 15.400	MAXIMUM	40.600	SUM	408.700
C.V. PCT	168.414	.95 C.I.	3.368	TO	7.678
VALID CASES	74	MISSING CASES	0		

   a. Formally state the hypotheses, and determine the decision rule.
   b. If construction activity has in fact increased by 5.5%, find the probability that we will fail to reject H_o. Illustrate by drawing the distribution of $\bar{X}$ if $\mu = 5.5\%$ and indicate the area β.
   c. Sketch the OC and power curves for this test.
   d. Use the decision rule to come to a conclusion. What type of error might have been made?

6. Using a sample of size 100 from a population whose standard deviation is 75, we will test these hypotheses at the 1% significance level:

$$H_o: \mu \geq 250$$

$$H_a: \mu < 250$$

   a. Find the critical value and state the decision rule.
   b. Find the probability of a Type II error if μ is: 220, 230, $\bar{X}^*$, 240.
   c. Use the values found in part (b) to sketch the operating characteristic and power curves of the test.
   d. Suppose α were 5%. What effect would this have on β and on the curves drawn in part (c)?
   e. We find that $\bar{X}$ is 229. What conclusion is drawn and what type of error might have been made?
   f. Perform the test again by comparing the z statistic with the critical normal deviate, and by comparing the probability that $\bar{X}$ would be less than its observed value with α.

7. Using the following statistics from the USNEWS file, test the claim that the change in factory worker income was less than 10% at the 5% significance level.

VARIABLE X6   CHANGE IN FACTORY WORKERS INCOME

MEAN	8.673	STD ERR	.650	STD DEV	5.551
VARIANCE	30.814	KURTOSIS	3.735	SKEWNESS	− .040
MINIMUM	− 10.900	MAXIMUM	28.300	SUM	633.100
C.V. PCT	64.007	.95 C.I.	7.377	TO	9.968

VALID CASES	73	MISSING CASES	1

  a. Formally state the hypotheses, and develop a decision rule.

  b. Find β if the true mean change in factory worker income is 9.0%.

  c. Sketch the OC and power curves of this test.

  d. If the sample size were increased, how would the curves of part (c) change?

  e. Use the decision rule to reach a conclusion.

  f. Repeat the test by comparing the normal deviate to the critical normal deviate, and by comparing $P(\bar{X} < 8.673|H_o)$ to α.

 8. Show that for any lower-tail test of hypotheses for μ, β is 0.5000 if $\mu = X^*$.

 9. Find $z_{\alpha/2}$ for $\alpha = 1\%$, 2%, 5%, and 10%.

10. Using a sample of 200 values from a population which has standard deviation 0.55, these hypotheses will be tested at the 5% significance level:

$$H_o: \mu = 7.6$$

$$H_a: \mu \neq 7.6$$

  a. Find the critical values and state the decision rule.

  b. Find the probability of Type II error if $\mu = 7.5$, $\bar{X}_1^*$, 7.55, $\bar{X}_2^*$, 7.7. (*Hint:* $\beta = P(\bar{X}_1^* < \bar{X} < \bar{X}_2^*|H_a$ with $\mu = \mu_a)$.)

  c. Use the values found in part (b) to draw the OC and power curves of this test.

  d. The observed value of $\bar{X}$ is 7.652. What conclusion is reached?

  e. Repeat the test by comparing the z statistic to the critical normal deviates $\pm z_{\alpha/2}$, and by comparing the probability that $\bar{X}$ would be further from μ_o than 7.652 to α.

11. The following statistics were produced from the USNEWS data file with the SPSS subprogram BREAKDOWN. Use them to test the null hypothesis that mean unemployment in all central cities is 5% against the alternative that it is not at the 10% significance level.

CRITERION VARIABLE X3 UNEMPLOYMENT
  BROKEN DOWN BY X1 CITY REGION

VARIABLE	CODE	MEAN	STD DEV	N VALUE LABEL
FOR ENTIRE POPULATION		5.691	1.647	70
X1	1.	5.836	1.567	25 EASTERN
X1	2.	5.446	1.792	28 CENTRAL
X1	3.	5.882	1.554	17 WESTERN

    TOTAL CASES = 74
   MISSING CASES = 4   OR   5.4 PCT

  a. State the hypotheses and develop a decision rule.

  b. Calculate the test statistic and use the decision rule to reach a conclusion.

c. Write an SPSS program which will produce this output.

12. The TRS-80 Statistical Analysis program T-Test was used to test the null hypothesis that the mean change in department store sales in U.S. cities was 10.0% against the alternative that it was not at the 5% significance level, using the values in the USNEWS file. After the 70 nonmissing values had been entered, the following display appeared:

T-TEST RESULTS

MU OR CONTROL VALUE   =   10.0

SAMPLE MEAN   =   10.180
STANDARD DEVIATION   =   6.684
SAMPLE SIZE (N)   =   70

T-RATIO   =   0.22531
DIFFERENCE (MEAN-MU)   =   0.180
DEGREES OF FREEDOM   =   69

WITH A SAMPLE OF 25 AT RATIO OF $(+ \text{ OR } -)0.22531$

COULD HAVE OCCURRED BY CHANCE MORE THAN 5 TIMES IN 100

WANT TO RUN ANOTHER SET OF DATA $(1 = \text{YES}, 2 = \text{NO})$?

a. Formally state the hypotheses being tested.

b. What conclusion is reached? How do you know?

13. The test in Exercise 12 was repeated for eastern U.S. cities, using the sample of 25 such in the USNEWS file. The displayed results were these:

T-TEST RESULTS

MU OR CONTROL VALUE   =   10.0

SAMPLE MEAN   =   8.32
STANDARD DEVIATION   =   5.78
SAMPLE SIZE (N)   =   25

T-RATIO   = $-$ 1.4533
DIFFERENCE (MEAN-MU)   = $-$ 1.68
DEGREES OF FREEDOM   =   24

WITH A SAMPLE OF 25 AT RATIO OF $(+ \text{ OR } -)1.4533$

COULD HAVE OCCURRED BY CHANCE MORE THAN 5 TIMES IN 100

WANT TO RUN ANOTHER SET OF DATA $(1 = \text{YES}, 2 = \text{NO})$?

a. What distribution applies to the problem?

b. What conclusion is reached? Why? What type of error might have been made?

14. A sample of size 20 is taken from a large population to test these hypotheses at the 5% significance level:

$$H_o: \mu = 50$$

$$H_a: \mu \neq 50$$

    a. State the decision rule for this test.

    b. The sample has mean 51.6 and standard deviation 9.1. What conclusion is reached?

15. The t distribution can also be used to compute confidence intervals for μ based on small samples.

    a. Beginning with this statement, develop formulas for the endpoints of such an interval:

    Given C, we can find t corresponding to $n - 1$ degrees of freedom so that

$$P\left(-t < \frac{\bar{X} - \mu}{s/\sqrt{n}} < t\right) = C$$

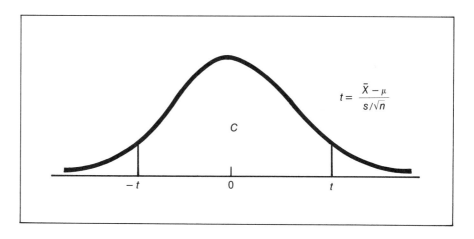

    b. Using the information in this printout, find a 90% confidence interval for the mean income of factory workers in all western U.S. cities.

CRITERION VARIABLE   X5   AVERAGE INCOME OF FACTORY WORKERS
BROKEN DOWN BY   X1   CITY REGION

VARIABLE	CODE	MEAN	STD DEV	N VALUE LABEL
FOR ENTIRE POPULATION		15065.918	2314.596	73
X1	1.	14504.852	2636.825	27 EASTERN
X1	2.	15723.667	2019.017	30 CENTRAL
X1	3.	14779.438	2064.741	16 WESTERN

    TOTAL CASES =   74
    MISSING CASES =   1   OR   1.4 PCT

    c. Write an SPSS program to produce this output.

16. We wish to test the null hypothesis that the average income of factory workers in all eastern U.S. cities is no more than \$14,000 against the alternative that it is greater than \$14,000, at the 5% significance level.

    a. State formally the hypotheses and the decision rule.

    b. Based on the statistics given in Exercise 15, draw a conclusion.

    c. Find a 90% confidence interval for mean income of factory workers in all eastern cities.

17. Use the information in this printout to test, at the 10% significance level, whether it can be convincingly claimed that construction activity in all eastern U.S. cities grew by less than 7%.

CRITERION VARIABLE X7		CHANGE IN CONSTRUCTION ACTIVITY		
BROKEN DOWN BY X1		CITY REGION		
VARIABLE	CODE	MEAN	STD DEV	N VALUE LABEL
FOR ENTIRE POPULATION		5.523	9.301	74
X1	1.	6.493	9.816	27 EASTERN
X1	2.	4.677	7.331	30 CENTRAL
X1	3.	5.476	11.739	17 WESTERN
TOTAL CASES =	74			

18. Use the information given in Exercise 17 to test whether mean construction activity in western cities grew by more than 4%. Use $\alpha = 5\%$.

19. Find a 95% confidence interval for the mean growth in construction activity in all western cities, using the information given in Exercise 17.

20. Show that if

$$P\left(\chi^2_{1-\alpha/2} < \frac{(n-1)S^2}{\sigma^2} < \chi^2_{\alpha/2}\right) = C,$$

then

$$P\left(\frac{(n-1)S^2}{\chi^2_{\alpha/2}} < \sigma^2 < \frac{(n-1)S^2}{\chi^2_{1-\alpha/2}}\right) = C.$$

21. Using this information from the USNEWS file, find a 90% confidence interval for the variance of the change in nonfarm employments in cities in the:

a. East.

b. West.

CRITERION VARIABLE X4		CHANGE IN NONFARM EMPLOYMENT		
BROKEN DOWN BY X1		CITY REGION		
VARIABLE	CODE	MEAN	STD DEV	N VALUE LABEL
FOR ENTIRE POPULATION		2.699	2.820	74
X1	1.	2.363	2.302	27 EASTERN
X1	2.	1.823	2.506	30 CENTRAL

X1	3.	4.776	3.172	17 WESTERN

TOTAL CASES = 74

22. Use the information given in Exercise 21 to test the null hypothesis that the variance of the change in nonfarm employment in eastern cities is 4 against the alternative that is not at the 10% significance level.

23. Using this information from the USNEWS file, can it be concluded at the 5% significance level that the variance of the growth in department store sales is less than 64 in eastern U.S. cities?

CRITERION VARIABLE X2		CHANGE IN DEPARTMENT STORE SALES		
BROKEN DOWN BY X1		CITY REGION		
VARIABLE	CODE	MEAN	STD DEV	N VALUE LABEL
FOR ENTIRE POPULATION		10.180	6.684	70
X1	1.	8.320	5.776	25 EASTERN
X1	2.	12.687	7.082	30 CENTRAL
X1	3.	8.267	5.963	15 WESTERN

TOTAL CASES = 74
MISSING CASES = 4 OR 5.4 PCT

24. Using the information in Exercise 23, can it be concluded at the 5% significance level that the variance in growth in department store sales is greater than 25 in the west?

25. Write an SPSS program to produce the outputs shown in Exercises 21 and 23.

26. A process fills soft-drink bottles with an average of 16 oz of soda, with standard deviation no more than 0.2 oz. If the standard deviation exceeds 0.2, the process must be adjusted; since this costs time and money, management hopes to adjust the process unnecessarily no more than 5% of the time. Periodically, a sample of 25 bottles is taken to see if adjustment is required. If one such sample has standard deviation 0.27 oz, should the process be adjusted?

27. We have found that the standard deviation of the 70 nonmissing values of X2 (change in department store sales) in the USNEWS file is 6.684. Use the result stated in Exercise 19 of Chapter 8 to construct a test of the null hypothesis that the variance of the population of all such values is no more than 6 against the alternative that the variance is greater than 6 at the 5% significance level. (*Note:* We must assume the population is normally distributed.)

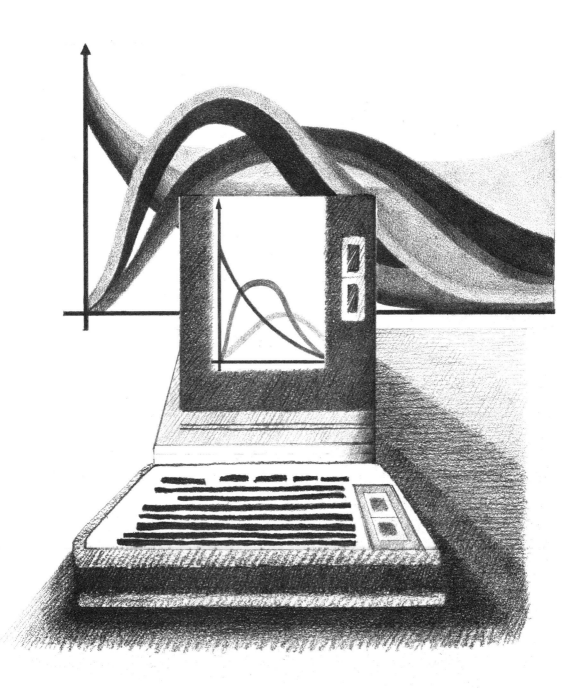

10

Two-Sample Hypothesis Tests

INTRODUCTION

In most experiments where samples are taken to estimate population parameters, the researcher is not so much concerned with differences of parameters from particular predetermined values (the tests in Chapter 9), as with differences among corresponding parameters of several populations. For example, the marketing strategist who has to choose between two packages for a new product wants to know which package is more effective in promoting sales. In evaluating a new treatment for a disease, the question is not whether the new treatment shortens expected recovery time below 14 days, but if the new treatment is more effective than the old. In each case, two samples will be taken and compared.

In this chapter we explore techniques for comparisons of two populations, beginning with hypothesis tests for the equality of two population variances.

Tests with Two Population Variances

A company that packages liquid products such as lubricants and solvents plans to install a new automatic filling machine. The company's representatives have narrowed the choice down to two competing devices. If one device is consistently more accurate than the other, this will be an important factor in their choice, so they visit the two factories where the equipment is made to witness demonstrations. They wish to find out if the variances of the two processes are significantly different.

The presumption is that the accuracies of the two processes are about the same, so these hypotheses are being tested:

$$H_o: \sigma_A^2 = \sigma_B^2$$

$$H_a: \sigma_A^2 \neq j_B^2$$

We now consider this problem in general.

Given two populations, A and B, we wish to test for a significant difference in their variances. Are the values of one more widely dispersed than the values of the other? To do this, we take *independent* samples of sizes n_A and n_B from the populations and examine the value of the F statistic, $F = s_A^2/s_B^2$.

If H_o is true and the two population variances are equal, then we would expect the value of $F = s_A^2/s_B^2$ to be near one; a value of F far from 1 would be evidence that $\sigma_A^2 \neq \sigma_B^2$. To find how far F must be from 1 to enable us to reject the null hypothesis, we must know the sample distribution of the F statistic when H_o is true. This is provided by the following result, from Sir Ronald Fisher:

If independent samples of sizes n_A and n_B are taken from two normally distributed populations with common variance σ^2, the sampling distribution of the F statistic, $F = s_A^2/s_B^2$, is an *F* **distribution** with $(n_A - 1)$ and $(n_B - 1)$ degrees of freedom.

Mathematically, the F random variable is the ratio of two chi-square random variables, each divided by its own number of degrees of freedom:

$$F = \frac{X_A^2/df_A}{X_B^2/df_B}.$$

The F distribution has a number of degrees of freedom associated with its numerator (with s_A^2 in the F statistic), and another with its denominator (with s_B^2). These two parameters determine and describe the F distribution.

The F random variable is the ratio of two values which are always nonnegative, so it can never take on a value less than 0. Like the chi-square, the F distribution is skewed to the right, and it becomes more symmetrical and bell-shaped as both numbers of degrees of freedom increase (Figure 10.1).

FIGURE 10.1.   *F* **distributions.**

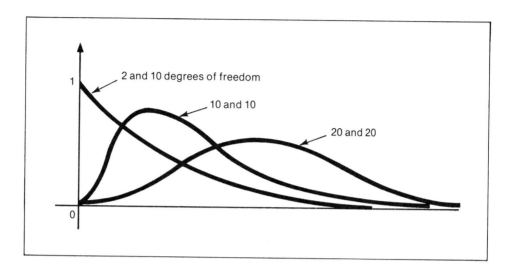

Given this information about the sampling distribution of the *F* statistic if H_o is true, we can find critical values for our test. Corresponding to our chosen significance level α, and to $n_A - 1$ and $n_B - 1$ degrees of freedom, Table A-7 provides us with the value $F_{\alpha/2}$. If H_o is true, the probability that $F = s_A^2/s_B^2$ will be greater than $F_{\alpha/2}$ is $\alpha/2$, so if this occurs, we will be convinced to reject H_o (Figure 10.2).

FIGURE 10.2.   **Critical values of the *F* distribution.**

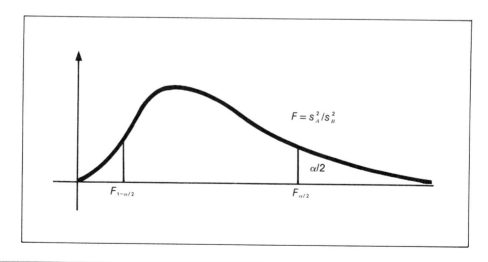

We will also be convinced to reject H_o if F is less than $F_{1-\alpha/2}$, but Table A-7 does not provide this value. However, while $F = s_A^2/s_B^2$ has an F distribution with $n_A - 1$ degrees of freedom in the numerator and $n_B - 1$ degrees of freedom in the denominator, the statistic F' $= s_A^2/s_B^2$ also has an F distribution, but with the numbers of degrees of freedom interchanged. For F' we can also find an upper critical value $F'_{\alpha/2}$ such that we will reject H_o if F' $> F'_{\alpha/2}$. But:

$$F' = s_B^2/s_A^2 > F'_{\alpha/2} \longleftrightarrow F = s_A^2/s_B^2 < \frac{1}{F'_{\alpha/2}}$$

That is, the lower critical value of F is the reciprocal of the upper critical value corresponding to interchanged numbers of degrees of freedom.

Suppose that the test in our example was performed at the 10% significance level, and that representatives of the company weighed a sample of 21 drums filled by machine A and an independent sample of 25 drums filled by machine B, finding that $s_a^2 = 4.3$ lb^2 and $s_B^2 = 2.2$ lb^2.

Corresponding to $21 - 1 = 20$ degrees of freedom in the numerator and $25 - 1 = 24$ degrees of freedom in the denominator, $F_{\alpha/2} = F_{0.05} = 2.03$ (see Figure 10.3.) Interchanging the numbers of degrees of freedom, $F'_{\alpha/2} = F'_{0.05} = 2.08$, so

$$F_{1-\alpha/2} = F_{0.95} = \frac{1}{2.08} \doteq 0.481.$$

The null hypothesis will be rejected if $F < 0.481$ or $F > 2.03$. The observed value is $F = s_A^2/s_B^2 = 4.3/2.2 = 1.95$, which is between the critical values, so H_o cannot be rejected. They cannot conclude at the 10% significance level that there is a significant difference in the accuracy of the two machines.

FIGURE 10.3.   *F* with 20 and 24 degrees of freedom.

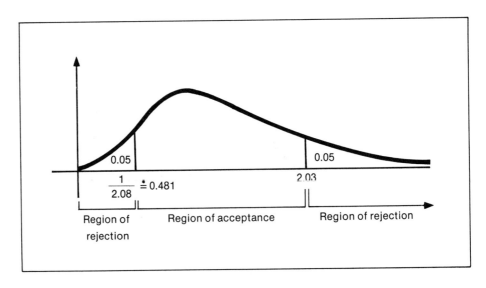

Upper-tail tests of the form

$$H_o: \sigma_A^2 \le \sigma_B^2$$

$$H_a: \sigma_A^2 > j_B^2$$

can be performed by finding the one critical value F_α, and any lower-tail test can be transformed into an upper-tail test by placing the suspected larger variance in the numerator of the F statistic.

Unlike the t and chi-square tests discussed so far, the F test for equality of variances is very sensitive to the assumption that both populations are normally distributed, particularly when the samples are small. The F test should not be used with small samples unless the populations can be shown to be normally distributed. The chi-square goodness-of-fit test described in Chapter 11 can be used for such a demonstration.

Independent Sample Tests for the Difference of Two Means

Most often when we compare two populations, we wish to demonstrate that their respective means are either equal or unequal; that is, we test these hypotheses:

$$H_a: \mu_A = \mu_B \qquad\qquad H_a: \mu_A - \mu_B = 0$$
or equivalently,
$$H_o: \mu_A \ne \mu_B \qquad\qquad H_a: \mu_A - \mu_B \ne 0$$

and this is often done by taking independent samples from the populations. Looking at the second of the two statements of hypotheses for such a test, it seems reasonable to use a test statistic related to $d = \bar{X}_A - \bar{X}_B$, where $\bar{X}_A$ and $\bar{X}_B$ are the means of the respective samples. The expected value of d is $\mu_A - \mu_B$, and we also have this important result:

When independent samples of sizes n_A and n_B are taken from two normally distributed populations A and B *with common variance* σ^2, the sampling distribution of the statistic $t = [d - (\mu_A - \mu_B)]/s_d$ is a t distribution with $n_A + n_B - 2$ degrees of freedom, where s_d is the standard deviation of the sampling distribution of d.

To describe the distribution of t, we must find s_d. Since the two samples, and therefore X_A and X_B, are independent, and both populations have variance σ^2,

$$\text{Var}(d) = \text{Var}(\bar{X}_A - \bar{X}_B) = \text{Var}(\bar{X}_A) + \text{Var}(\bar{X}_B) = \frac{\sigma^2}{n_A} + \frac{\sigma^2}{n_B}.$$

Then

$$\sigma_d = \left(\frac{\sigma^2}{n_A} + \frac{\sigma^2}{n_B}\right)^{1/2} = \sigma\left(\frac{1}{n_A} + \frac{1}{n_B}\right)^{1/2}.$$

s_A^2 and s_B^2 are both unbiased estimators of σ^2, so we can *pool* them to provide a better estimate of the common population variance:

$$\frac{(n_A - 1)s_A^2 + (n_B - 1)s_B^2}{n_A + n_B - 2} \text{ is an unbiased estimator of } \sigma^2.$$

Then

$$S_d = \left[\frac{(n_A - 1)s_A^2 + (n_B - 1)s_B^2}{n_A + n_B - 2} \right]^{\frac{1}{2}} \left(\frac{1}{n_A} + \frac{1}{n_B} \right)^{\frac{1}{2}}.$$

Further, if H_o is true, then $\mu_A = \mu_B$, and our test statistic is simply

$$t = \frac{d - (\mu_A \times \mu_B)}{S_d} = \frac{d}{S_d}.$$

Once again, corresponding to our choice of significance level α and to $n_A + n_B - 2$ degrees of freedom, we can find $t_{\alpha/2}$ so that (Figure 10.4)

$$P(t > t_{\alpha/2}) = \alpha/2 \text{ and } P(t < -t_{\alpha/2}) = \alpha/2.$$

The critical values are $\pm t_{\alpha/2}$. We will reject the null hypothesis and conclude that the population means are different if the t statistic is not between them. (Keep in mind that this development is valid only if the population *variances* are equal.)

Suppose that we want to determine if mean change in nonfarm employment in eastern cities is different from that in the central region, based on the data in the USNEWS file. Our hypotheses, which we will test at the 5% significance level, are:

$$H_o: \mu_E = \mu_C$$

$$H_a: \mu_E \neq \mu_C ,$$

and the SPSS subprogram **BREAKDOWN** can be used to provide these statistics:

FIGURE 10.4. $t = d/s_d$ **under** H_o.

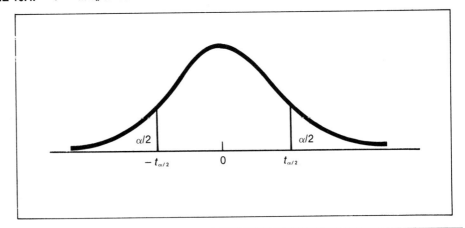

CRITERION VARIABLE X4		CHANGE IN NONFARM EMPLOYMENT		
BROKEN DOWN BY X1		CITY REGION		
VARIABLE	CODE	MEAN	STD DEV	N VALUE LABEL
FOR ENTIRE POPULATION		2.699	2.820	74
X1	1.	2.363	2.302	27 EASTERN
X1	2.	1.823	2.506	30 CENTRAL
X1	3.	4.776	3.172	17 WESTERN

TOTAL CASES   =   74

To employ the method derived above, the variances of the two populations must be equal; that is the case here, as can be verified by an *F* test.

From Table A–5, corresponding to $\alpha = 5\%$ and $27 + 30 - 2 = 55$ degrees of freedom, $t_{\alpha/2} = t_{0.025} \doteq 2.004$, so we will reject H_o and conclude that the two levels of unemployment are unequal if the value of the *t* statistic is less than -2.004 or greater than 2.004.

To find the value of the *t* statistic, we must first find s_d, based on the pooled variance estimate of the common population variance σ^2.

$$s_d = \left[\frac{(27-1)\,2.302^2 + (30-1)\,2.506^2}{27 + 30 - 2} \right]^{1/2} \left(\frac{1}{27} + \frac{1}{30} \right)^{1/2} = 0.6398.$$

Then

$$t = \frac{d}{s_d} = \frac{\bar{X}_E - \bar{X}_C}{s_d} = \frac{2.363 - 1.823}{0.6398} = 1.69.$$

t is between the critical values; it falls in the region of acceptance, so we do not reject H_o. We do not conclude that the two mean changes in nonfarm employment are different.

When the variances of the two populations are not equal, the sampling distribution of the *t* statistic, though symmetrical, is not a *t* distribution, and the technique developed above does not apply. The question of describing the distribution is called the *Fisher-Behrens problem*, and several statisticians (Lehman, Fisher and Behrens, and others) have proposed solutions.

One such solution, implemented in an SPSS subprogram to be discussed shortly, uses an approximation to the *t* statistic,

$$t = \frac{d - (\mu_A - \mu_B)}{(s_A^2/n_A + s_B^2/n_B)^{1/2}},$$

whose sampling distribution approximates a *t* distribution with this number of degrees of freedom:

$$\frac{[s_A^2/n_A + s_B^2/n_B]^2}{(s_A^2\,n_A)^2/(n_A - 1) + (s_B^2/n_B)^2/(n_B - 1)}$$

This number is not usually an integer, but reasonable accuracy is obtained by rounding to the nearest integer.

As when the population variances are equal, we find the critical values $\pm\, t_{\alpha/2}$ corresponding to α and the approximate number of degrees of freedom, and compare the approximate *t* statistic to the critical values in order to draw a conclusion.

For example, suppose we wish to determine if the average change in factory worker income in the East is different from that in the West, at the 5% significance level, based on the information in the USNEWS file. That is, we wish to test these hypotheses:

$$H_o: \mu_E = \mu_W$$

$$H_a: \mu_E \neq \mu_W$$

The appropriate statistics of the values of the variable X5 from the file, calculated by the SPSS subprogram BREAKDOWN, are these:

CRITERION VARIABLE   X6   CHANGE IN FACTORY WORKERS INCOME
        BROKEN DOWN BY   X1   CITY REGION

VARIABLE	CODE	MEAN	STD DEV	N VALUE LABEL
FOR ENTIRE POPULATION		8.673	5.551	73
X1	1.	9.015	3.648	27 EASTERN
X1	2.	8.423	6.178	30 CENTRAL
X1	3.	8.562	7.121	16 WESTERN

TOTAL CASES = 74
MISSING CASES = 1 OR   1.4 PCT

Critical values of the F statistic with 26 numerator and 15 denominator degrees of freedom are approximately 2.30 and 0.474 at the 10% significance level. The value of the F statistic is $3.648^2/7.121^2 = 0.252$, less than the lower critical value, so we can conclude that the population variances are unequal.

Returning to consideration of the means, the number of degrees of freedom of the approximate t statistic is

$$\frac{\left(\frac{3.648^2}{27} + \frac{7.121^2}{16}\right)^2}{\frac{\left(\frac{3.648^2}{27}\right)^2}{26} + \frac{\left(\frac{7.121^2}{16}\right)^2}{15}} = \frac{13.412}{0.009 + 0.670} - 19.75 \doteq 20.$$

At the 5% significance level, the critical values are ± 2.086. The value of the approximate t statistic is

$$t = \frac{d}{(s_E^2/n_E + s_W^2/n_W)^{1/2}} = \frac{9.015 - 8.562}{\left(\frac{3.648^2}{27} + \frac{7.121^2}{16}\right)^{1/2}} = \frac{0.453}{1.914} = 0.237 .$$

This is between the critical values, so we do not reject the null hypothesis; we do not conclude that the change in factory worker income in the East is different from that in the West.

We know that as the number of degrees of freedom increases, the t distribution approaches the standard normal. Therefore, with large sample sizes, the critical normal deviate $z_{\alpha/2}$ can be used to approximate $t_{\alpha/2}$. This can be done when the variances are equal or unequal.

From the above discussion and examples, we can also see how to perform one-tail tests, and any of these tests, including those when population variances are equal, can be performed by comparing the area under the graph of the sampling distribution of the t statistic beyond the observed value to the significance level of the test, as was done in Chapter 9 with one-sample tests. This is, in fact, what is done in an SPSS subprogram that performs t-tests for the difference of two population means based on independent samples.

The T-Test Subprogram for Independent Samples

When our file of cases contains sample values of a variable from two populations, we can perform a t-test for the difference of the means of those populations with the SPSS subprogram T-TEST. In the specification field of the card which calls this subprogram, we specify the groups into which the cases are to be divided, and the variables upon which the test is to be performed, as shown in this example, which performs the test described in the previous section for the mean difference in the changes in factory worker income (X6) in eastern and western cities.

```
? 10   GET FILE;USNEWS
? 20   T-TEST;GROUPS = X1(1,3)/VARIABLES = X6
```

The command word T-TEST occupies the command field of the statement. In the specification field, the cases in the file are grouped according to the values of X1, region. The cases for which X1 is 1, the eastern cities, comprise GROUP 1, and the western cities, for which X1 is 3, are GROUP 2. Following the slash, we specify that this test is to be performed on the values of the variable X6, and the program produces this output:

VARIABLE	NUMBER OF CASES	MEAN	STANDARD DEVIATION	STANDARD ERROR
X6				
CHANGE IN FACTORY WORKERS INCOME				
GROUP 1	27	9.0148	3.648	.702
GROUP 2	16	8.5625	7.121	1.780

		POOLED VARIANCE ESTIMATE			SEPARATE VARIANCE ESTIMATE		
F VALUE	2-TAIL PROB	T VALUE	DEGREES OF FREEDOM	2-TAIL PROB.	T VALUE	DEGREES OF FREEDOM	2-TAIL PROB.
3.81	.003	.28	41	.784	.24	19.75	.816

Two hypothesis tests are performed. The first is an F test for the equality of the two population variances. The value of the F statistic, calculated with the larger variance in the numerator, is 3.81, and the probability, if the two population variances are equal, of obtaining a ratio of variances more extreme than this is 0.003, the *two-tail probability*. This is less than a likely significance level for such a test, so we conclude that the two variances are unequal (Figure 10.5).

To perform the t-test, therefore, we look to the *separate variance estimate*. The approximate t statistic, calculated as shown in the previous section, is 0.24, and corresponds to 19.75

FIGURE 10.5.   **Critical values of the *F* statistic.**

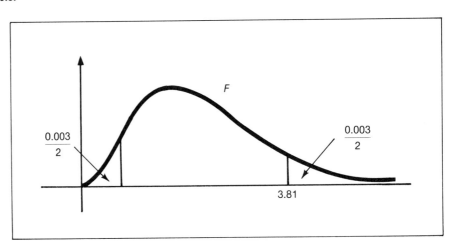

degrees of freedom. If the null hypothesis is true and the two population means are equal, the probability that the *t* statistic would be at least as extreme as observed (the two-tail problem) is 0.816. This is more than the significance level of the test (5%), so we do not reject the null hypothesis (Figure 10.6).

Had the two-tail probability of the *F* statistic been greater than the significance level for the test of equality of variances, we would have drawn our conclusion from the *pooled variance estimate*, as in this example from p. 212, in which we test for inequality of mean change in nonfarm employment (X4) between eastern and central cities:

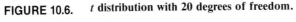

FIGURE 10.6.   *t* **distribution with 20 degrees of freedom.**

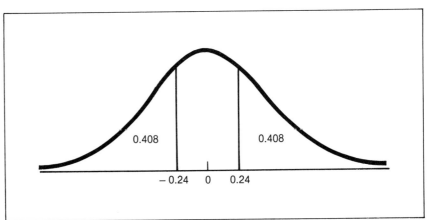

```
10.        GET FILE
10.005     USNEWS
20.        T-TEST
20.005     GROUPS = X1 (1,2)/VARIABLES = X4
```

GROUP 1, for which X1 is 1, is composed of the eastern cities in the USNEWS file; GROUP 2, with X1 = 2, is the central cities; and the variable of interest is X4.

VARIABLE	NUMBER OF CASES	MEAN	STANDARD DEVIATION	STANDARD ERROR
X4				
CHANGE IN NONFARM EMPLOYMENT				
GROUP 1	27	2.3630	2.302	.443
GROUP 2	30	1.8233	2.506	.458

		POOLED VARIANCE ESTIMATE			SEPARATE VARIANCE ESTIMATE		
F VALUE	2-TAIL PROB.	T VALUE	DEGREES OF FREEDOM	2-TAIL PROB.	T VALUE	DEGREES OF FREEDOM	2-TAIL PROB.
1.19	.664	.84	55	.403	.85	54.97	.400

The two-tail probability of the *F value* is 0.664; there is no significant difference between the two variances. Using the *pooled variance estimate*, the *T value* (the value of the *t* statistic) is 0.85, and its two-tail probability is 0.400. This is more than α, so H_o is not rejected. We cannot conclude that the mean changes in nonfarm employment are different.

If the cases of the data file are ordered so that all those of GROUP 1 precede those of GROUP 2 at the beginning of the file, the groups can be specified in this way:

GROUPS = 27, 30

Here the first 27 cases form GROUP 1, the next 30, GROUP 2. Also, more than one variable can be specified in one T-TEST. For example,

VARIABLES = X2 to X4, X7

The T-TEST subprogram has no optional statistics.

Matched-Pair Tests for the Difference of Two Means

So far, when testing hypotheses concerning the means of two populations, we have used independent samples; the values in the first sample have no effect on those in the second. Suppose, however, that we wish to assess the effectiveness of a training program by comparing pre-training and post-training test scores for a single group of people. An individual who scored well before the training could be expected to score well after; the two samples are *dependent*, and rather than examine the two samples separately, we use as our sample *matched pairs* of values, the pairs of scores for each individual in the experiment.

Samples of matched pairs occur in two situations. In the first, of which the experiment described above is an example, measurements are made on the elements of a single sample group before and after the application of an experimental treatment, or after each of two different treatments. In the second, members of two groups are paired, and different treatments applied to the two groups. In a study of the effects of smoking on the incidence of heart disease, individuals might be matched by such characteristics as age, sex, weight, occupation, diet, family medical history, and environment, but with one member of each pair a smoker. Differences between members of a pair in subsequent medical history can then be more reasonably ascribed to smoking, since other factors were made identical.

Consider a matched pair of samples (X_{A_i}, X_{B_i}) from two populations A and B. Again, we test these hypotheses:

$$H_o: \mu_A = \mu d_1$$

$$H_a: \mu_A \neq \mu_B$$

We do not, however, find $\bar{X}_A$ and $\bar{X}_B$. Instead, for each pair we calculate the **paired difference** $d_i = X_{A_i} - X_{B_i}$. These values are really our sample. We find the **mean-paired difference** $\bar{d}$ and the standard deviation s_d in the usual ways:

$$\bar{d} = \frac{1}{n} \sum_{i=1}^{n} d_i; \; s_d = \left[\frac{1}{n-1} \sum_{i=1}^{n} (d_i - \bar{d})^2 \right]^{1/2}$$

These can be thought of as statistics of a sample from the population of all possible paired differences. The mean-paired difference $\bar{d}$ is an unbiased estimator of $\mu_A - \mu_B$, and the sampling distribution of the quantity

$$t = \frac{\bar{d} - (\mu_A - \mu_B)}{s_d/\sqrt{n}}$$

is a t distribution with $n - 1$ degrees of freedom.

If H_o is true, $\mu_A - \mu_B = 0$, and $t = \bar{d}/(s_d/\sqrt{n})$. We will be convinced to reject the null hypothesis if t is less than $-t_{\alpha/2}$ or greater than $t_{\alpha/2}$, where α is the significance level of the test (Figure 10.7).

To draw another example from the USNEWS file, consider testing to see if there is any difference in American cities between the mean change in nonfarm employment and the mean change in construction activity at $\alpha = 5\%$. The relevant variables are X4 and X7, and we use the pairs of values associated with each city in the file. For Atlanta, the pair is (3.2, 0.5), and $d = 3.2 - 0.5 = 2.7$.

The mean paired difference $\bar{d}$ is -2.8243 with standard deviation 7.982 and the critical values, corresponding to $\alpha = 5\%$ and $74 - 1 = 73$ degrees of freedom, are approximately ± 1.99.

The value of the t statistic is $-2.8243/(7.982/\sqrt{74}) = -3.043$, which is not between the critical values; we reject the null hypothesis and conclude that there is a difference between the mean change in nonfarm employment and the mean change in construction activity in U.S. cities.

Again, when the sample size is large, values from the standard normal distribution can be used to approximate the t values; here, $z_{0.025} = 1.96$ is very close to the t deviate used, and would have provided an acceptable test.

One-tailed tests are performed by comparing t to $-t_\alpha$ or to t_α.

FIGURE 10.7. $t = d/(s_{d/\sqrt{n}})$ **if** H_o **is true.**

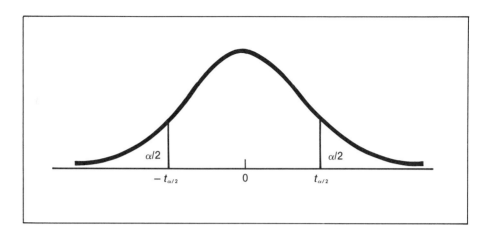

The T-Test Subprogram for Matched-Pair Tests

When using the SPSS subprogram T-TEST to perform matched-pair tests, the pairs must be formed *within* cases. We specify which variables are to be paired in the specification field, as shown in this example, which performs the test described in the previous section:

```
10.       GET FILE
10.005    USNEWS
20.       T-TEST
20.005    PAIRS = X4 WITH X7
```

For each case, a pair is formed of the values of X4 and X7, and this output is produced:

VARIABLE	NUMBER OF CASES	MEAN	STANDARD DEVIATION	STANDARD ERROR
X4				
CHANGE IN NONFARM EMPLOYMENT				
	74	2.6986	2.820	.328
	74	5.5230	9.301	1.081
X7				
CHANGE IN CONSTRUCTION ACTIVITY				

(DIFFERENCE) MEAN	STANDARD DEVIATION	STANDARD ERROR	CORR.	2-TAIL PROB.	T VALUE	DEGREES OF FREEDOM	2-TAIL PROB.
− 2.8243	7.982	.928	.586	.000	− 3.04	73	.003

Variable labels and statistics of the two variables are printed. (DIFFERENCE) MEAN is $\bar{d}$, STANDARD DEVIATION is s_d, and STANDARD ERROR is the standard deviation of the sampling distribution of $\bar{d}$, $sd/\sqrt{n}$. CORR. (correlation) is a measure of the degree of relationship between the values of X4 and X7. It will be discussed in detail in Chapter 12; for now we observe that this value should be positive, indicating some degree of direct relationship between the variables. (If it were negative, indicating an inverse relationship, we would gain nothing by using matched pairs.) The first 2-TAIL PROB. is the significance of the correlation. Here, there is a significant relationship between X4 and X7. T VALUE is the value of the t statistic, and its two-tailed probability is the second 2-TAIL PROB. This is 0.003, less than $\alpha = 0.05$, so the t statistic must be beyond the critical values, and we reject H_o.

To perform a one-tailed matched-pair test using T-TEST, check the sign of the t statistic and compare half its 2-TAIL PROB. to α.

Several t-tests can be performed with a single command, by placing all the appropriate specifications in the specification field. This command will perform the three examples shown in this chapter:

```
T-TEST     GROUPS = X1(1,3)/VARIABLES = X6/
           GROUPS = X1(1,2)/VARIABLES = X4/
           PAIRS = X4 WITH X7
```

T-Tests for the Difference of Population Proportions

The t distribution can also be applied to testing the difference of two population proportions. For example, is the proportion of recalled automobiles produced by one manufacturer different from that of another? The hypotheses in such a situation are these:

$$H_o: P_A = P_B$$

$$H_a: P_A \neq P_B,$$

where P_A and P_B are the proportions of "successes" in two populations A and B. We take independent samples of sizes n_A and n_B, and find the sample proportions $\hat{P}_A$ and $\hat{P}_B$. We know that $\hat{P}_A$ and $\hat{P}_B$ are unbiased estimators of P_A and P_B, respectively, so $E(\hat{P}_A - \hat{P}_B) = P_A - P_B$. Since the samples are independent, so are $\hat{P}_A$ and $\hat{P}_B$, and the standard deviation of their difference is approximated by

$$\left[\frac{\hat{P}_A(1 - \hat{P}_A)}{n_A} + \frac{\hat{P}_B(1 - \hat{P}_B)}{n_B}\right]^{1/2}.$$

We test with the statistic

$$t = \frac{\hat{P}_A - \hat{P}_B}{\left[\frac{\hat{P}_A(1 - \hat{P}_A)}{n_A} + \frac{\hat{P}_B(1 - P_B)}{n_B}\right]^{1/2}}$$

which has, when H_o is true, a t distribution whose number of degrees of freedom is the maximum of $n_A - 1$ and $n_B - 1$.

Recall that when the elements of a sample which have the characteristic of interest are indicated by 1, and those which do not by 0, the sample mean $\bar{X}$ is also the sample proportion p. With such a coding, then, tests of two proportions may be performed with the SPSS procedure T-TEST.

EXERCISES

1. Corresponding to 10 degrees of freedom in the numerator of the F distribution and 12 degrees of freedom in the denominator, find $F_{0.01}$, $F_{0.05}$, $F_{0.95}$, and $F_{0.99}$.

2. A sample of size 12 from population A has mean 12.7 and variance 23.2; a sample of size 15 from population B has mean 16.1 and variance 39.2. Assuming that both populations are normally distributed, can it be concluded at the 10% significance level that their variances are unequal?

3 Based on these statistics generated by the SPSS subprogram BREAKDOWN, can it be concluded that:

   a. the variance of change in factory workers' income is different in eastern and central cities, at the 10% significance level?

   b. the variance of change in factory workers' income is greater in the West than in the East at the 5% significance level?

| CRITERION VARIABLE X6 | | CHANGE IN FACTORY WORKERS INCOME | | |
| BROKEN DOWN BY X1 | | CITY REGION | | |
VARIABLE	CODE	MEAN	STD DEV	N VALUE LABEL
FOR ENTIRE POPULATION		8.673	5.551	73
X1	1.	9.015	3.648	27 EASTERN
X1	2.	8.423	6.178	30 CENTRAL
X1	3.	8.562	7.121	16 WESTERN

TOTAL CASES = 74
MISSING CASES = 1 OR   1.4 PCT.

4. Based on the data in the USNEWS file, can it be concluded that:

   a. the variance of the change in construction activity (X7) in the East is different from that in the West, at the 2% significance level?

   b. The variance of the change in construction activity in the central region is less than that in the western, at the 5% significance level?

   (*Hint:* Use the computer to generate the necessary statistics.)

5. In Exercises 3 and 4, what assumption was made about the populations from which the samples were taken?

6. Let $d = \bar{X}_A - \bar{X}_B$, where $\bar{X}_A$ and $\bar{X}_B$ are the means of independent samples from populations A and B. Show that d is an unbiased estimator of $\mu_A - \mu_B$.

7. A sample of size 21 from population A has mean 56 and standard deviation 12. A sample of size 26 from population B has mean 63 and standard deviation 14. Assuming that the variances of their respective populations are equal, test the claim that the population means are unequal at the 5% significance level.

8. Consider these data from the USNEWS file.

   a. Assuming that the variances of X4 are equal for the central and western regions, test the claim that their means are different at the 10% significance level.

   b. Test the claim that the mean value of X4 is greater for the western region than for the central at the 1% significance level.

CRITERION VARIABLE	X4		CHANGE IN NONFARM EMPLOYMENT		
BROKEN DOWN BY	X1		CITY REGION		
VARIABLE	CODE	MEAN	STD DEV	N VALUE LABEL	
FOR ENTIRE POPULATION		2.699	2.820	74	
X1	1.	2.363	2.302	27 EASTERN	
X1	2.	1.823	2.506	30 CENTRAL	
X1	3.	4.776	3.172	17 WESTERN	

   TOTAL CASES = 74

9. Consider the data given in Exercise 8. Assuming that the variances of X4 for the eastern and western regions are *unequal*,

   a. test, at the 5% significance level, the claim that the respective means are different.

   b. test the claim that the mean of X4 for western cities is greater than the mean for eastern cities at the 1% significance level.

10. A sample of 15 widgets made by machine A has mean length 12.50 cm with standard deviation 0.06 cm, while a sample of 10 widgets made by machine B has mean length 12.45 cm with standard deviation 0.03 cm.

    a. Use the F test to verify, at the 10% significance level, that the respective population variances are unequal. Is the assumption of normality reasonable here?

    b. At the 5% significance level, test the claim that the mean length of widgets produced by machine A is different from the mean length of those produced by machine B.

11. The following is the output of an SPSS T-TEST run on the values of X5 for the eastern (GROUP 1) and central (GROUP 2) cities in the USNEWS file.

VARIABLE	NUMBER OF CASES	MEAN	STANDARD DEVIATION	STANDARD ERROR
X5				
AVERAGE INCOME OF FACTORY WORKERS				
GROUP 1	27	14504.8519	2636.825	507.457
GROUP 2	30	15723.6667	2019.017	368.620

		POOLED VARIANCE ESTIMATE			SEPARATE VARIANCE ESTIMATE		
F VALUE	2-TAIL PROB.	T VALUE	DEGREES OF FREEDOM	2-TAIL PROB.	T VALUE	DEGREES OF FREEDOM	2-TAIL PROB.
1.71	.164	− 1.97	55	.054	− 1.94	48.56	.058

   a. Describe the two hypothesis tests being performed here.

   b. At the 10% significance level, are the variances of average income in the eastern and central regions unequal? What importance does this have for the test of means?

   c. At the 5% significance level, are the two mean incomes different?

   d. Can it be concluded at the 10% significance level that mean income of factory workers is larger in central cities than in eastern?

   e. Write the SPSS command which caused this output.

12. In this printout of a T-TEST run on the values of X7 from the USNEWS file, the central cities are GROUP 1, and the western cities are GROUP 2.

VARIABLE	NUMBER OF CASES	MEAN	STANDARD DEVIATION	STANDARD ERROR
X7				
CHANGE IN CONSTRUCTION ACTIVITY				
GROUP 1	30	4.6767	7.331	1.339
GROUP 2	17	5.4765	11.739	2.847

		POOLED VARIANCE ESTIMATE			SEPARATE VARIANCE ESTIMATE		
F VALUE	2-TAIL PROB.	T VALUE	DEGREES OF FREEDOM	2-TAIL PROB.	T VALUE	DEGREES OF FREEDOM	2-TAIL PROB.
2.56	.027	− .29	45	.775	− .25	23.23	.802

   a. At the 5% significance level, are the respective population variances equal?

   b. Are changes in construction activity in these two regions unequal at the 5% significance level?

   c. Write an SPSS program to produce this output.

13. Write and run a T-TEST program to test at the 5% significance level the claim that the average income of factory workers is greater in the West than in the East. What conclusions do you reach, and why?

14. Write and run a T-TEST program to test at the 5% significance level the claim that mean change in department store sales is greater in the central region than in the eastern. What conclusions do you reach?

15. Write and run a T-TEST program to test the claim that the mean change in nonfarm employment is less in the East than in the West at the 2% significance level. What conclusions do you reach?

16. Let $\bar{d} = (1/n)\sum_{i=1}^{n} d_i$ be the sum of the matched-pair differences $d_i = X_{A_i} - X_{B_i}$. Show that $\bar{d}$ is an unbiased estimator of $\mu_A - \mu_B$.

17. A randomly chosen group of individuals is given a standardized reading test before and after taking a speed-reading course. Their scores are these:

Individual	1	2	3	4	5	6	7	8	9	10
Before	86	83	86	70	66	90	70	85	77	86
After	82	79	91	63	68	86	81	90	85	94

   a. Assuming that the population of score differences is normally distributed, do these results indicate, at the 5% significance level, that reading ability has been increased by the course?

   b. If it is known that the variance of the population of score differences is $\sigma^2 = 48.4$, what conclusion is reached?

18. Two tire-tread materials are being compared. Pairs of tires of the two materials are mounted on test cars and their lifetimes measured. The results are these:

Pair	1	2	3	4	5	6	7	8	9	10	11	12
Life of Material A(1000s of miles)	42.6	43.1	41.9	45.2	43.6	41.0	42.6	43.7	42.2	46.3	42.4	41.8
Life of Material B(1000s of miles)	40.3	42.7	42.0	42.1	43.4	39.5	41.5	43.4	41.4	43.1	41.6	40.1

At the 5% significance level, can it be concluded that there is a difference in the mean lifetimes of the two materials?

19. The following is the output of a matched-pair T-TEST using the values of X2 and X4 from the USNEWS file:

VARIABLE	NUMBER OF CASES	MEAN	STANDARD DEVIATION	STANDARD ERROR
X4				
CHANGE IN NONFARM EMPLOYMENT				
	70	2.4971	2.538	.303
	70	10.1800	6.684	.799

X2
CHANGE IN DEPARTMENT STORE SALES

(DIFFERENCE) MEAN	STANDARD DEVIATION	STANDARD ERROR	CORR.	2-TAIL PROB.	T VALUE	DEGREES OF FREEDOM	2-TAIL PROB.
− 7.6829	6.767	.809	.157	.194	− 9.50	69	.000

    a. Is a matched-pair T-TEST appropriate here? Why?

    b. At the 5% significance level, is there a difference in the change in department store sales and the change in nonfarm employment in U.S. cities?

    c. Can a stronger conclusion be made?

    d . Write the SPSS command that caused this output.

20. Write and run a program to perform the test described in Exercise 18. What conclusions do you reach?

21. Write and run a program to perform the test described in Exercise 17(a). Are your earlier conclusions verified?

22. Write and run a program which uses the USNEWS file to test the claim that there is a difference in the mean change in department store sales and the mean change in factory workers' income in U.S. cities. Use $\alpha = 5\%$.

23. Write and run a program which uses the USNEWS file to test the claim that mean nonfarm employment has risen more than factory workers' income in U.S. cities. Use $\alpha = 2\%$.

24. In Chicago, of 200 Republicans surveyed, 42 support George Leroy Tyrebiter for mayor, while he has the support of 87 of 300 Democrats. At the 5% significance level, can it be concluded that Tyrebiter has more support among Democrats than among Republicans? (*Note:* For large numbers of degrees of freedom, $t_\alpha \doteq z_\alpha$.)

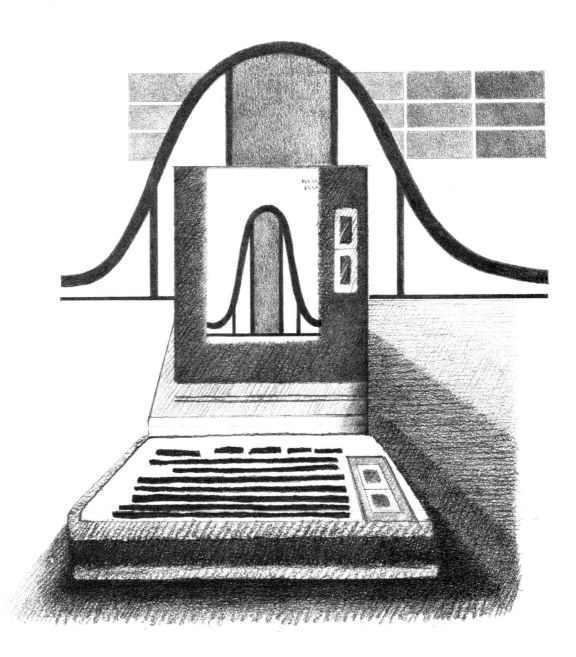

Applications of Chi-Square Statistics

INTRODUCTION

In preceding chapters we have described hypothesis testing, and have performed a variety of tests involving population means, variances, and proportions. In this chapter we increase the number of situations that can be examined with hypothesis tests by considering several in which the test statistic has a chi-square distribution.

Chi-Square Tests of Multinomial Data

We have seen that the binomial probability distribution $b(n,p)$ corresponds to selecting n objects, with replacement, from a population of which the proportion p are "successes." The probability of drawing exactly r "successes" in the n independent trials is

$$f(r) = \binom{n}{r} p^r (1-p)^{n-r}, \qquad r = 0, 1, 2, \ldots, n.$$

The **multinomial** distribution is an extension of the binomial: From a population containing proportion p_1 of objects of type 1, p_2 of objects of type 2, $\ldots$, p_k of objects of type k, we select n objects at random and with replacement. The probability that we select n_1 objects of type 1, n_2 objects of type 2, $\ldots$, n_k objects of type k is

$$\frac{n!}{n_1! n_2! \ldots n_k!} \, p_1^{n_1} p_2^{n_2} \cdots p_k^{nk}$$

where $\sum_{i=1}^k p_i = 1$.

Calculations of probabilities for the multinomial distribution are tedious when n is large, but as a binomial probability may be approximated with a normal, a multinomial may be approximated with a chi-square.

In a data set, a *multinomial variable* is one whose values, not necessarily numeric, can be classified into three or more categories; that is, whose values are *nominal*. If we take a sample from a population of multinomial data—for example, ask a random group of voters to name their first choice for president—the probabilities with which various numbers of responses from the categories occur in the sample are described by the multinomial distribution, while the probabilities of various numbers of elements of type i occurring in the sample are given by the binomial distribution $b(n,p_i)$.

Given a population of multinomial values, we will want to test hypotheses concerning the proportions of various values (types of objects) in the population. That is, we wish to test whether a given set of values for the proportions accurately describes the population. Our hypotheses are:

$$H_o: p_1 = p_{1_o}; p_2 = p_{2_o}; \ldots; p_k = p_{k_o}$$

H_a: at least one of the proportions is incorrect

We take a random sample of size n from the population, and compute a test statistic. Keep in mind that a test statistic should tend to indicate the truth or falsity of the hypotheses, and should have a known distribution when H_o is true. In this case, we base the test statistic on the expected number of elements in the sample from each category if H_o is true and the postulated proportions hold.

If category i has population proportion p_{io}, the distribution of the number X_i of sample elements from that category is $b(n,p_{io})$, with expected value

$$E_i = E(X_i) = np_{io}.$$

If O_i is the number of elements from category i observed in the sample, then for large sample sizes the statistic

$$X^2 = \sum_{i=1}^{k} \frac{(O_i - E_i)^2}{E_i}$$

has an approximate chi-square distribution whose number of degrees of freedom is $k - 1$, one less than the number of categories. Note that the number of degrees of freedom of the chi-square distribution in this application does not depend on the sample size. (See Figure 11.1.)

   If the observed frequencies are near the expected frequencies supporting H_o, the value of the chi-square statistic will be near 0; if the O_i are far from the E_i, indicating rejection of H_o, the value of the statistic will be larger. Critical values of the chi-square statistic are found in Table A-6, and we reject H_o when X^2 is greater than the appropriate critical value. This is always an upper-tail test.

Example 1 In 1968 Richard Nixon, Hubert Humphrey, George Wallace, and several others were candidates for the presidency. Voters could be grouped according to their first choice for that office into the categories N, H, W, and O. Suppose that a pollster who had been studying the campaign wished to test these hypotheses:

$$H_o\colon p_N = 0.45;\ p_H = 0.40;\ P_W = 0.10;\ p_0 = 0.05$$

$$H_a\colon \text{At least one of the above is wrong.}$$

at the 1% significance level. If a sample of 600 voters contained 250 who supported Nixon, 240 for Humphrey, 70 for Wallace, and 40 for others, what conclusion did he reach?

FIGURE 11.1.   Chi-square with $k - 1$ degrees of freedom.

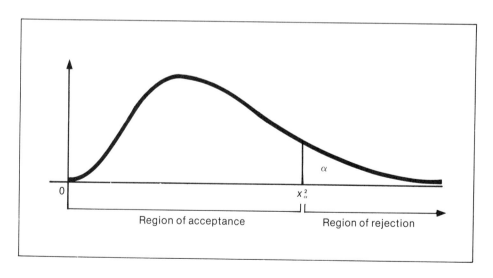

Solution

First, we construct a table of the observed and expected counts. For Nixon,

$$E_N = E(X_N) = np_{N0} = 600 \times 0.45 = 270.$$

Continuing in this way, we complete the table:

	N	H	W	O
Observed	250	240	70	40
Expected	270	240	60	30

The value of the chi-square statistic is

$$X^2 = \sum_{i=1}^{k} \frac{(O_i - E_i)^2}{E_i} = \frac{(250 - 270)^2}{270} + \frac{(240 - 240)^2}{240} + \frac{(70 - 60)^2}{60}$$
$$+ \frac{(40 - 30)^2}{30}$$
$$= 1.48 + 0 + 1.67 + 3.33 = 6.48.$$

Corresponding to $\alpha = 1\%$ and to $4 - 1 = 3$ degrees of freedom, the critical value is 11.345. $X^2 = 6.48 < 11.345$, so we do not reject the null hypothesis. From the sample information, the pollster could not conclude that his hypothesized proportions were wrong.

Some implementations of SPSS include the subprogram NPAR TESTS, which can perform the calculations associated with chi-square tests of observed vs. expected frequencies. In this procedure, each value of the variable being considered is treated as a separate class; when the variable has interval measure, it is necessary to group its values using a RECODE statement. The subprogram also requires that the expected frequencies be provided.

Example 2 Suppose it is claimed that of U.S. cities, one-third experienced increases of construction activity of 10% or less, one-third had increases of from 10% to 20%, and one-third had increases of more than 20%. At the 5% significance level, can we reject this claim using the variable X7 in the USNEWS file?

Solution

In a sample of size 74, the expected frequencies are all 24.67. The following program will perform the test.

```
10.       GET FILE
10.005    USNEWS2
20.       RECODE
20.005    X7 (LO THRU 10 = 5)(10.01 THRU 20 = 15)
20.010    (20.01 THRU HI = 25)
30.       NPAR TESTS
30.005    CHI-SQUARE = X7/EXPECTED = 24.67,24.67,24.67
```

We begin by grouping the values with RECODE, then specify the chi-square test by naming the variable and giving the expected frequencies. The following output results:

```
— — — — — —  CHI-SQUARE TEST

        X7      CHANGE IN CONSTRUCTION ACTIVITY

      VALUE        5.0         15.0         25.0
      COUNT        54.          13.          7.
    EXPECTED      24.67        24.67        24.67

           CHI-SQUARE                    D.F.      SIGNIFICANCE
             53.054                        2           .000
```

The significance of the chi-square statistic, whose number of degrees of freedom is one less than the number of classes, is less than $\alpha = 5\%$, so we reject H_o. The suggested proportions are not correct.

Chi-Square Tests of Independence

We are often concerned with finding if two variables in a research study are connected in some way, if a particular value of one increases the likelihood of a particular value of the other. If this is true, we say that the variables are **dependent**, and if the two variables do not interact, we say that they are **independent**, as earlier with events and random variables. The two variables in a study can be thought of as random variables, and are independent if for all possible values of each, $P(V1 = v_1 \text{ and } V2 = v_2) = P(V1 = v_1) \times P(V2 = v_2)$.

When the two variables being considered are qualitative, we can use a chi-square statistic to perform this test of their independence:

H_o: The variables are independent.

H_a: The variables are dependent.

In hoping to reject the null hypothesis, we seek to demonstrate that there is a relationship between the two variables.

Suppose that we wish to find if there is a relationship between city region and unemployment in U.S. cities, based on the data in the USNEWS file. In the file, the variable X3 is numeric, but we can obtain a qualitative measure of unemployment by grouping its values into two classes; 0 to 4.99% will be *low* unemployment, while values of 5% and above will be *high*.

We construct a *contingency table* showing how many cases have each of the possible combinations of attributes of the two variables (Table 11.1). Each of the 70 members of the sample with nonmissing values of X3 will be counted in exactly one of the six *cells* of the table; Boise, a western city with high unemployment, is counted in the lower right cell, contributing one to the total there of 12. Row and column totals are recorded in the margins, and the total number of cases is in the extreme lower right.

TABLE 11.1. **Contingency table.**

	EASTERN	REGION CENTRAL	WESTERN	ROW TOTALS
Low (0 to 4.99%)	8	11	5	24
High (5% and up)	17	17	12	46
Column totals	25	28	17	70

Before we can perform the test of independence, we require some notation. In a contingency table, rows are numbered from top to bottom, columns from left to right, and cells are indicated by their row and column numbers, in that order. The high unemployment row is row 2, the eastern column is column 1, and the cell of eastern cities with high unemployment is cell 21. R_i is the total frequency of row i, C_j is the total frequency of column j, and the sample size, as always, is n.

Since we are testing the independence of two variables, we can rewrite the hypotheses in this way:

$$H_o: P_{ij} = P_i \times P_j \text{ for all } i \text{ and } j$$

$$H_a: P_{ij} \neq P_i \times P_j \text{ for some } i \text{ and } j$$

where P_{ij} is the probability that a case falls in cell ij; P_i is the probability that a case falls in row i, and P_j is the probability that a case falls in column j. The null hypothesis in this form restates the independence of the variables in terms of independence of events.

Our test statistic is again calculated from a comparison of expected and observed frequencies, the frequencies of the cells in the contingency table. The observed frequencies O_{ij} have been recorded, and the expected frequencies E_{ij}, based on the assumption that the null hypothesis is true, are found from this formula:

$$E_{ij} = \text{expected frequency of cell } ij$$

$$= P_{ij} \times n$$

$$= P_i \times P_j \times n$$

$$= \frac{R_i}{n} \times \frac{C_j}{n} \times n = \frac{R_i C_j}{n}.$$

(Note that the sums of the expected frequencies in each row and column are equal to the marginal frequencies.)

For example, the expected frequency of cell 12 is

$$E_{12} = \frac{R_1 C_2}{n} = \frac{24 \times 28}{70} = 9.6.$$

The test statistic is found by summing a quantity similar to that used in the multinomial test over all the cells in the contingency table:

$$X^2 = \sum_i \sum_j \frac{(O_{ij} - E_{ij})^2}{E_{ij}} \, .$$

When the null hypothesis is true, this statistic has an approximate chi-square distribution with $(R-1) \times (C-1)$ degrees of freedom, where R is the number of rows in the contingency table and C is the number of columns. Again, the chi-square distribution is an approximation to a multinomial, and the sample size does not determine the number of degrees of freedom.

If the variables being tested are independent, the observed frequencies should be near those expected under independence, and X^2 should be near 0. Observed frequencies far from the E_{ij} tend to support H_a, and will cause a larger value of the chi-square statistic. The critical value can be found in Table A-6, corresponding to the significance level α and to $(R-1) \times (C-1)$ degrees of freedom.

In our example, the value of the chi-square statistic is 0.548; if we choose 5% as the significance level of the test, the critical value corresponding to $(2-1)(3-1) = 2$ degrees of freedom is 5.991. X^2 is less than the critical value, so we cannot reject H_o. We cannot conclude from this test that there is a relationship between region and unemployment in the United States.

Note that tests of independence are always upper-tail tests.

Tests of Independence and Subprogram Crosstabs

The astute reader will have observed that the table produced by the SPSS subprogram CROSSTABS is a contingency table. Statistic number 1 of this subprogram is the chi-square statistic with its associated significance. The following program will perform the test described above:

```
10.       GET FILE
10.005    USNEWS2
20.       RECODE
20.005    X3(LO THRU 4.99 = 4)(5.00 THRU 15 = 7)
30.       VALUE LABELS
30.005    X3 (4) LOW (7) HIGH
40.       CROSSTABS
40.005    TABLES = X3 BY X1
45.       STATISTICS
45.005    1
```

The RECODE statement groups the values of X3 into two classes, and the VALUE LABELS command labels the classes. The output of the program is this:

X3   UNEMPLOYMENT
BY X1   CITY REGION

X1

COUNT ROW PCT COL PCT TOT PCT		EASTERN 1.	CENTRAL 2.	WESTERN 3.	ROW TOTAL
X3					
LOW	4.	8 33.3 32.0 11.4	11 45.8 39.3 15.7	5 20.8 29.4 7.1	24 34.3
HIGH	7.	17 37.0 68.0 24.3	17 37.0 60.7 24.3	12 26.1 70.6 17.1	46 65.7
COLUMN TOTAL		25 35.7	28 40.0	17 24.3	70 100.0

RAW CHI SQ =   .54790   WITH   2 D.F., SIG. =   .7604

MISSING OBSERVATIONS —   4

The value of the chi-square statistic is the same as that found earlier. SIG. is the significance of this value, the area to the right of X^2 and under the graph of its sampling distribution when H_o is true. This significance is greater than 5%, our chosen α, so we cannot reject the null hypotheses.

GUIDELINES FOR USING THE CHI-SQUARE TEST OF INDEPENDENCE

To obtain meaningful results, the following guidelines should be followed when using the chi-square test of independence:

1. For contingency tables with two rows and two columns, the average expected cell frequency should be at least 7.5.

2. For contingency tables with equal expected cell frequencies throughout, the expected cell frequencies should be at least 2 when $\alpha = 5\%$, at least 4 when $\alpha = 1\%$.

3. For contingency tables with unequal expected cell frequencies, the average expected cell frequency should at at least 6 when $\alpha = 5\%$, at least 10 when $\alpha = 1\%$.

4. The number of expected cell frequencies equal to 0 should be minimized, not by collapsing rows or columns, but by taking sufficiently large samples.

Many mathematicians simply state that in any chi-square test of independence, every expected cell frequency should be at least 5.

In our example, the average expected cell frequency is 11.67, which justifies the use of the chi-square test by rule (3).

Chi-Square Tests of Goodness-of-Fit

We can use a chi-square test based on the comparison of expected and observed frequencies to test whether a sample could have come from a population with a specified distribution. Such a test is called a **goodness-of-fit** test (sometimes a **one-sample chi-square** test) and can be used to test the normality of a population involved in a t-test, or to determine if the values produced by a random number generator are, in fact, selected from a uniform distribution. In general, these hypotheses are tested:

H_o: The sample is from a population having a specified distribution.

H_a: The population is from some other distribution.

Consider the values of X5 in the USNEWS data file. We will test at the 5% significance level the null hypothesis that the values of X5 were taken from a normally distributed population with mean 15,000 and standard deviation 2300 against the alternative hypothesis that the population has some other distribution.

We begin by dividing the range of population values into cells and finding the number of sample values we should expect in each cell if the null hypothesis is true. If the population from which the values of X5 were taken is normally distributed with $\mu = 15000$ and $\sigma = 2300$, it seems reasonable to use cells of width 2000 from 12,000 to 18,000, with an unbounded cell at each end of this range. The probability that a single sample value would fall in the cell between 12,000 and 14,000 is

$$P(12{,}000 < X < 14{,}000) = P\left(\frac{12000 - 15000}{2300} < \frac{X - 15000}{2300} < \frac{14000 - 15000}{2300} \right)$$

$$= P(-1.30 < Z < -0.43) = 0.4032 - 0.1664 = 0.2368.$$

We are taking a sample of size 73 (one value is missing), so the number of values which will fall in the cell from 12,000 to 14,000 has the binomial distribution $b(73, 0.2368)$ and expected value $73 \times 0.2368 = 17.2684$.

In general, the expected frequency of cell i is:

$E_i = n \times P$(a randomly selected population value falls in cell i).

Similar calculations provide us with the expected frequencies of the other cells (Figure 11.2).

After recording the observed frequencies, we calculate the chi-square statistic in the usual way:

$$X^2 = \sum_{i=1}^{k} \frac{(O_i - E_i)^2}{E_i}$$

FIGURE 11.2   **Expected and observed frequencies in the goodness-of-fit test.**

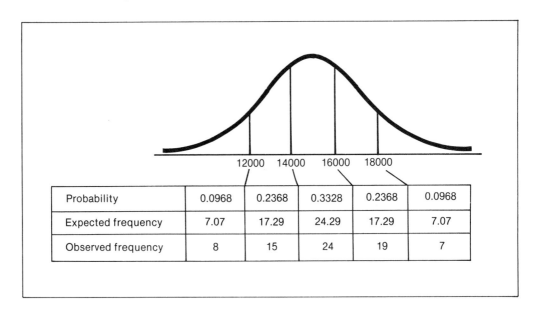

	12000	14000	16000	18000	
Probability	0.0968	0.2368	0.3328	0.2368	0.0968
Expected frequency	7.07	17.29	24.29	17.29	7.07
Observed frequency	8	15	24	19	7

where E_i = expected frequency of cell i, O_i = observed frequency of cell i, and k = number of cells.

When the null hypothesis is true, the statistic has a chi-square sampling distribution with $k-1$ degrees of freedom. If parameters of the population are being estimated from the sample, then the number of degrees of freedom is [$k-1$ − (number of estimated parameters)], and when the number of degrees of freedom is 1, **Yate's correction** should be used in finding X^2:

$$X^2 = \sum_{i=1}^{k} \frac{(O_i - E_i - \frac{1}{2})^2}{E_i} .$$

In our example, corresponding to $5 - 1 = 4$ degrees of freedom and $\alpha = 5\%$, the critical value is $X^2_{0.05} = 9.488$. The value of the chi-square statistic is $X^2 = 0.599$; since this is less than the critical value, we cannot reject the null hypothesis. We cannot conclude that the values of X5 did not come from the normal distribution described above.

Again, having calculated the expected frequencies, the SPSS subprogram NPAR TESTS can be used to complete the test.

```
10.      GET FILE
10.005   USNEWS2
20.      RECODE
20.005   X5 (LO THRU 12000 = 10000)
20.010   (12000.1 THRU 14000 = 13000)
20.015   (14000.1 THRU 16000 = 15000)
20.020   (16000.1 THRU 18000 = 17000)
```

```
20.025   (18000.1 THRU 25000  =  20000)
30.      NPAR TESTS
30.005   CHI-SQUARE = X5/EXPECTED  =  7.07,17.29,24.29,17.29,7.07
```

Note that the expected values are given in ascending order of the corresponding cells of values of X5. The output is this:

— — — — — CHI-SQUARE TEST

X5					AVERAGE INCOME OF FACTORY WORKERS
VALUE	10000.0	13000.0	15000.0	17000.0	20000.0
COUNT	8.	15.	24.	19.	7.
EXPECTED	7.07	17.29	24.29	17.29	7.07
CHI-SQUARE			D.F.		SIGNIFICANCE
.599			4		.963

The significance of the chi-square statistic is greater than $\alpha = 5\%$, so we do not reject H_o.

Chi-Square Analysis of Missing Data

In many data-collection situations, particularly those involving human beings as the experimental units, missing values carry as much information as complete responses. If many people refuse to answer questions on a particular topic, this suggests that the topic is sensitive and perhaps important, and may deserve further study.

Chi-square tests of independence can be used to analyze the missing values in a file in an exploratory way, by detecting significant dependence between the missing values of one variable and responses or nonresponses to another. If such a dependence is found, further qualitative (nonstatistical) analysis might explain the connection.

Data files can be broken down by variables, by cases, and by subfiles (subsets of the file). Exploratory chi-square analysis of missing data includes tests of independence for variable vs variable and for subfile vs variable. The contingency tables for such tests will have the forms shown in Table 11.2.

TABLE 11.2.   Tests of missing values.

		VARIABLE i				*VARIABLE i*	
		RESPONSE	*NONRESPONSE*			*RESPONSE*	*NONRESONSE*
VARIABLE j	Response			*SUBFILE k*	In subfile		
	Non-response				Not in subfile		

Such tests detect significant relationships between nonresponses to variable i and nonresponses to variable j, and between nonresponses to variable i and membership in subfile k. These relationships could then be explored in more detail.

EXERCISES

1. A large population of objects is one-half type As, one-third type Bs, and one-sixth type Cs. If a sample of 10 objects is randomly selected, what is the proability that it contains 6 type As, 3 type Bs, and 1 type C?

2. A machine is designed to mix candy in the following proportions: 20% chocolate, 25% lemon, 15% orange, 30% grape, and 10% butterscotch. In a bag of 200 pieces of candy, 35 are chocolate, 55 lemon, 25 orange, 60 grape, and 25 butterscotch. At the 1% significance level, should it be concluded that the machine is not mixing candy in the specified proportions?

3. The manager of a mail-order firm believes that 10% of their orders come from the West, 20% from the South, 40% from the East, 20% from the Midwest, and 10% from foreign countries. A sample of 150 orders has these destinations:

West	20
South	33
East	14
Midwest	48
Foreign	10
Total	150

    At the 5% significance level, is the manager correct about the proportions?

4. a. In your opinion, what proportions of your class would agree with each of the responses to this question?

    "Check the one statement with which you most agree."

    _____On the whole, wars do the world some good.

    _____Peace and war are both essential to progress.

    _____War is hardly ever necessary in the modern world.

    _____There is no conceivable justification for war.

    b. Administer the question to the class, and test your proposed proportions at the 5% significance level.

5. It is claimed that the average income of factory workers is less than $12,000 in 25% of American cities; between $12,000 and $16,000 in half; and greater than $16,000 in the remaining 25%. Test this claim, using data in the USNEWS file, at the 5% and 1% significance levels.

6. Use the USNEWS file to test the claim that the proportions of American cities having unemployment less than 4%, from 4% to 7%, from 7% to 10%, and above 10% are each $\frac{1}{4}$. Use the significance level $\alpha = 5\%$.

7. In 1960 John F. Kennedy became the first Catholic president when he defeated Richard Nixon. A polling organization at the time wishes to know if there is a relationship between religious affiliation of voters and presidential preference; they survey 300 voters and obtain these results:

| | VOTED FOR: | | |
RELIGION	KENNEDY	NIXON	TOTALS
Catholic	70	30	100
Protestant	70	80	150
Other or none	10	40	50
Totals	150	150	300

At the 5% significance level, does the organization conclude that there is a relationship between a voter's religion and his or her presidential preference?

8. A paperback book has been bound in three differently colored covers. At a bookstore, 200 people buy copies of the book and provide the data for this table. At the 5% significance level, is there a relationship between the sex of the customer and the color of the book selected?

| | COLOR OF BOOK JACKET | | |
	BLUE	RED	YELLOW
Men	43	30	14
Women	19	66	28

9. Administer this question to the members of your class, and record the response (yes or no) and the respondent's sex.

"Do you favor the legalization of marijuana?"

At the 5% significance level, is there a relationship between support for the legalization of marijuana and a person's sex among the members of your class?

10. Using the data in the USNEWS file, test this claim at the 5% significance level: "Change in construction activity in U.S. cities is dependent on region." (*Hint:* Group the values of X7 with a RECODE.)

11. Using the data in the USNEWS file, test this claim at the 5% significance level: "In American cities, the average income of factory workers is dependent on region."

12. Administer the following two questions to your class:

a. Where did you buy your texts for this term?

_____Campus bookstore

_____Off-campus bookstore

_____Some other source

b. What percentage of your total expenditures for supplies was spent at the campus bookstore?

Use a chi-square test of independence to discover if there is a relationship between responses to the two questions at the 5% significance level.

13. Ask the following question of a random sample of students, teachers, and local community members unconnected with your school:

Do you agree or disagree that varsity athletics are harmful to the purposes of a college or university?

_____Agree

_____Disagree

_____No opinion

At the 5% significance level, is the response to this question independent of the respondents' position relative to your school?

14. The following values were randomly selected from a large population.

96	101	110	114	92
99	99	78	105	109
69	87	97	103	101
126	115	98	106	93
98	117	86	89	108
110	112	73	90	100

Test the hypothesis that the population was normally distributed with mean 100 and standard deviation 10. Use the 2% significance level.

15. Test the null hypothesis that these values were randomly selected from the uniform distribution $U[0,10]$ against the alternative that they were selected from some other distribution at the 5% significance level.

1.4	6.2	5.7	3.1	0.3	6.2	2.3
6.3	9.1	1.4	2.9	5.6	3.4	4.6
2.7	3.4	4.0	7.9	4.6	3.1	3.5
1.0	0.2	0.6	8.0	7.1	4.6	8.1
5.4	7.1	1.5	4.2	3.2	9.0	2.4
2.4	4.3	2.0	1.1	3.0	8.9	2.5
1.0	5.3	8.0	2.8	0.5	3.5	6.1
0.7	3.6	4.9	6.4	1.4	0.6	1.5

16. At the 5% significance level, test the hypothesis that the values of X3 in the USNEWS data file came from a normally distributed population with mean 5.5 and standard deviation 1.6.

17. This table gives the number of deaths in highway accidents per day in Trinidad in 1975.*

---

*Richards, Winston A., "The Poisson and Exponential Models," *The MAYTC Journal,* Spring 1978: 113-117.

NUMBER OF DEATHS	NUMBER OF DAYS
0	239
1	97
2	19
3	5
4	5

Using a chi-square goodness-of-fit test, test the hypothesis that the number of deaths per day possesses a Poisson distribution. Use $\alpha = 5\%$. *(Hint:* Since you must estimate the rate of the Poisson distribution with the mean of the observed numbers of deaths, the number of degrees of freedom associated with the chi-square statistic is $5 - 1 - 1 = 3$.)

18. The article which provided the data for Exercise 17 also contained this table for the length of time in days between highway deaths in Trinidad in 1975:

TIME IN DAYS BETWEEN HIGHWAY DEATHS	FREQUENCY
0	50
1	48
2	24
3	14
4	15
5	5
6	6
7	4
8	0
9	2
10	0
11	0
12	1

At the 5% significance level, test the hypotheses that these values came from an exponential distribution.

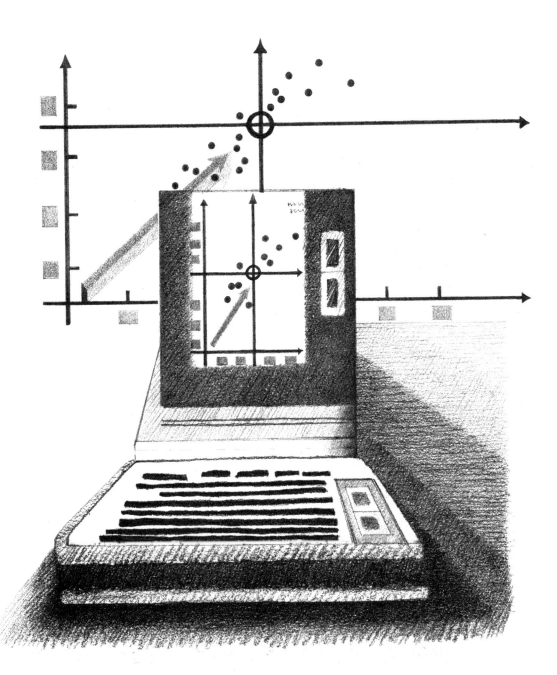

12

Correlational
Analysis

INTRODUCTION

In Chapter 11 we saw that the chi-square test of independence can be used to detect a relationship between two variables in a data file. When such a test reveals a dependence, however, it tells us nothing about the nature of the relationship. Is it strong or weak? Do large values of one variable generally correspond to large or small values of the other? In this chapter we will develop numerical measures of the **correlation** *between the values of two variables.*

Scatter Diagrams

We have seen how useful histograms can be in visualizing the distribution of the values of a variable. Similarly, we can create a graphical display of the values of two variables that illustrates the nature and degree of any relationship they may share. Such displays are called **scatter diagrams,** and are formed by plotting in the $X-Y$ plane points whose coordinates are the paired values of the two variables. When one variable is thought to be dependent on the other, the values of the dependent variable are plotted along the vertical (Y) axis, while the values of the independent variable are plotted on the horizontal (X) axis.

For example, Table 12.1 gives values of floor area and natural gas used in January for 15 midwestern homes heated with natural gas.

TABLE 12.1.   Values of floor area and natural gas in January for 15 midwestern homes.

FLOOR AREA (Thousands of square feet)	NATURAL GAS USED IN JANUARY (Thousands of cubic feet)
1.4	31.6
1.7	40.3
2.1	52.0
1.9	41.5
2.4	50.7
1.8	43.8
1.7	44.0
1.9	48.2
1.3	30.5
1.2	30.7
2.0	55.2
1.9	47.6
2.2	57.1
1.4	36.2
1.6	34.5

We build the scatter diagram of Figure 12.1 by plotting the pairs of values given in Table 12.1. Since floor area can be expected to influence gas use, the amount of gas is the dependent variable and is plotted on the vertical axis.

From Figure 12.1 we see that natural gas use is higher for larger houses and that the data points seem to cluster around a straight line. Such a relationship is said to be *positive* or *direct* and *linear*. Many kinds of relationships are possible, as Figure 12.2 illustrates.

The strength of a relationship can also be inferred from a scatter diagram. When the data points are closely bunched around a line or curve, the relationship between the two variables is strong; in a weak relationship, the points are scattered (Figure 12.3).

In SPSS, scatter diagrams may be generated with the command SCATTERGRAM, which has this general form:

SCATTERGRAM   variable list WITH variable list

The subprogram plots a scatter diagram pairing each variable in the first list with each variable in the second, in turn. The variables in the first list are always considered dependent and are plotted vertically.

FIGURE 12.1.   The pairs of values given in Table 12.1.

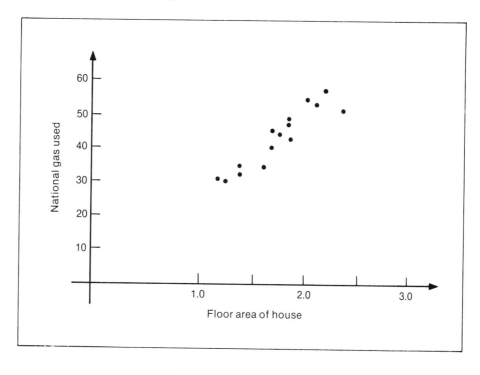

The following program will produce the scatter diagram of house floor area and natural gas use shown on p.000:

```
FILE NAME; REGEX
VARIABLE LIST; NUM, AREA, GAS
INPUT FORMAT; FREEFIELD
N OR CASES; 15
SCATTERGRAM; GAS WITH AREA
```

The Pearson Correlation Coefficient

While a scatter diagram will illustrate the relationship (or lack of one) between two variables, we still seek *numerical* measures of correlation. The most widely used of these is applied to interval- or ratio-level data, like the house floor area and gas use figures of the last section.

Let us indicate the house areas as X_i and the gas volumes as $Y_i (i = 1, 2, 3, \ldots, 15)$. Using the FREQUENCIES or CONDESCRIPTIVE subprograms, we can find their means, $\bar{X} = 1.767$ and $\bar{Y} = 42.927$. Using these means, we create new pairs of values:

$$X_i' = (X_i - \bar{X}) \text{ and } Y_i' = (Y_i - \bar{Y}).$$

FIGURE 12.2.   Relationships between two variables.

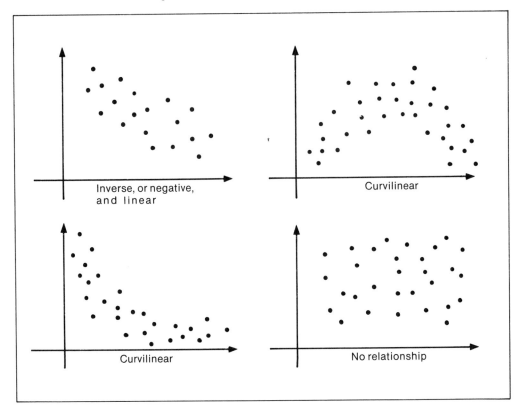

Any relationship between the original X_i and Y_i will be preserved by the X_i' and Y_i', but the scatter diagram is transformed in a useful and interesting way: the axes are moved so that the origin is at the point $(\bar{X}, \bar{Y})$. See Figure 12.4.

Most of the points of the original scatter diagram are now in the first and third quadrants relative to the new axes. If a generally negative relationship existed between the two variables, most of the points would be in the second and fourth quadrants; if no relationship existed, the points would be scattered in all four quadrants of the new diagram. From these distinctions we build our measure of correlation.

We begin by calculating the *Sum of the Cross Products of the Deviations* (SCPD)* of the values from their means

$$\text{SCPD} = \sum_{i=1}^{n} X_i'Y_i' = \sum_{i=1}^{n} (X_i - \bar{X})(Y_i - \bar{Y}).$$

The cross product $X_i'Y_i'$ is positive for points in the first and third quadrants, negative for those in the second and fourth. When a positive relationship exists, this sum will be positive;

---

*This quantity is commonly called the **covariance** of X and Y, and is a measure of the *joint* variation of the variables.

FIGURE 12.3. Diagram showing the strength of relationships between two variables.

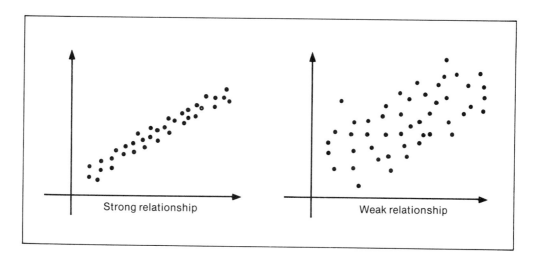

FIGURE 12.4. Transformation of the scatter diagram when $X_i' = X_i - \bar{X}$ and $Y_i' = Y_i - \bar{y}_i.$

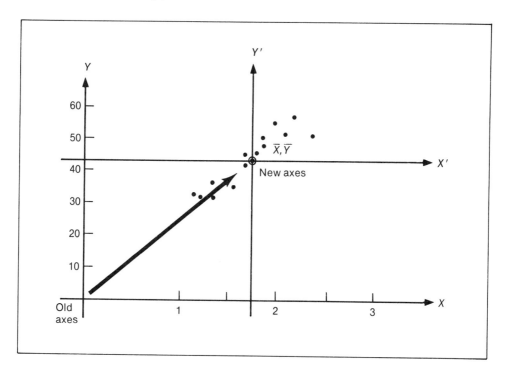

when negative, negative. When no relationship exists, the sum will be near 0. In our example,

$$SCPD = (1.4 - 1.767)(31.6 - 42.927) + \ldots + (1.6 - 1.767)(34.5 - 42.927)$$

$$= 39.013.$$

This positive value corresponds to the positive relationship between the X_i and the Y_i.

While the SCPD is a statistic of the data pairs (X_i, Y_i) which indicates the direction of the relationship of the two variables, it does not tell us about their *degree* of correlation. Is the value found above large or small? Further, the SCPD is in the units of the original measurements; this prevents us from comparing such values for two different pairs of variables.

These difficulties can be resolved by computing the *Sum of the Cross Products of the z-Scores*, the SCPZ:

$$SCPZ = \sum_{i=1}^{n} \frac{X_i - \bar{X}}{s_X} \times \frac{Y_i - \bar{Y}}{s_Y} = \frac{1}{s_X s_Y} \times SCPD.$$

Dividing the SCPD by $s_X s_Y$ causes the resulting measure to be unitless, and the SCPZ has definite maximum and minimum values.

If there is a perfect positive linear relationship between the X_i and the Y_i, then $z_{X_i} = z_{Y_i}$ for every data point, so that

$$SCPZ = \sum_{i=1}^{n} z_{X_i} z_{Y_i}$$

$$= \sum_{i=1}^{n} z_{X_i}^2$$

$$= \sum_{i=1}^{n} \frac{(X_i - \bar{X})^2}{s_X^2}$$

$$= \frac{\sum_{i=1}^{n} (X_i - \bar{X})^2}{\sum_{i=1}^{n} (X_i - \bar{X})^2 / n - 1} = (n - 1).$$

Similarly, with a perfect negative linear relationship between the two variables, $z_{X_i} = -z_{X_i}$ for every data point, and $SCPZ = -(n - 1)$.

Thus, $-(n - 1) \leq SCPZ \leq (n - 1)$; in our example, the value of this statistic must fall between -14 and $+14$. From the FREQUENCIES subprogram, $s_X = 0.344$ and $s_Y = 8.877$. The sum of the cross products of the z-scores is

$$SCPZ = \frac{SCPD}{s_X s_Y} = \frac{39.013}{0.344 \times 8.877} = 12.776.$$

This indicates a strong relationship between the floor area of a house and the amount of gas it uses.

Still, this measure of correlation is influenced not only by the relationship between the values in the data pairs, but also by the sample size. Division of the SCPZ by $(n-1)$ overcomes this deficiency and gives us this definition:

DEFINITION

Given n data pairs (X_i, Y_i), the **Pearson product-moment correlation coefficient*** (or simply the *correlation coefficient*) R is defined to be

$$R = \sum_{i=1}^{n} \frac{(X_i - \bar{X})(Y_i - \bar{Y})}{(n-1)\, s_X s_Y} \; .$$

Since $R = \text{SCPZ}/(n-1)$, it will always take on a value between -1 and $+1$; values near -1 indicate a strong negative relationship, values near $+1$, a strong positive relationship. If R is near 0, there is little correlation between the two variables.

In our example, $R = 12.776/14 = 0.913$, indicating a strong positive relationship between house area and gas use.

The Pearson correlation coefficient can be found in SPSS by using the PEARSON CORR subprogram. For example,

PEARSON CORR; GAS WITH AREA

The correlation coefficient is also statistic number 1 of the SCATTERGRAM subprogram, and the authors recommend finding it in that way so as to obtain the illustration as well. The scatter diagram is helpful in checking that three assumptions on which the calculation of the Pearson correlation coefficient is founded are satisfied. They are:

1. A linear relationship exists between the two variables.

2. The variables are continuously measured.

3. For all values of X within the range of the study, the corresponding Y values have the same dispersion. This condition is called **homoscedasticity.**

In the graph on the left side of Figure 12.5, the distributions of Y values corresponding to each X value are equally dispersed, *homoscedastic*. This condition is not satisfied by the data pairs depicted on the right, where dispersion increases as X increases.

A program to display the scatter diagram and compute R for our example is:

---

*Named for Karl Pearson, who developed this measure in 1896.

FIGURE 12.5.

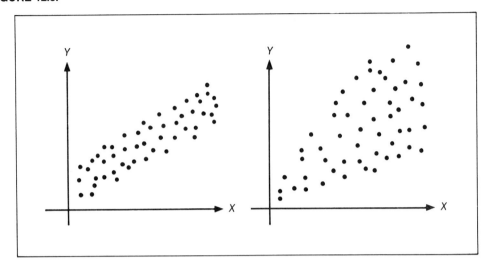

```
10.       FILE NAME
10.005    REGEX
20.       VARIABLE LIST
20.005    NUM,AREA,GAS
30.       INPUT FORMAT
30.005    FREEFIELD
40.       N OF CASES
40.005    15
50.       SCATTERGRAM
50.005    GAS WITH AREA
60.       STATISTICS
60.005    1
```

The data had been stored in the computer in the raw data file REGEX: the variable name NUM is associated with the line numbers of this file. The program produces the output on the following page:

Estimating the Population Correlation Coefficient

The data pairs (X_i, Y_i) which we collect can be thought of as a sample from the population of all possible such data pairs. The computed correlation coefficient R is an estimate of the true population correlation coefficient ϱ (rho), which indicates the true degree and kind (negative or positive) of relationship between the two variables.

The sampling distribution of R depends on the true value of ϱ. When ϱ is zero, the dis-

— — — SCATTERGRAM — — —

GIVEN   2   VARIABLES,   MAXIMUM CM ALLOWS FOR 7308 CASES

(DOWN)   GAS
(ACROSS)   AREA

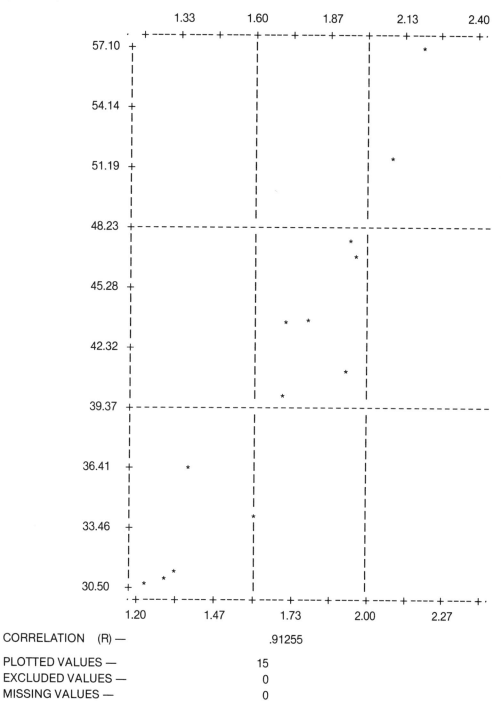

CORRELATION (R) —   .91255

PLOTTED VALUES —   15
EXCLUDED VALUES —   0
MISSING VALUES —   0

tribution of R is symmetrical, but as ϱ departs from zero, the distribution of R becomes increasingly skewed, positively as ϱ approaches -1, negatively as ϱ nears $+1$. For sample sizes (numbers of data pairs) greater than 30, the distribution of R is approximately normal with mean ϱ and standard deviation $\sigma_R = (1 - \varrho^2)/(n - 1)$, when ϱ is near zero.

R.A. Fisher,* however, identified a function of R whose sampling distribution is normal in general. The result is this:

Let R be the Pearson correlation coefficient to be computed from n data pairs, with $n > 30$. The sampling distribution of the quantity

$$z_R = \ln_e \left[(1 + R)/(1 - R)\right]^{\frac{1}{2}}$$

is approximately normal with mean 0 and standard deviation $s_{z_R} = 1/(n - 3)^{\frac{1}{2}}$.

For $n < 30$, Fisher demonstrated that the statistic

$$t = R\left[(n - 2)/(1 - R^2)\right]^{\frac{1}{2}}$$

has a t distribution with $n - 2$ degrees of freedom when $\varrho = 0$.

With this knowledge of the sampling distribution of R, we can test hypotheses about the population correlation coefficient. In our example, suppose we wish to test the null hypothesis that house area and January natural gas use are not related against the alternative that they are at the 5% significance level. Formally,

$$H_o\colon \varrho = 0$$

$$H_a\colon \varrho \neq 0$$

There are 15 cases in our file, so if H_o is true, the quantity $t = R[(n-2)/(1-R^2)]^{\frac{1}{2}}$ has a t distribution with $15 - 2 = 13$ degrees of freedom. Corresponding to $\alpha = 5\%$, the critical values are ± 2.160, and we will reject the null hypothesis if the t statistic does not fall between them (Figure 12.6).

FIGURE 12.6. Critical values for the test of ϱ.

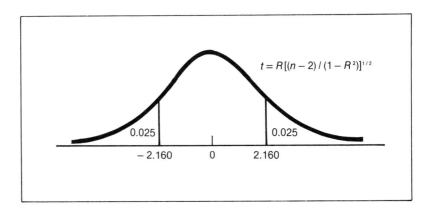

---

*R.A. Fisher, *Statistical Methods for Research Workers*, 10th ed., 1948.

From the SCATTERGRAM subprogram, we found that $R = 0.91255$, so

$$t = R\left(\frac{n-2}{1-R^2}\right)^{1/2} = 0.91255 \left(\frac{15-2}{1-0.91255^2}\right)^{1/2} = 8.045.$$

This value is outside the critical values, so we reject the null hypothesis and conclude that there is a significant correlation between house area and natural gas use.

Suppose that R had been 0.40. Then

$$t = 0.40 \left(\frac{15-2}{1-0.40^2}\right)^{1/2} = 1.56.$$

This value falls between the critical values, and we would not be able to conclude that a significant correlation exists.

The SCATTERGRAM subprogram's statistic number 3 is the two-tail significance of R in a test of this form; we may compare this significance with our chosen α. For the above example, this value, which is also calculated by PEARSON CORR, is 0.00001, much less than $\alpha = 0.05$.

The reader should be aware that the presence of a significant correlation between the values of two variables does not necessarily imply that a cause-and-effect mechanism links them. They might both be influenced by some third quantity, or their apparent connection might be a numerical coincidence. A significant correlation tells us only that some physical interaction *might* be present, and that its possibility should be investigated.

The Point-Biserial Correlation Coefficient

The Pearson product-moment correlation coefficient measures the correlation between two continuous interval- or ratio-level variables. Many other situations are possible. To examine the interaction of a continuous variable with one which is **dichotomous**, which can take on only two distinct values, we find the **point-biserial correlation coefficient**.

Many correlation coefficients are called Pearson correlation coefficients because their formulas are algebraically equivalent to that of the Pearson product-moment coefficient. The point-biserial is one of these. To compute it, the formula of the last section may be used, or it may be found in this way:

DEFINITION

Let X be a continuous variable in a data set, and let Y be a dichotomous variable with values y_1 and y_2. The point-biserial correlation coefficient is:

$$R_{pbi} = \frac{M_1 - M_2}{s_X} \sqrt{qp} \,,$$

where M_1 = the mean of X for those cases with $Y = y_1$, M_2 = the mean of X for those cases with $Y = y_2$, s_X = the standard deviation of all the X values, p = the proportion of cases with $Y = y_1$, and $q = 1 - p$ = the proportion of cases with $Y = y_2$.

The values of the dichotomous variable Y do not enter into the calculation; the values of Y are used only to categorize the values of X.

Since the formula above is equivalent to that of the product-moment correlation coefficient, the point-biserial coefficient may be found using the SCATTERGRAM or PEARSON CORR subprograms. In this example program, we return to the USNEWS file to investigate the correlation of region with growth of department store sales for central and eastern cities. (We must exclude one of the three values of X1 so that X1 will be dichotomous.)

```
10.      GET FILE
10.005  USNEWS2
20.      SELECT IF
20.005  (X1 EQ 1 OR X1 EQ 2)
30.      PEARSON CORP
30.005  X2 WITH X1
```

In line 20 we select only those cases which represent eastern and central cities. (Cases may be temporarily selected for only one task with the command *SELECT IF.) The output of the program is this:

```
— — — PEARSON CORR — — —

              X1

X2           .3215
             (  55)
             P = .008
```

(COEFFICIENT / CASES / SIGNIFICANCE) (99.0000 MEANS UNCOMPUTABLE)

The value of the coefficient is positive, indicating a direct relationship.

Since the point-biserial correlation coefficient is algebraically equivalent to the Pearson product-moment correlation coefficient, we may use the significance of the coefficient as calculated by either SCATTERGRAM or PEARSON CORR to perform this hypothesis test:

$$H_o: \varrho_{pbi} = 0$$

$$H_a: \varrho_{pbi} \neq 0$$

In our example, the sugnificance of R_{pbi} is 0.008, smaller than any reasonable significance level, so we reject the null hypothesis. We conclude that there is a significant correlation between region and growth of department store sales between the eastern and central regions of the United States.

The Biserial Correlation Coefficient

When one of two continuously distributed variables is *treated* as a dichotomy, the relationship between the variables is often measured with the **biserial coefficient,** our first example of a correlation coefficient which is not a Pearson coefficient, not algebraically equivalent to the Pearson product-moment correlation coefficient. The need for a measure of correlation for such situations is clear: many continuously valued variables are more easily reported as dichotomies. For example, household incomes are essentially continuously distributed, but people are more likely to respond when asked simply if their income is less than or greater than $15,000.

Like the point-biserial coefficient, computation of the biserial correlation coefficient R_{bi} is based on dividing the observed values of the continuous variable into two groups according to the values of the dichotomous variable, and comparing the means of the continuous variable for the two groups:

DEFINITION

Let X and Y be interval- or ratio-level variables, Y *normally distributed*, with the values of Y grouped to form a dichotomy with values y_1 and y_2. The **biserial correlation coefficient** R_{bi} is:

$$R_{bi} = \frac{M_2 - M_1}{S_X} \times \frac{pq}{f(z_y)},$$

where M_1 = the mean of X for those cases where $Y = y_1$, M_2 = the mean of X for those cases where $Y = y_2$, p = the proportion of cases for which $Y = y_1$, $q = 1 - p$ = the proportion of cases for which $Y = y_2$, and $f(z_y)$ = the ordinate of the point on the standard normal curve which divides the area under the curve into the complementary areas p and q.

Figure 12.7 illustrates finding the value $f(z_y)$. The proportions p and q locate z_y through the standard normal table (Table A-4); then the standard normal probability function is evaluated at z_y.

FIGURE 12.7   The computation of $f(z_y)$.

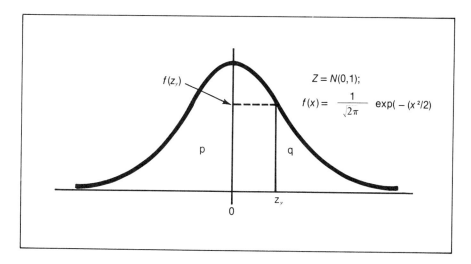

No subprogram in SPSS computes the biserial coefficient directly, but the values M_1, M_2, and s_X can be found using BREAKDOWN, or by an appropriate choice of *SELECT IF commands preceding several CONDESCRIPTIVES. Also, the point-biserial coefficient, calculated by SPSS, is related to the biserial coefficient in this way:

$$\frac{\sqrt{pq}}{f(z_y)} \quad R_{pbi} = \quad \frac{\sqrt{pq}}{f(z_y)} \quad \times \quad \frac{M_1 - M_2}{s_X} \sqrt{pq}$$

$$= \frac{M_1 - M_2}{s_X} \times \frac{pq}{f(z_y)} = R_{bi} .$$

We can use this relationship to find R_{bi} from the value of R_{pbi} computed by SCAT-TERGRAM or PEARSON CORR.

Suppose that the values of unemployment (X3) in the USNEWS study had been reported in only two categories: 0 to 5% and above 5%. The following program finds the point-biserial correlation coefficient of the now dichotomous X3 with X5 (average factory workers' income) from which we will compute the biserial coefficient.

```
10.     GET FILE
10.005 USNEWS2
20.     RECODE
20.005 X3(LO THRU 5 = 1) (5.01 THRU 25 = 2)
30.     PEARSON CORR
30.005 X3 WITH X5
```

The RECODE statement categorizes the values of X3, and this output is produced:

```
                X5

X3              .2231
              (   69)
              P = .033
```

(COEFFICIENT / CASES / SIGNIFICANCE) (99.0000 MEANS UNCOMPUTABLE)

Corresponding to $p = 25/69 = 0.362$ (the proportion of cities for which X3 is 5% or less and values are present for X3 and X5) and $q = 44/69 = 0.638$, the value of z_y is -0.35, since $P(z < -0.35) = 0.362$.

$$f(z_y) = \frac{1}{\sqrt{2\pi}} \exp(-z_y^2/2) = \frac{1}{\sqrt{2\pi}} \exp[-(-0.35)^2/2]$$

$$= 0.3989 \times 0.9406 \doteq 0.375.$$

The value of the biserial correlation coefficient for these two variables is

$$R_{bi} = \frac{(0.362 \times 0.638)^{1/2}}{0.375} \times 0.2231 = 0.286.$$

Many mathematicians question the use of the biserial coefficient because its sampling distribution is difficult to describe and use. In particular, the standard error of R_{bi} is

$$s_{R_{bi}} = \frac{\sqrt{pq} / f(z_y) - R_{bi}^2}{\sqrt{n}} .$$

The shape of the sampling distribution of R_{bi} depends on the proportions of observations in each category of the dichotomy and even the above expression for $S_{R_{bi}}$ is incorrect if the underlying population of the dichotomous variable is not normally distributed. If the normality assumption is violated, values of R_{bi} larger than one are possible.

Finally, because of these difficulties with the sampling distribution of R_{bi}, we are unable to conveniently perform hypothesis tests concerning the population biserial coefficient.

The Tetrachoric Correlation Coefficient

When we wish to measure the correlation of two artificially dichotomized variables, the **tetrachoric correlation coefficient** is often used. With this statistic, which is not a Pearson coefficient, it is assumed that both of the dichotomous variables being examined represent continuous, normally distributed variables which share a linear relationship.

The definition of the tetrachoric coefficient is very complicated, but a well-known approximation uses the cross-tabulation of the two dichotomous variables.

TABLE 12.2.

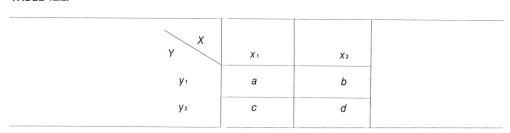

Y \ X	x_1	x_2	
y_1	a	b	
y_2	c	d	

a, b, c, and d are the numbers of cases in each cell of the cross-tabulation; from them we obtain this approximation to the tetrachoric coefficient:

$$R_{tet} \doteq \cos \left[\frac{\pi}{1 + \left(\frac{ad}{bc} \right)^{1/2}} \right] .$$

The *cosine-pi formula* provides a good approximation to R_{tet} when the dichotomies of both variables are formed by dividing the values near their medians.

The standard error of R_{tet} is more easily established:

$$S_{tet} = \frac{abcd}{f(z_x)f(z_y)\, n^2 \sqrt{n}} ,$$

where $f(z_x)$ = the ordinate of the point on the normal curve which divides the area under the curve into two complementary areas $(a + c)/n$ and $(b + d)/n$, and $f(z_y)$ = the same four areas $(a + b)/n$ and $(c + d)/n$. (See Figure 12.8.)

The values a, b, c, and d can be found with the CROSSTABS subprogram in SPSS, but some implementations of the package support the subprogram TETRACHORIC, which computes R_{tet} and its two-tailed significance.

FIGURE 12.8. Computing $f(z_x)$.

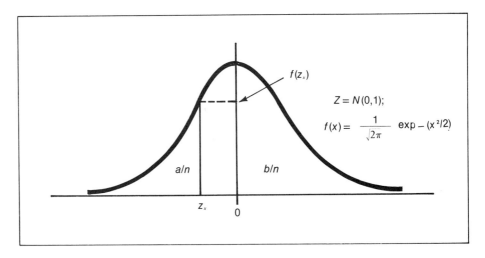

We can manufacture an example of such a computation from the USNEWS file by forcing dichotomies on the two variables X3 (unemployment) and X6 (change in factory workers' income). In the following program, both are grouped into two categories, and the subprogram TETRACHORIC is used to find R_{tet} and its significance.

```
10.     GET FILE
10.005 USNEWS
20.     RECODE
20.005 X3 (LO THRU 5 = 1) (5.01 THRU 15 = 2)
25.     RECODE
25.005 X6 (LO THRU 5 = 1) (5.01 THRU 30 = 2)
30.     TETRACHORIC
30.005 VARIABLES = X3,X6(1,2)/CORRELATIONS = X3 WITH X6
```

In the TETRACHORIC command, we must specify the variables that we will consider, the two values of each which form the dichotomy, and the correlation coefficients to be computed. The output is this:

```
― ― ― TETRACHORIC ― ― ―

               X6

X3            ― .5666
              (   69)
              S =   .008
```

The value of R_{tet} is negative and reasonably large, and its significance is small; we can reject H_o: $\varrho_{tet} = 0$, and conclude that an inverse linear relationship exists between the original values of X3 and X6.

The Phi and Cramer's V Correlation Coefficients

The Pearson correlation coefficient is called the **phi coefficient** when it is calculated for two genuinely dichotomous variables. If we construct a cross-tabulation of the variables, with cell frequencies *a, b, c,* and *d* as shown in Table 12.3,

TABLE 12.3.

Y \ X	X_1	X_2
y_1	a	b
y_2	c	d

the value of the phi coefficient can be found very easily from this formula:

$$\phi = \frac{bc - ad}{[(a+b)(c+d)(a+c)(b+d)]^{1/2}} \; .$$

Suppose that we wish to find if a relationship exists between having finished high school and having or not having a criminal record. If of 15 people who did not finish high school, 8 have criminal records, and the same is true of 6 out of 23 who did finish, we obtain the cross-tabulation shown in Table 12.4.

TABLE 12.4.   Criminal Record

	HIGH SCHOOL YES	NO
Criminal Record — Yes	6	8
No	17	7

and the value of the phi coefficient is

$$\phi = \frac{8 \times 17 - 6 \times 7}{[(6+8)(17+7)(6+17)(8+7)]^{1/2}} \; .$$

$$= \frac{94}{[115920]^{1/2}} \doteq 0.276.$$

Though ϕ is a Pearson coefficient, and finding its value is a simple process, calculation of its standard error is far too complicated to be useful. Also, the two-by-two cross-tabulation places restrictions on ϕ which do not usually apply to the Pearson product-moment correlation R. ϕ will vary from -1 to $+1$ only when all four marginal totals are equal; in general, the largest and smallest possible values of ϕ are functions of the marginal proportions of the cross-tabulations so that the sampling distribution of ϕ is difficult to identify and apply.

Its presence as a statistic of a cross-tabulation, however, suggests a link between the phi coefficient, a measure of the correlation between the two variables, and the chi-square statistic X^2 of the test of independence. This link takes a very simple form:

$$X^2 = n\,\phi^2 \text{ or } \phi = (X^2/n)^{1/2} \ .$$

If the chi-square statistic shows a significant dependence between the two variables, then the phi coefficient is also significant.

When either of the two variables has more than two classes (attributes), the *contingency coefficient C*, also related to the chi-square statistic, is sometimes used:

$$C = \left(\frac{X^2}{x+X^2}\right)^{1/2} \ .$$

This statistic, like ϕ, is a function of the marginal frequencies. A coefficient which has the value 1 when a perfect relationship exists between the two variables is *Cramer's V* or *Cramer's statistic*:

$$Cr = \left(\frac{X^2}{n(L-1)}\right)^{1/2} ,$$

where $L =$ the minimum of the number of rows and the number of columns.

The significance of Cr is again the significance of the corresponding chi-square statistic.

Cramer's V (or the phi coefficient for a two-by-two table) and the contingency coefficient are statistics 2 and 3 of the SPSS subprogram CROSSTABS. The following program finds these statistics and X^2 in examining the relationship between city region (X1) and a dichotomized X3 (unemployment; see line 20) in the USNEWS file.

The significance of the statistics is large, so we cannot conclude from these data that there is a relationship between region and unemployment. (Note that because the dichotomy of X3 here is artificial rather than inherent, Cramer's V and the contingency coefficient are not, in fact, appropriate.)

CPO335

```
10.     GET FILE
10.005 USNEWS
20.     RECODE
20.005 X3(LO THRU 5 = 1) (5.01 THRU 15 = 2)
30.     CROSSTABS
30.005 TABLES = X3 BY X1
35.     STATISTICS
35.005 1,2,3
```

| X3 | | UNEMPLOYMENT | | | |
| BY X1 | | CITY REGION | | | |

		X1			
COUNT ROW PCT COL PCT TOT PCT		EASTERN 1.	CENTRAL 2.	WESTERN 3.	ROW TOTAL
X3					
	1.	8 32.0 32.0 11.4	12 48.0 42.9 17.1	5 20.0 29.4 7.1	25 35.7
	2.	17 37.8 68.0 24.3	16 35.6 57.1 22.9	12 26.7 70.6 17.1	45 64.3
COLUMN TOTAL		25 35.7	28 40.0	17 24.3	70 100.0

RAW CHI SQ =   1.06656 WITH   2 D.F., SIG. =   .5867

CRAMER'S V =   .12344

CONTINGENCY COEFFICIENT =   .12251

MISSING OBSERVATIONS —   4

Partial Correlation Coefficients

In many situations, an apparent correlation between two variables X and Y occurs when both are related to some third variable Z. In such cases we would like to investigate the relationship between X and Y while *controlling for* the effect of Z. For example, in the USNEWS data file, a small but significant correlation exists between X3 (unemployment) and X5 (average income of factory workers); both are related to X4 (change in nonfarm employment). Could X4 be influencing both X3 and X5 to produce the appearance of a correlation? We measure the correlation of X3 and X5 while removing the effect of X4 on their relationship.

In general, we calculate the *first-order partial correlation coefficient* of X and Y while controlling for Z, $R_{XY \cdot Z}$, in this way:

$$R_{XY \cdot Z} = \frac{R_{XY} - R_{XZ} R_{YZ}}{[(1 - R_{XZ}^2)(1 - R_{YZ}^2)]^{1/2}} .$$

This value identifies the part of the correlation of X and Y which is not an artifact of their relationship with Z.

In our example, computed from the 69 cases which have no missing values of X3, X4, or X5, $R_{X3,X5} = 0.2728$, $R_{X3,X4} = -0.3793$, and $R_{X5,X4} = -0.1986$, so that

$$R_{X3,X5.X4} = \frac{R_{X3,X5} - R_{X3,X4}R_{X5,X4}}{[(1 - R^2_{X3,X4})(1 - R^2_{X5,X4})]^{1/2}} = \frac{0.2728 - (-0.3793)(-0.1986)}{[(1 - (-0.3793)^2)(1 - (-0.1986)^2)]^{1/2}}$$

$$= \frac{0.1975}{[0.8224]^{1/2}} = 0.2178$$

We can conclude that some of the correlation between X3 and X5 can be associated with X4, but even when controlling for X4, there is a small positive correlation between X3 and X5.

Second-order partial correlation coefficients measure the interaction of two variables while controlling for the effects of *two* others; the partial correlation of X and Y while controlling for Z and W is

$$R_{XY.ZW} = \frac{R_{XY.W} - R_{XZ.W}R_{YZ.W}}{[(1 - R^2_{XZ.W})(1 - R^2_{YZ.W})]^{1/2}} .$$

Second- and higher-order partial correlation coefficients are difficult to interpret, and are rarely used.

Zero-order partial correlation coefficients are the familiar Pearson correlations, when the effects of other variables are not controlled.

The significance of first-order partial correlation coefficients can be investigated with the following statistic, whose sampling distribution is a t distribution with $n - 3$ degrees of freedom, when the population partial correlation is zero:

$$t = \frac{R_{XY.Z}}{[(1 - R^2_{XY.Z})/(n - 3)]^{1/2}} .$$

For example, we have found that $R_{X3,X5.X4} = 0.2178$, so the value of the associated t statistic is

$$t = \frac{R_{X3,X5.X4}}{[1 - R^2_{X3,X5.X4})/(69 - 3)]^{1/2}} = \frac{0.2178}{[(1 - 0.2178^2)/66]^{1/2}} \doteq 1.81.$$

Corresponding to 66 degrees of freedom (recall that $n = 69$), $t_{0.95} \doteq 1.658$ and $t_{0.975} \doteq 2.000$, so this positive correlation is significant in a one-tail test at the 5% significance level but not at $\alpha = 2.5\%$.

In SPSS, partial correlation coefficients are calculated by the subprogram PARTIAL CORR. This example finds the partial correlation coefficient of X3 and X5 while controlling for X4.

```
10.     GET FILE
10.005 USNEWS
20.     PARTIAL CORR
20.005 X3 WITH X5 BY X4 (1)
30.     STATISTICS
30.005 ALL
```

In the specification field of the PARTIAL CORR command, we give the two variables whose partial correlation is to be found, separated by the keyword WITH. Following the keyword BY, we give the variable whose effects are to be controlled, and, in parentheses, the order of the partial correlation coefficient. The other statistics which appear in the output, including the zero-order partial correlations, are caused by the statement STATISTICS; ALL.

VARIABLE	MEAN	STANDARD DEV	CASES
X3	5.6899	1.6589	69
X5	14998.8406	2358.3787	69
X4	2.7667	2.8259	69

ZERO ORDER PARTIALS

	X3	X5	X4
X3	1.0000 (0) P = * * * * *	.2728 (67) P = .012	− .3793 (67) P = .001
X5	.2728 (67) P = .012	1.0000 (0) P = * * * * *	− .1986 (67) P = .051
X4	− .3793 (67) P = .001	− .1986 (67) P = .051	1.0000 (0) P = * * * * *

(COEFFICIENT   /   (D.F.)   /   SIGNIFICANCE)
(A VALUE OF 99.0000 IS PRINTED IF A COEFFICIENT CANNOT BE COMPUTED)

CONTROLLING FOR..   X4

	X5
X3	.2178 (66) P = .037

(COEFFICIENT   /   (D.F.)   /   SIGNIFICANCE)
A VALUE OF 99.0000 IS PRINTED IF A COEFFICIENT CANNOT BE COMPUTED.

In the printout we see that the correlation of X3 and X5 is reduced slightly to 0.2178 when controlling for X4, and the partial correlation is significant. This might be interpreted to mean that the correlation of X3 and X5 is not caused by the influence on both of X4.

RECOMMENDATIONS WHEN USING CORRELATION COEFFICIENTS

When investigating relationships between variables with statistics such as those described in this chapter, the following points should be kept in mind:

1. Pearson coefficients—the point-biserial, phi, Cramer's V—are preferable, since their sampling distributions are well known and hypotheses related to them can be tested.

2. A large value of a correlation coefficient is not, by itself, meaningful; the significance of the coefficient is a function of sample size.

3. The presence of a significant correlation between two variables does not necessarily imply a cause-and-effect relationship.

4. The value of any correlation coefficient is meaningless if the assumptions on which it is based are not satisfied by the data being analyzed.

EXERCISES

1. The following are the curb weights in pounds, horsepowers, and EPA mileage estimates of 12 1979-model American automobiles:

WEIGHT	HORSEPOWER	MPG
3205	125	14
2945	80	20
4380	125	21
2040	74	29
3295	165	16
3520	195	16
2195	77	25
3205	195	16
1760	66	28
2515	130	21
3205	97	17
2480	93	28

   a. Draw the scatter diagram associated with weight and mileage. What type of relationship, if any, is suggested?
   b. Find the value of the Pearson product-moment correlation coefficient R for weight and mileage.
   c. At the 5% significance level, can we conclude from these data that a linear relationship exists between the weight of a car and its mileage? If so, describe that relationship.

2. For the data given in Exercise 1,

   a. Draw the scatter diagram associated with weight and horsepower. What type of relationship is suggested?
   b. Find the Pearson product-moment correlation coefficient R for weight and horsepower.

   c. At the 5% significance level, can we conclude that a linear relationship exists between the weight and horsepower of an automobile?

3. Consider the data given in Exercise 1.

   a. Draw the scatter diagram associated with the values of horsepower and mileage. What type of relationship is suggested, if any?

   b. Find $R_{\text{Horsepower, mpg}}$.

   c. At $\alpha = 5\%$, is the value found in part (b) significant? What conclusion can we draw?

4. For the players on your school's varsity basketball team, obtain the number of points scored and the number of fouls committed in the last three games.

   a. Compute the Pearson correlation coefficient R for the two variables.

   b. Interpret the result of part (a). Should the coach encourage more fouls? What explanation would you suggest for the value of R?

5. Collect the college and high school grade point averages (GPA) of the members of your class, making sure that everyone uses the same scale of measurement.

   a. Plot a scatter diagram of the value pairs. Which variable do you choose to plot on the vertical axis?

   b. Compute R for these data.

   c. At the 5% significance level, is there a relationship between high school and college GPAs? Interpret this result.

6. Repeat Exercise 5 comparing these two values for each member of your class: College GPA and entrance exam score (ACT or SAT). If you had to predict success in college from high school GPA or entrance exam score, which would you use? Why?

7. Use your computer system to perform the following analyses of data in the USNEWS file:

   a. Plot a scatter diagram of the values of X3 and X4 (unemployment and change in nonfarm employment).

   b. Find R for X3 and X4.

   c. At the 5% significance level, is there a relationship between X3 and X4? If so, describe and interpret that relationship.

8. Repeat Exercise 7 investigating X4 and X7.

9. Let X and Y be two variables. Show algebraically that:

   a. $s_{x+y}^2 = s_x^2 + s_y^2 + 2R_{xy}s_x s_y$

   b. $s_{x-y}^2 = s_x^2 + s_y^2 - 2R_{xy}s_x s_y$

$$\left[Hint:\ s_{x+y}^2 = \frac{1}{n-1} \sum_{i=1}^{n} ((X_i + Y_i) - (\bar{X} + \bar{Y}))^2 = \frac{1}{n-1} \sum_{i=1}^{n} ((X_i - \bar{X}) + (Y_i - \bar{Y}))^2. \right]$$

10. Let X and Y be two variables, and a, b, c, d be constants, with a and c positive.

   a. Show that $R_{aX + b, cY + d} = R_{X, Y}$.

   b. Interpret the result in part (a).

11. A sample of 50 data pairs (X_i, Y_i) produces a correlation coefficient of $R = 0.4140$. At the 5% significance level, can we conclue that $\varrho > 0$?

12. Let $X, Y,$ and Z be three variables in a data file. We might wish to investigate the relationships of both X and Y to Z. Do they both have the same correlation with Z? These hypotheses,

$$H_o: p_{xz} = p_{yz}$$

$$H_a: p_{xz} \neq p_{yz},$$

can be tested using the test statistic

$$t = \frac{(R_{xz} - R_{yz}) \, [(n-3)(1+R_{xy})]^{1/2}}{[2(1 - R_{xy}^2 - R_{yz}^2 - R_{xy}^2 + 2R_{xy}R_{xz}R_{yz})]^{1/2}}$$

whose sampling distribution, if $\varrho_{xz} = \varrho_{yz}$, is a t distribution with $n - 3$ degrees of freedom. Use this result to test, at the 5% significance level, whether the correlation between X2 and X4 is equal to that between X4 and X7, using the USNEWS data file.

13. For each of the following groups of data pairs, determine what type of relationship exists between X and Y, and try to find R:

a.

X	1	2	2	2	2	3	3	3	4	4	5
Y	2	2	2	2	2	2	2	2	2	2	2

b.

X	-1	$-.75$	$-.5$	$-.25$	0	$-.5$
Y	0	.60	.80	.95	1	.81

c.

X	.25	.50	1	.75	.75
Y	.94	.83	.02	.58	.62

14. Using the USNEWS data file, calculate R_{pbi} for X2 and X1 for cities in the eastern and central regions with the formula for R_{pbi}.

15. Show that the formula for R_{xy} is algebraically equivalent in the point-biserial situation to that of R_{pbi} if s_x is computed using n in the denominator rather than $n - 1$. (*Hint*: Let the two values of Y be 0 and 1.)

16. Show that R_{pbi} is a function of p and q. Under what conditions is R_{pbi} unable to achieve unity? (*Hint*: Create examples with different marginal frequencies but the same sample size.)

17. In exam construction, educators often employ the point-biserial correlation coefficient to measure the degree to which a particular question discriminates between those who do well on the exam and those who do poorly. This is called *item analysis*. The following table gives the total exam scores for 14 students, and whether or not each answered question 5 correctly.

INDIVIDUAL	TOTAL SCORE	QUESTION 5
1	70	0
2	75	1
3	65	0
4	50	0
5	50	1
6	30	0
7	25	0
8	40	0
9	90	1
10	50	0
11	45	0
12	85	1
13	77	1
14	92	1

    a. Find M_1, the mean score of those who answered question 5 correctly.

    b. Find M_o, the mean score of those who missed question 5.

    c. Find p and q; what do they represent?

    d. Find R_{pbi}; what does it represent? (Be sure to test $H_o : \varrho_{pbi} = 0$ against $H_a: \varrho_{pbi} \neq 0$ before interpreting R_{pbi}.)

    e. Would a scatter diagram be useful in a situation like this? Why or why not?

18. For the data given with Exercise 1, group the values of automobile weight into two classes, those values below 3000, and those above. Now find the biserial correlation coefficient of weight with mileage. What might this value mean, and what assumption do we make in using it?

19. Consider the data given with Exercise 1. Group the values of horsepower into two classes by dividing them at 100, and find the biserial correlation coefficient of horsepower with mileage. What might this value mean, and what assumptions do we make in using it?

20. Find and interpret the value of R_{bi} for X5 and X3 from the USNEWS file with the values of X5 grouped into those below \$15,226 and those above.

21. **a.** When is R_{pbi} an appropriate measure of correlation?

    b. When should R_{pbi} be converted to R_{bi}?

22. Reconsider the data gathered in Exercise 5. Divide the values of high school GPA into two classes at their mean and compute R_{bi}. What does this value indicate?

23. When reporting a biserial correlation coefficient, what assumption and limitations should a statistician also mention?

24. Develop a graph showing the ratio of R_{pbi} to R_{bi} for various values of p. What does the graph indicate?

25. Force dichotomies on the data gathered for Exercise 4 by dividing the values of each variable into two classes. Then calculate and interpret the tetrachoric correlation coefficient R_{tet}.

26. Repeat Exercise 25 using the data gathered for Exercise 5. What assumptions are being made about the two variables)?

27. **a.** In the USNEWS file, force dichotomies on the values of X5 and X7, then calculate and interpret R_{tet}.

    b. Repeat part (a) using an SPSS program and the subprogram TETRACHORIC.

28. What limitations and cautions should be reported along with R_{tet}?

29. Show that

$$\phi = \frac{ad - bc}{[(a+b)(a+c)(b+d)(c+d)]^{1/2}}$$

is algebraically equivalent to R_{xy}, and is therefore a Pearson coefficient.

30. Show algebraically that $X^2 = n\phi^2$. (*Hint*: Begin by setting up a 2 by 2 contingency table and writing out the expression for ϕ and for

$$X^2 = \sum_{ij} \frac{(O_{ij} - E_{ij})^2}{E_{ij}} \,.$$

Note that n is the sum of the four cell frequencies.)

31. How does ϕ change with p_x and p_y, where p_x = proportion of cases with $X = X_1$ and p_y = proportion of cases with $Y = y_1$? What are the largest and smallest possible values of ϕ?

33. For the members of your class, use the ϕ coefficient to find if there is a link between owning a car and living off campus.

34. **a.** For the USNEWS file, use zero-order partial correlations to calculate $R_{X2,X4.X7}$.

    b. Use the statistic $t = \dfrac{R_{X2,X4.X7}}{[(1 - R^2_{X2,X4.X7})/(n-3]^{1/2}}$ to test

these hypotheses: H_o: $\varrho_{X2,X4.X7} = 0$

$$H_a:\ \varrho_{X2,X4.X7} \neq 0$$

Is the value of $R_{X2,X4.X7}$ significant at $\alpha = 5\%$?

35. Using the data given with Exercise 1, find the partial correlation of horsepower with mileage while controlling for weight. Is this value significant at $\alpha = 10\%$? What does it mean?

36. Consider the following data for a random group of 13 boys:

AGE (YEARS)	HEIGHT (INCHES)	WEIGHT (POUNDS)
10	48	60
10	58	75
10	56	80
12	59	96
12	62	120
14	68	140
14	70	172
14	71	182
14	73	175
15	68	190
15	74	205
17	70	160
17	72	180

a. Why should a partial correlation coefficient be used to measure the linear relationship between height and weight?

b. Find $R_{\text{Weight, Height, Age}}$.

c. Is the value found in part (b) significant?

37. For the USNEWS file, use partial correlations to examine the relationship between X4 (change in nonfarm employment) and X7 (change in construction activity) while controlling for X5 (average factory workers' income). What interpretation might be placed on your results?

38. Consider the values in the USNEWS file as a population rather than as a sample; then $R_{X3,X5}$ becomes $\varrho_{X3,X5}$.

a. Obtain four 10% random samples from the USNEWS file using the SAMPLE command, and for each, find $R_{X3,X5}$.

b. Compare the four sample values of $R_{X3,X5}$. Why are they different?

c. Would the mean of the four values of $R_{X3,X5}$ be a better estimate of $P_{X3,X5}$ than any of the individual values? Find that mean. If the sample sizes are different, compute the weighted mean.

d. The values of $R_{X3,X5}$ may also be averaged by finding the mean of the Fisher coefficients $z_{R_{X3,X5}}$ and transforming the mean z_R back into an estimate of $\varrho_{X3,X5}$. Do this, again using a weighted mean if the sample sizes are different. Which value is nearer $\varrho_{X3,X5}$, this value or the one found in part (c)? Which averaging technique appears to provide the better estimate of $\varrho_{X3,X5}$?

[*Hint:* To obtain an estimate of $\varrho_{X3,X5}$ from the mean of the Fisher coefficients, you must solve the equation .

$$z_R = \ln\left(\frac{1+R}{1-R}\right)^{1/2} \text{ for } R\ ]$$

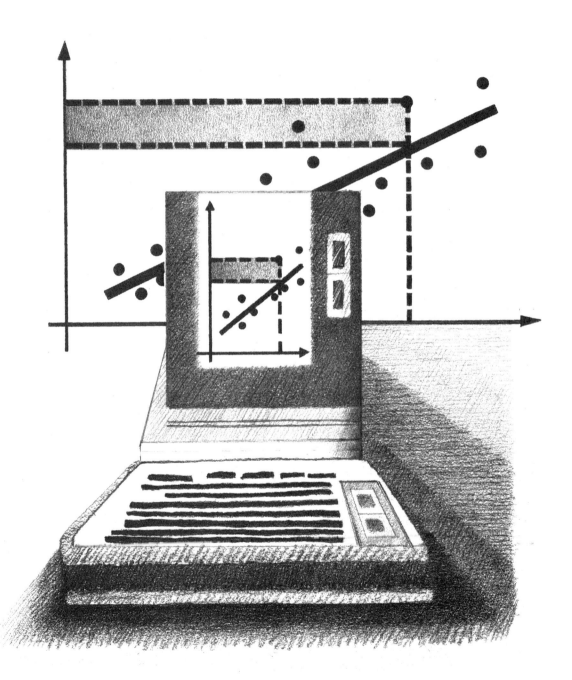

13

Regression

INTRODUCTION

In Chapter 12 we developed a number of statistics which measure the degree of correspondence or relationship between two variables. These statistics provide information about the degree, and, sometimes, the direction of the relationship. However, we can go beyond correlation analysis to predict, with known accuracy, the value of one of the variables when the value of the other is known. This prediction process is called regression.

The word regression comes from the first study that used the concept, which was published by Sir Francis Galton in 1885.* In it he examined the relationship between the heights of parents and their children and found that the heights of children of unusually tall or short parents tended toward the average height of the population; that is, their heights regressed *toward the average height. Regression has come to mean the prediction of the value of a variable from the values of one or more other variables.*

Consider the simplest case, in which we have collected value pairs for two variables, X and Y, and we wish to predict the value of Y which might correspond to a given value of X.

We have seen that a scatter diagram graphically illustrates the relationship between two variables, and a linear relationship exists if the data points of the scatter diagram appear to cluster around a straight line. We now develop methods to identify the line which comes "closest" to all the data points, that "best" summarizes the apparent relationship between X and Y. This line is called the regression line, *and its equation is the* regression equation. *With this equation, we can predict the value Y_0 corresponding to a given value X_0. (See Figure 13.1.)*

*Sir Francis Galton, *Regression toward Mediocrity in Hereditary Stature*, 1885.

FIGURE 13.1. Regression line and its equation.

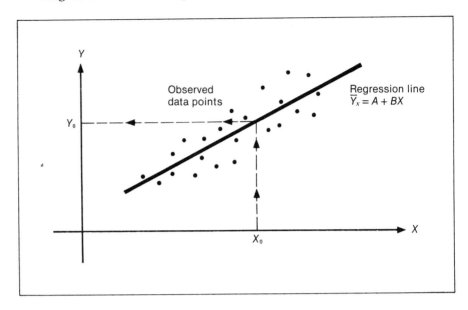

In regression, we develop an equation for the purpose of prediction. When we predict the value of a single dependent variable Y from one independent variable X, and the relationship between them is assumed to be linear, the equation is of the form

$$\bar{Y}_X = A + BX,$$

where $\bar{Y}_x$ is the predicted value of Y corresponding to X, B is the slope of the regression line, and A is the y-intercept of the regression line. This is called *simple linear regression.*

We may wish to incorporate more than one independent variable into our prediction of the dependent variable Y. When the relationship between each independent variable and Y is linear, this process is called *multiple linear regression*, and the regression equation is of the form

$$Y' = b_1 X_1 + b_2 X_2 + \ldots + b_k X_k + b_o,$$

where $b_1, b_2, \ldots, b_k$ are the coefficients of the independent variables $X_1, X_2, \ldots, X_k$, and b_o is the constant term.

Regression analyses serve three purposes in statistical research: *description, statistical control,* and *prediction.* In some situations, previously collected data are examined to discover the relationship between some dependent quantity and one or more independent variables. Here description is the goal.

In a study of the relationship between two variables, the researcher can use regression analysis to statistically control the influence of some other quantity. For example, in a study of two teaching methods, tests might be administered after each course of study. Regression equations can be found to relate the test scores to students' IQ's, and these equations can be used to adjust the scores, inflating those of students of low IQ, deflating those of students of high IQ. Remaining differences in test scores are then due to differences in the two teaching methods.

Finally, a regression equation can be used to predict the value of a variable from the value of some other quantity, as in predicting college grade point averages (GPA) from high school GPA.

Simple Linear Regression

We begin with the dependent variable related linearly to one independent variable. For example, suppose that over the past 12 quarters, a small business has spent the amounts shown in Table 13.1 on advertising and has achieved the corresponding sales; both are given in thousands of dollars.

TABLE 13.1.   **Advertising costs and corresponding sales of a small business.**

QUARTER	ADVERTISING (in $1000s)	SALES (in $1000s)
1	3.2	30
2	3.9	42
3	4.2	36
4	4.2	46
5	4.7	43
6	4.6	51
7	5.2	55
8	4.7	55
9	5.8	60
10	5.6	54
11	5.6	59
12	6.3	60

The scatter diagram of Figure 13.2 illustrates that a linear relationship appears to hold between the two variables; we now seek an equation with which to describe the relationship, the equation of a line "close" to all the data points.

In general, when the dependent variable is linearly related to one independent variable, the model which describes the relationship has this form:

$$Y_i = \beta_o + \beta_1 X_i + \epsilon_i,$$

where Y_i is the ith observation of the dependent variable, X_i is the ith observation of the independent variable, β_o and β_1 are the parameters of the model, and ϵ_i is the random error of Y_i.

We will estimate β_o and β_1, and thereby arrive at an equation relating X and Y which will allow us to predict Y, given a value of X. In developing these two estimates, we make the following four assumptions:

FIGURE 13.2.   Scatter diagram showing a linear relationship between two variables.

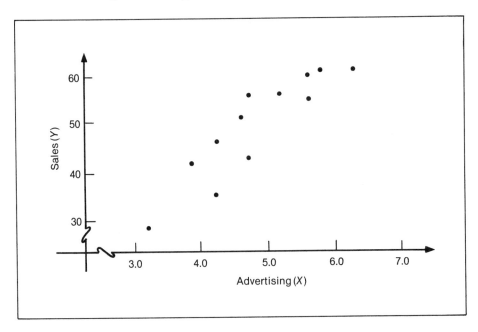

1. X is not a random variable; that is, as the independent variable, X is thought of as influencing the value of Y, and is often under the control of the researcher.

2. $E(\epsilon_i) = 0$. The expected value of each error is 0. Therefore,

$$E(Y_i) = E(\beta_o + \beta_1 X_i + \epsilon_i)$$

$$= E(\beta_o) + E(\beta_1 X_i) + E(\epsilon_i)$$

$$= \beta_o + \beta_1 X_i.$$

3. The random errors are independent; the magnitude of ϵ_i has no effect on ϵ_j when $i \neq j$.

4. The random errors have common variance σ^2.

$$\text{Then } \sigma^2(Y_i) = \sigma^2(\beta_o + \beta_1 X_i + \epsilon_i)$$

$$= \sigma^2(\epsilon_i)$$

$$= \sigma^2.$$

That is, each Y_i comes from a probability distribution with mean $E(Y_i) = \beta_o + \beta_1 X_i$ and variance σ^2, and the Y_i are independent of each other.

The line corresponding to the equation $Y = \beta_o + \beta_1 X$ is sometimes called the *true regression line*, since it represents the true relationship between the independent and the dependent variables.

Estimating β_o and β_1 is the same as estimating the true regression line. It seems reasonable that the line which best matches the sample data points in the scatter diagram will also be the best estimate of the true regression line. Therefore, we seek A and B, estimates of β_o and β_1, respectively, so that the line $\bar{Y}_x = A + BX$ matches as closely as possible the data points of the scatter diagram (Figure 13.3).

We want to find A and B to minimize the errors between the Y_i and the corresponding values indicated by the regression line, $\bar{Y}_i = A + BX_i$. Some of these errors are positive and some negative; if we sum these errors they will tend to cancel out, so that this sum is not a useful measure of the total error:

$$\sum_{i=1}^{n} (Y_i - \bar{Y}_i) = \sum_{i=1}^{n} (Y_i - (A + BX_i)).$$

To provide a useful measure of the total error between the regression line and the observed data points, we square each of the individual error terms indicated above, then sum over all the data points. This gives us the *error sum of squares*, ESS:

$$\text{ESS} = \sum_{i=1}^{n} (Y_i - \bar{Y}_i)^2 = \sum_{i=1}^{n} (Y_i - (A + BX_i))^2 = \sum_{i=1}^{n} (Y_i - A - BX_i)^2$$

$$= \sum_{i=1}^{n} (Y_i^2 - 2AY_i - 2BX_iY_i + A^2 + 2ABX_i + B^2X_i^2).$$

We find the values of A and B which minimize the error sum of squares. This is the *method of least squares*, and the line found in this way, whose equation is $\bar{Y}_x = A + BX$, is the *least squares regression line*.

FIGURE 13.3. Matching the regression line to the data points.

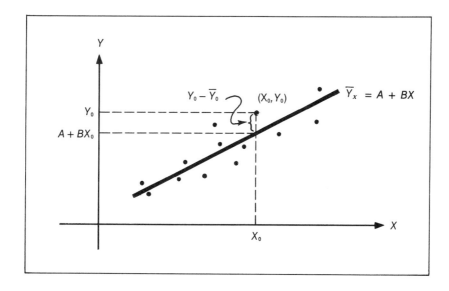

In calculus, the location of a minimum or maximum value of a function of one variable is found by setting the derivative of the function equal to zero and solving the resulting equation. Since the ESS is a function of both A and B, finding the location of its minimum value requires the simultaneous solution of the system of two equations obtained by setting equal to zero the two partial derivatives of ESS with respect to A and to B. Keep in mind that when taking the partial derivative of a function with respect to one variable, all other variables are treated as constants, and that the derivative of a sum or difference is the sum or difference of the derivatives.

We take the partial derivative of ESS with respect to A:

$$\frac{\delta \text{ESS}}{\delta A} = \sum_{i=1}^{n} (0 - 2Y_i - 0 - 2A + 2BX_i + 0)$$

$$= \sum_{i=1}^{n} (-2Y_i + 2A + 2BX_i)$$

$$= -2 \sum_{i=1}^{n} Y_i + 2nA + 2B \sum_{i=1}^{n} X_i.$$

Next, the partial derivative of ESS with respect to B:

$$\frac{\delta \text{ESS}}{\delta B} = \sum_{i=1}^{n} (0 - 0 - 2X_iY_i + 0 + 2AX_i + 2BX_i^2)$$

$$= \sum_{i=1}^{n} (-2X_iY_i + 2AX_i + 2BX_i^2)$$

$$= -2 \sum_{i=1}^{n} X_iY_i + 2A \sum_{i=1}^{n} X_i + 2B \sum_{i=1}^{n} X_i^2.$$

Setting the two partial derivatives equal to zero, we derive the *normal equations*:

$$\frac{\delta \text{ESS}}{\delta A} = 0 \qquad\qquad \frac{\delta \text{ESS}}{\delta B} = 0$$

$$\rightarrow -2 \sum_{i=1}^{n} Y_i + 2nA + 2B \sum_{i=1}^{n} = 0; \quad \rightarrow -2 \sum_{i=1}^{n} X_iY_i + 2A \sum_{i=1}^{n} X_i + 2B \sum_{i=1}^{n} X_i^2 = 0$$

$$\rightarrow \sum_{i=1}^{n} Y_i = nA + B \sum_{i=1}^{n} X_i \quad ; \quad \rightarrow \sum_{i=1}^{n} X_iY_i = A \sum_{i=1}^{n} + B \sum_{i=1}^{n} X_i^2.$$

We have two linear equations in the two unknowns A and B. Solving them simultaneously (as you are asked to do in Exercise 1), we obtain these formulas for the regression coefficients:

$$B = \frac{n \left(\sum\limits_{i=1}^{n} X_i Y_i \right) - \left(\sum\limits_{i=1}^{n} X_i \right) \left(\sum\limits_{i=1}^{n} Y_i \right)}{n \sum\limits_{i=1}^{n} X_i^2 - \left(\sum\limits_{i=1}^{n} X_i \right)^2},$$

$$A = \bar{Y} - B\bar{X}.$$

Second-order partial derivatives can be used to show that the error sum of squares attains a minimum value with these expressions for A and B, and that the line whose equation is $\bar{Y}_X = A + BX$ therefore comes closest to all the data points, as measured by the ESS.

In our advertising and sales example, the following sums and means can be computed:

$$\sum_{i=1}^{12} X_i = 58.0 \qquad \sum_{i=1}^{12} X_i^2 = 289.160 \qquad \bar{X} = 4.033,$$

$$\sum_{i=1}^{12} Y_i = 591.0 \qquad \sum_{i=1}^{12} X_i Y_i = 2944.20 \qquad \bar{Y} = 49.250.$$

The regression coefficients are then:

$$B = \frac{12 \times 2944.20 - 58.0 \times 591.0}{12 \times 289.16 - (58.0)^2} \doteq 9.936$$

$$A = 49.250 - 9.936 \times 4.833 \doteq 1.229,$$

and the least squares regression equation is

$$\bar{Y}_X = 1.229 + 9.936X.$$

The regression equation can now be used to predict the value of $\bar{Y}_X$ which corresponds to a given value of X; that is, to estimate the expected value of Y for a given X. For example, if the business spends $5,300 on advertising in a given quarter, then they can expect sales of $53,890 in that quarter:

$$\bar{Y}_{5.3} = 1.229 + 9.936 \times 5.3 \doteq 53.89.$$

The computed regression equation is an estimate of the true regression equation, and the coefficients A and B are estimates of β_o and β_1, respectively. By assuming that the random errors ϵ_i are normally distributed, the Y_i and therefore A and B are random variables, and we can describe the sampling distributions of A and B, which are statistics of the sample of data pairs. First, however, we consider a related quantity.

We assume that the random error terms ϵ_i, and therefore the Y_i, have equal variance σ^2. We can estimate σ^2 by using the squared deviations of the Y_i, not from their mean $\bar{Y}$, but from the regression line. The deviations are $(Y_i - \bar{Y}_i) = (Y_i - (A + BX_i))$, and the estimator of σ^2 is:

$$\frac{\sum\limits_{i=1}^{n} (Y_i - \bar{Y}_i)^2}{(n-2)} = \frac{\sum\limits_{i=1}^{n} (Y_i - A + BX_i)^2}{(n-2)} = \frac{ESS}{n-2}.$$

Note that the numerator is the error sum of squares which we minimized to find A and B, and that we divide by $(n-2)$ in order that this quantity be an unbiased estimator of σ^2; A and B are being used to estimate β_o and β_1.

More often used than this variance estimator is its square root s_e, the *standard error of estimate:*

$$s_e = \left[\frac{\sum\limits_{i=1}^{n} (Y_i - \bar{Y}_i)^2}{(n-2)} \right]^{1/2} = \left[\frac{\sum\limits_{i=1}^{n} Y_i^2 - A \sum\limits_{i=1}^{n} Y_i - B \sum\limits_{i=1}^{n} X_i Y_i}{(n-2)} \right]^{1/2}.$$

s_e estimates the standard deviation of the Y distribution corresponding to each possible X value, and is more commonly calculated by the second formula shown above. It is a measure of the dispersion of the data points around the regression line. (See Figure 13.4.)

In our example, we need only find $\sum_{i=1}^{12} Y_i^2$ to compute the standard error of estimate:

$$\sum\limits_{i=1}^{12} Y_i^2 = 30173.0 .$$

Then

$$s_e = \frac{30173 - 1.229 \times 591 - 9.936 \times 2944.2}{12 - 2} \doteq 4.4.$$

FIGURE 13.4   The standard error of estimate s_e.

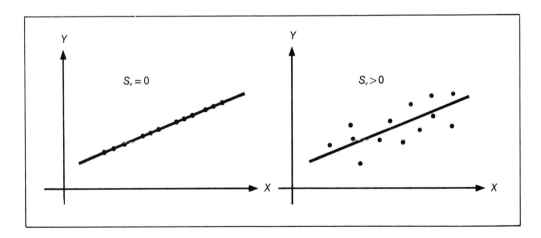

We can now describe the sampling distribution of B, the slope of the regression line.

If the ϵ_i are normally distributed, then since β_o, β_1, and X_i are all constants, the $\bar{Y}_i = \beta_o + \beta_1 X_i + \epsilon_i$ are also normally distributed. B can be written in this way:

$$B = \frac{\sum_{i=1}^{n} (X_i - \bar{X})Y_i}{\sum_{i=1}^{n} (X_i - \bar{X})^2} .$$

So if we let

$$k_i = \frac{(X_i - \bar{X})}{\sum_{i=1}^{n} (X_i - \bar{X})^2} , \text{ then } B = \sum_{i=1}^{n} k_i Y_i.$$

But each k_i is a constant; B is a linear combination of independent normal random variables (the Y_i), and must itself be normally distributed, as we pointed out in Chapter 7 (p. 147).

Further,

$$E(B) = E\left(\sum_{i=1}^{n} k_i Y_i\right) = \sum_{i=1}^{n} k_i E(Y_i) = \sum_{i=1}^{n} k_i(\beta_o + \beta_1 X_i)$$

$$= \beta_o\left(\sum_{i=1}^{n} k_i\right) + \beta_1 \left(\sum_{i=1}^{n} k_i X_i.\right)$$

But

$$\sum_{i=1}^{n} k_i = \left(\sum_{j=1}^{n} \frac{X_i - \bar{X}}{\sum_{i=1}^{n} (X_i - \bar{X})^2}\right) = \frac{\sum_{i=1}^{n} (X_i - \bar{X})}{\sum_{i=1}^{n} (X_i - \bar{X})^2} = 0,$$

and

$$\sum_{i=1}^{n} k_i X_i = \left(\sum_{j=1}^{n} \frac{(X_i - \bar{X})X_i}{\sum_{i=1}^{n} (X_i - \bar{X})^2}\right)$$

$$= \frac{\sum_{i=1}^{n} (X_i - \bar{X})X_i}{\sum_{i=1}^{n} (X_i - \bar{X})^2}$$

$$= \frac{\sum_{i=1}^{n} (X_i - \bar{X})(X_i - \bar{X})}{\sum_{i=1}^{n} (X_i - \bar{X})^2} = 1,$$

since

$$\sum_{i=1}^{n} (X_i - \bar{X})X_i = \sum_{i=1}^{n} (X_i - \bar{X})^2. \text{ (See Exercise 2.)}$$

Then $E(B) = \beta_o \times 0 + \beta_1 \times 1 = \beta_1$

Also,

$$\sigma^2(B) = \sigma^2(\sum_{i=1}^{n} k_i Y_i) = \sum_{i=1}^{n} k_i^2 \sigma^2(Y_i) = \sum_{i=1}^{n} k_i^2 \sigma^2$$

$$= \sigma^2 \sum_{i=1}^{n} k_i^2 = \sigma^2 \sum_{i=1}^{n} \left[\frac{(X_i - \bar{X})}{\sum_{i=1}^{n} (X_i - \bar{X})^2} \right]^2$$

$$= \sigma^2 \frac{\sum_{i=1}^{n} (X_i - \bar{X})^2}{\left[\sum_{i=1}^{n} (X_i - \bar{X})^2 \right]^2} = \frac{\sigma^2}{\sum_{i=1}^{n} (X_i - \bar{X})^2},$$

where σ^2 is the common variance of the Y_i.

To summarize, the sampling distribution of B is normal with mean β_1 and variance

$$\sigma_B^2 = \frac{\sigma^2}{\sum_{i=1}^{n} (X_i - \bar{X})^2}.$$

Not only is B an unbiased estimator of β_1, but it can also be shown that as an estimator of β_1, B is sufficient and of minimum variance.

The statistic

$$z = \frac{B - \beta_{1_o}}{\sigma_B} \text{ or } t = \frac{B - \beta_{1_o}}{\sigma_B} \text{ (with } n - 2 \text{ degrees of freedom)}$$

can be used, depending on the sample size, to test the hypothesis $H_o: \beta_1 = \beta_{1_o}$. Since

$$\sigma_B = \frac{\sigma}{\left[\sum_{i=1}^{n} (X_i - \bar{X})^2 \right]^{1/2}}$$

and s_e is an estimate of σ, the statistic is usually calculated according to this formula:

$$z \text{ or } t = \frac{B - \beta_{1_o}}{s_e} \left(\sum_{i=1}^{n} X_i^2 - n\bar{X}^2 \right)^{1/2}.$$

When the slope of the regression line is zero, no relationship between the two variables is indicated. Thus, the hypothesis test most often performed with B is this:

$$H_o: \beta_1 = 0$$

$$H_a: \beta_1 \neq 0.$$

The null hypothesis says that there is no linear relationship between the variables; in attempting to accept the alternative, we hope to show that such a (numerical) relationship exists.

For example, to perform such a test with our advertising and sales data at the 5% significance level, the critical values corresponding to $12 - 2 = 10$ degrees of freedom are $\pm t_{.025} = \pm 2.228$. We have found that $B = 9.936$ and $S_e = 4.4$, so the value of the t statistic is

$$t = \frac{B - 0}{S_e} \left(\sum_{i=1}^{n} X_i^2 - n\bar{X}^2 \right)^{1/2} = \frac{9.936}{4.4} \left(289.16 - 12 \times 4.833^2 \right)^{1/2} = 6.72.$$

This value is beyond the critical values, so we can reject the null hypothesis and conclude, based on our sample data, that there is a linear relationship between advertising and sales for the business. Again, be aware that the existence of a significant *numerical* relationship does not necessarily imply a causal, physical relationship between the quantities represented. The existence of a causal relationship must be investigated through qualitative rather than quantitative methods.

Of the two regression coefficients, β_1 is the more important, since it indicates the expected change in Y corresponding to a unit change in X. Inferences about β_o, based on its estimator A, can also be useful, however $\bar{X}$ and $\bar{Y}$ are constants, and $A = \bar{Y} - B\bar{X}$, so since the sampling distribution of B is normal (or a t distribution for small sample sizes), that of A is also. It can further be shown that

$$E(A) = \beta_o, \; j_A^2 = \sigma^2 \left[\frac{1}{n} + \frac{\bar{X}^2}{\sum_{i=1}^{n} (X_i - \bar{X})^2} \right],$$

and that A is a sufficient, minimum variance estimator of β_o. The test statistic z or t (with $n - 2$ degrees of freedom) for tests on β_o is

$$z \text{ or } t = \frac{A - \beta_{oo}}{S_e \left(\frac{1}{n} + \frac{\bar{X}^2}{\sum_{i=1}^{n} X_i^2 - n\bar{X}^2} \right)^{1/2}}.$$

The fact that A and B are "good" estimators of β_o and β_1 supports the claim that the least squares regression line is a good estimator of the true regression line. Still, how accurate are estimates of Y corresponding to a particular value of X? To investigate this question, we examine the sampling distribution of $\bar{Y}_o$, the estimated expected value of Y corresponding to some value of X, X_o. (The true mean of the distribution of Y values corresponding to X_o is the *conditional mean* $\mu_{X_o} = \beta_o + \beta_1 X_o$. It is this quantity of which $\bar{Y}_o$ is an estimate.)

First,

$$E(\bar{Y}_o) = E(A + BX_o) = E(A) + E(BX_o)$$

$$= \beta_o + X_o E(B) = \beta_o + \beta_1 X_o = \mu_{X_o}.$$

Second,

$$\sigma^2(\bar{Y}_o) = \sigma^2(A + BX_o) = \sigma^2((\bar{Y} - B\bar{Y}) + BX_o)$$
$$= \sigma^2(\bar{Y} + B(X_o - \bar{X}))$$
$$= \sigma^2(\bar{Y}) + (X_o - \bar{X})^2\, \sigma^2(B),$$

since $\bar{Y}$ and B are independent. But $\bar{Y}$ is simply the mean of the Y_i; if σ^2 is the common variance of the Y distributions, then $\sigma^2(\bar{Y}) = \sigma^2/n$, so that

$$\sigma^2(\bar{Y}_o) = \frac{\sigma^2}{n} + \frac{(\bar{X} - X_o)^2\sigma^2}{\sum_{i=1}^{n}(X_i - \bar{X})^2},$$

since

$$\sigma^2(B) = \frac{\sigma^2}{\sum_{i=1}^{n}(X_i - \bar{X})^2}.$$

$\sigma(\bar{Y}_o)$ is called the *mean-square error* (MSE), and is often written this way:

$$MSE = \left[\frac{\sigma^2}{n} + \frac{(X_i - X_o)^2\sigma^2}{\sum_{i=1}^{n}X_i^2 - n\bar{X}^2}\right]^{1/2} = \sigma\left[\frac{1}{n} + \frac{(\bar{X} - X_o)^2}{\sum_{i=1}^{n}X_i^2 - n\bar{X}^2}\right]^{1/2}.$$

Again, σ can be estimated by the standard error of estimate s_e.

Finally, the sampling distributions for large n of both A and B are normal; so is the sampling distribution of $\bar{Y}_{X_o} = A + BX_o$. We can now construct confidence intervals for the conditional mean μ_{X_o} corresponding to X_o, and perform tests of the hypothesis $H_o: \mu_{X_o} = \mu_{X_{oo}}$ using, depending on the sample size, the statistic

$$z = \frac{\bar{Y}_{X_o} - \mu_{X_o}}{MSE} \text{ or } t = \frac{\bar{Y}_{X_o} - \mu_{X_o}}{MSE} \text{ (with } n - 2 \text{ degrees of freedom).}$$

Note that since the accuracy of estimates of conditional means depends on the MSE, we would like this quantity to be as small as possible, which can be accomplished by making n large.

SPSS AND SIMPLE LINEAR REGRESSION

The astute reader will have noticed long since that all the statistics discussed above and their significances are calculated by the SPSS subprogram SCATTERGRAM. The following program plots the scatter diagram and performs those calculations for the example we have used throughout this chapter. Again, the variable NUM is the line number required in constructing a raw data file. The only unexplained value on the printout is R SQUARED, which we will discuss in the next section.

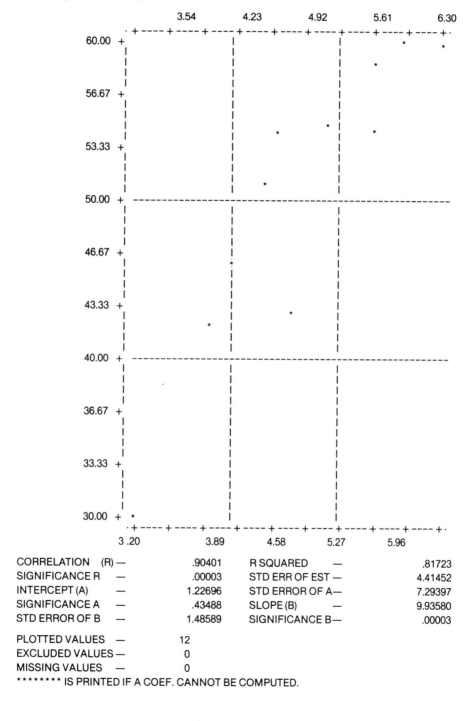

— — — SCATTERGRAM — — —

GIVEN   2   VARIABLES,   MAXIMUM CM ALLOWS FOR 7648 CASES

(DOWN) Y   SALES
(ACROSS) X   ADVERTISING

CORRELATION (R) —	.90401	R SQUARED —	.81723
SIGNIFICANCE R —	.00003	STD ERR OF EST —	4.41452
INTERCEPT (A) —	1.22696	STD ERROR OF A—	7.29397
SIGNIFICANCE A —	.43488	SLOPE (B) —	9.93580
STD ERROR OF B —	1.48589	SIGNIFICANCE B—	.00003

PLOTTED VALUES —	12
EXCLUDED VALUES —	0
MISSING VALUES —	0

* * * * * * * * IS PRINTED IF A COEF. CANNOT BE COMPUTED.

```
10.     FILE NAME
10.005 ADSALES
20.     VARIABLE LIST
20.005 NUM,X,Y
30.     INPUT FORMAT
30.005 FREEFIELD
40.     N OF CASES
40.005 12
50.     VAR LABELS
50.005 X,ADVERTISING/Y,SALES
60.     SCATTERGRAM
60.005 Y WITH X
70.     STATISTICS
70.005 ALL
```

In the output, the values of the statistics agree with those found in our earlier work; differences are due to rounding. The significance of A is given for the test

$$H_o: \beta_o = 0$$

$$H_a: \beta_o \neq 0.$$

The Coefficient of Determination

From the data pairs graphed on the scatter diagram we can derive another measure of the predictive accuracy of the regression line by considering the difference between each Y_i value and the mean of the Y_i, $\bar{Y}$.

As Figure 13.5 shows, the difference between the observed value Y_i and the mean $\bar{Y}$ of the Y_i values is the sum of two differences:

$$Y_i - \bar{Y} = (Y_i - (A + BX_i)) + ((A + BX_i) - \bar{Y}).$$

Y_i differs from $\bar{Y}$ because it is partly determined by its corresponding X value $((A + BX_i) - \bar{Y})$, and because of the variability of the Y's $(Y_i - (A + BX_i))$. Part of the variation of Y_i from $\bar{Y}$ can be explained by the regression line, and part cannot.

We can obtain an equation that considers the variations of all the Y values represented in the scatter diagram by squaring both sides of the above statement and summing over all the data points. Note the distinction between $\bar{Y}$, the mean of the Y values, and $\bar{Y}_i$, the estimated mean of the Y distribution corresponding to X_i.

$$\sum_{i=1}^{n} (Y_i - \bar{Y})^2 = \sum_{i=1}^{n} [(Y_i - \bar{Y}_i) + (\bar{Y}_i - \bar{Y})]^2$$

$$= \sum_{i=1}^{n} (Y_i - \bar{Y}_i)^2 + 2\sum_{i=1}^{n} (Y_i - \bar{Y}_i)(\bar{Y}_i - \bar{Y}) + \sum_{i=1}^{n} (\bar{Y}_i - \bar{Y})^2$$

$$= \sum_{i=1}^{n} (Y_i - \bar{Y}_i)^2 + 2\sum_{i=1}^{n} \bar{Y}_i(Y_i - \bar{Y}_i) - 2\bar{Y} \sum_{i=1}^{n} (Y_i - \bar{Y}_i) + \sum_{i=1}^{n} (Y_i - \bar{Y})^2.$$

But

$$\sum_{i=1}^{n} \bar{Y_i}(Y_i - \bar{Y_i}) = 0, \text{ and } \sum_{i=1}^{n} (Y_i - \bar{Y_i}) = 0, \text{ so}$$

$$\sum_{i=1}^{n} (Y_i - \bar{Y})^2 = \sum_{i=1}^{n} (Y_i - \bar{Y_i})^2 + \sum_{i=1}^{n} (Y_i - \bar{Y})^2.$$

The sum on the left is called the *total sum of squares*, and is a measure of the total variability of the observed Y values about $\bar{Y}$. [Note that $\sum_{i=1}^{n}(Y_i - \bar{Y})^2 = (n - 1) \text{Var}(Y)$.] The sum on the right, the *regression sum of squares*, is a measure of the variability of the Y values that can be explained by the regression line, which can be thought of as being due to the influence of the X values. And the center sum, the *residual sum of squares*, measures the unexplained variation that is not due to the X values.

Consider the regression sum of squares (RSS):

$$\text{RSS} = \sum_{i=1}^{n} (\bar{Y_i} - \bar{Y})^2 = \sum_{i=1}^{n} (A + BX_i - \bar{Y})^2$$

$$= \sum_{i=1}^{n} (\bar{Y} - B\bar{X} + BX_i - \bar{Y})^2$$

$$= \sum_{i=1}^{n} (BX_i - B\bar{X})^2$$

$$= B^2 \sum_{i=1}^{n} (X_i - \bar{X})^2$$

$$= \left[\frac{\sum_{i=1}^{n} (X_i - \bar{X})(Y_i - \bar{Y})}{\sum_{i=1}^{n} (X_i - \bar{X})^2} \right]^2 \sum_{i=1}^{n} (X_i - \bar{X})^2$$

$$= \frac{\left[\sum_{i=1}^{n} (X_i - \bar{X})(Y_i - \bar{Y}) \right]^2}{\sum_{i=1}^{n} (X_i - \bar{X})^2}$$

$$= R^2 \sum_{i=1}^{n} (Y_i - \bar{Y})^2,$$

where R is the Pearson correlation coefficient.

Substituting this quantity into the original statement relating the three sums of squares and dividing the resulting expression by the total sum of squares initiates this development:

FIGURE 13.5. The components of the difference $Y_i - \bar{Y}$.

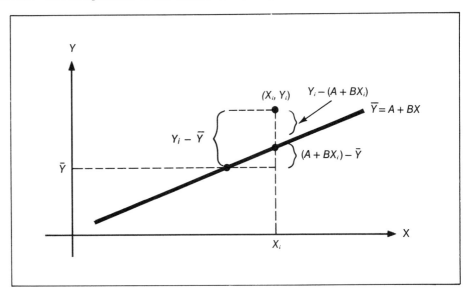

$$\sum_{i=1}^{n} (Y_i - \bar{Y})^2 = \sum_{i=1}^{n} (Y_i - \bar{Y}_i)^2 + R^2 \sum_{i=1}^{n} (Y_i - \bar{Y})^2$$

$$\rightarrow 1 = \frac{\sum_{i=1}^{n} (Y_i - \bar{Y}_i)^2}{\sum_{i=1}^{n} (Y_i - \bar{Y})^2} + R^2$$

$$\rightarrow R^2 = 1 - \frac{\sum_{i=1}^{n} (Y_i - \bar{Y}_i)^2}{\sum_{i=1}^{n} (Y_i - \bar{Y})^2} .$$

The quantity R^2, which is the square of the Pearson product-moment correlation coefficient R, is called the *coefficient of determination*. It is the proportion of the variability of the Y values which can be related to variation in the X values; R^2 measures the reduction in the sum of squares of deviations of the Y values from $\bar{Y}$ which can be achieved by considering their linear relationship to the X values. R^2 is often calculated using this formula:

$$R^2 = \frac{A\left(\sum_{i=1}^{n} Y_i\right) + B\left(\sum_{i=1}^{n} X_i Y_i\right) - n\bar{Y}^2}{\left(\sum_{i=1}^{n} Y_i^2\right) - n\bar{Y}^2} .$$

On the SPSS SCATTERGRAM printout, the coefficient of determination is identified as R SQUARED. Inferences involving R^2 may be drawn using the statistic $t = [R(n - 2)/(1 - R^2)]^{1/2}$ as discussed in Chapter 12 (p.250), and the significance of R^2 is the same as that of R.

In our example, the coefficient of determination is 0.87123, indicating that 87% of the variation of the Y values from their mean can be related to variation in the X values.

Observations about Linear Regression

Several observations should be made about linear regression:

1. Valid predictions of $\bar{Y}_i$ are based on the assumption that future observations of X and Y will follow the pattern observed in the past.

2. Do not attempt to predict Y from values of X beyond those examined when you are constructing the regression equation. There is no guarantee that the linear relationship observed in the study persists outside its bounds.

3. In practice, the value of X from which Y is estimated is often itself an estimate. This introduces another element of uncertainty.

4. In the z_X, z_Y plane, the regression equation may be written as $z_Y = Rz_X$.

5. Since $B = (S_Y/S_X)R$, when B is zero, so must be R. Thus when R is zero, the regression line is horizontal, and the significances of R, R^2, and B are always the same.

Multiple Linear Regression

In many situations, a dependent variable will be influenced by several independent variables rather than just one. For example, the mileage of a car will be influenced by its weight, engine displacement, horsepower, rolling resistance, axle ratio, and other factors.

When the relationships between the independent variables and the dependent variable are linear, we examine such a situation through **multiple linear regression analysis**, and we measure the collective relationship between the dependent variable and all the independent variables with R^2, the square of the multiple correlation coefficient R. The multiple correlation coefficient is the Pearson correlation coefficient of the observed values of the dependent variable with those predicted by the regression model.

Assuming, as we did with simple linear regression, that the error terms are normally distributed, the multiple linear regression model has this form:

$$Y_i = \sum_{j=1}^{p} \beta_j X_{ij} + \beta_o + \epsilon_i,$$

where $X_{i1}, X_{i2}, \ldots, X_{ip}$ are the values of the p independent variables (or *predictors*) for the ith case; Y_i is the observed value of the dependent variable; $\beta_1, \beta_2, \ldots, \beta_p$ are the parameters associated with the predictors; β_o is the constant term; ϵ_i is a normally distributed error term with mean 0 and variance σ^2; and i is the number of the case. The estimate of Y_i based on the sample is

$$E(Y_i) = \beta_o + \sum_{j=1}^{p} \beta_j X_{ij}.$$

We wish to construct an estimate of the true model based upon sample data; that is, we wish to estimate the regression coefficients (sometimes called *weights*) $\beta_o, \beta_1, \ldots, \beta_p$. The method used with simple linear regression to find a least squares solution can be extended to this larger situation, but such an approach creates enormously complex computations. A similar result can be achieved by using a matrix algebra notation which illustrates the concepts involved while avoiding tedious manipulations. It too provides a least squares solution.

We begin by defining the matrices

$$Y = \begin{bmatrix} Y_1 \\ Y_2 \\ \cdot \\ \cdot \\ \cdot \\ Y_p \end{bmatrix}, \beta = \begin{bmatrix} \beta_o \\ \beta_1 \\ \cdot \\ \cdot \\ \cdot \\ \beta_p \end{bmatrix}, \epsilon = \begin{bmatrix} \epsilon_1 \\ \epsilon_2 \\ \cdot \\ \cdot \\ \cdot \\ \epsilon_p \end{bmatrix}, \text{ and}$$

$$X = \begin{bmatrix} 1 & X_{1,1} & X_{1,2} & \ldots & X_{1,p} \\ 1 & X_{2,1} & X_{2,2} & \ldots & X_{2,p} \\ \cdot & & & & \cdot \\ \cdot & & & & \cdot \\ \cdot & & & & \cdot \\ 1 & X_{n,1} & X_{n,2} & \ldots & X_{n,p} \end{bmatrix},$$

in which the various quantities are as defined previously. The general linear model can then be written as

$$\mathbf{Y} = X\beta\} + \epsilon,$$

where $\mathbf{Y}$ is the vector of dependent variable observations; X is the $n \times (p + 1)$ matrix of independent variable observations (n cases by p predictors and the constant term); β is the vector of beta coefficients; and ϵ is the vector of independent normally and identically distributed errors.

Then $E(Y) = X\beta$, since for each ϵ_i, $E(\epsilon_i) = 0$. Now, let

$$\mathbf{b} = \begin{bmatrix} b_o \\ b_1 \\ \cdot \\ \cdot \\ \cdot \\ b_p \end{bmatrix} \text{ be the estimator vector for } \beta = \begin{bmatrix} \beta_0 \\ \beta_1 \\ \cdot \\ \cdot \\ \cdot \\ \beta_p \end{bmatrix}.$$

Then $\mathbf{Y}' = E(\mathbf{Y}) = \mathbf{Xb}$. Using matrix algebra, we solve for $\mathbf{b}$, multiplying on the left first by the transpose of $\mathbf{X}$, then by $(\mathbf{X}^{TR}\mathbf{X})^{-1}$. Recall that matrix multiplication is associative but not commutative.

$$X\mathbf{b} = \mathbf{Y}' \rightarrow (X^{TR}X)\mathbf{b} = X^{TR}\mathbf{Y}' \rightarrow \mathbf{b} = (X^{TR}_{X)^{-1}(X}{}^{TR}\mathbf{Y}'),$$

where X^{Tr} is the transpose of X and $(X^{TR}X)^{-1}$ is the inverse of the $(p+1) \times (p + 1)$ matrix $(X^{TR}X)$.*

---

* If a_{ij} are the elements of the $m \times n$ matrix A, then a_{ji} are the elements of the $n \times m$ matrix A^{Tr}. For example, if

$$A = \begin{pmatrix} 3 & -1 & 6 \\ 5 & 0 & 2 \end{pmatrix}, \text{ then } A^{Tr} = \begin{pmatrix} 3 & 5 \\ -1 & 0 \\ 6 & 2 \end{pmatrix}$$

As in simple linear regression, each element of the estimator vector **b** is an unbiased, sufficient, minimum variance, consistent estimator of the corresponding element of β.

Rather than solving the matrix equation developed above, many statistical software packages employ iterative techniques to find **b**. Such approaches are systematic, repetitive, easily programmed, and well-suited to the capabilities of a computer.

The Kelley-Salisbury technique is one of these. In it, successive approximations to β_z in the equation

$$R\beta_z = \mathbf{V}$$

are found until a satisfactory solution is obtained. **V** is the vector of *validities*, Pearson correlations between each of the predictor variables and the dependent variable, and R is the *intercorrelation matrix* of the predictors, the matrix of correlations of the independent variables with each other.

The vector β_z found by this method is not the same β discussed earlier; it is the vector of coefficients in the equation

$$z_Y = \sum_{j=1}^{p} \beta_{zj}\, z_{X_{ij}},$$

the *normalized* solution of the multiple regression problem. We can use the facts that

$$z_{Y'} = \frac{Y' - \bar{Y}'}{S_{Y'}} \text{ and } z_{X_i} = \frac{X_i - \bar{X}_i}{S_{X_i}} \text{ for } i = 1,2,\ldots,p,$$

to find the vector **b** of weights in this form of the solution:

$$\mathbf{Y}' = b_o + \sum_{j=1}^{p} b_j X_{ij}.$$

Consider the following example, taken from a text by Walker and Lev* and based on a sample of $n = 36$ cases. Given the intercorrelation matrix for a dependent variable Y and three predictors X_1, X_2, X_3, we seek the least squares multiple linear regression equation. The correlations of all the variables are given in Table 13.2.

TABLE 13.2.   Intercorrelation matrix.

	Y	X_1	X_2	X_3
Y	1.000	0.357	0.620	0.518
X_1	0.357	1.000	0.321	0.477
X_2	0.620	0.321	1.000	0.539
X_3	0.518	0.477	0.539	1.000

---

*Helen M. Walker and Joseph Lev, *Statistical Inference* New York: Holt, Rinehart and Winston, Inc., 1953.

The equation employed in the Kelley-Salisbury technique is this:

$$
\begin{array}{cccc}
R & & \beta_z & V
\end{array}
$$

$$
\begin{bmatrix} 1.000 & 0.321 & 0.477 \\ 0.321 & 1.000 & 0.539 \\ 0.477 & 0.539 & 1.000 \end{bmatrix} \times \begin{bmatrix} \beta_{z1} \\ \beta_{z2} \\ \beta_{z3} \end{bmatrix} = \begin{bmatrix} 0.357 \\ 0.620 \\ 0.518 \end{bmatrix}
$$

In general, the equation $\beta_z = R^{-1}V$ has a solution only if all the predictors are independent of one another. The iterative technique which is the heart of the Kelley-Salisbury method, however, will find a solution even when the predictors are statistically dependent.

We begin by choosing arbitrary values for the vector β_z of beta weights. We then perform one matrix multiplication, $R\beta_z = V_T$, where V_T is a vector of trial validities. The difference of largest magnitude between elements of V and elements of V_T is then added to the corresponding trial beta weight in the vector β_z.

In our example, we set the weights equal to the validities. Our first multiplication is

$$
R\beta_z = \begin{bmatrix} 1.000 & 0.321 & 0.477 \\ 0.321 & 1.000 & 0.539 \\ 0.477 & 0.539 & 1.000 \end{bmatrix} \times \begin{bmatrix} 0.357 \\ 0.620 \\ 0.518 \end{bmatrix} = \begin{bmatrix} 0.803 \\ 1.014 \\ 1.022 \end{bmatrix} = V_T.
$$

Then

$$
V - V_T = \begin{bmatrix} 0.357 \\ 0.620 \\ 0.518 \end{bmatrix} - \begin{bmatrix} 0.803 \\ 1.014 \\ 1.022 \end{bmatrix} = \begin{bmatrix} -0.446 \\ -0.394 \\ -0.504 \end{bmatrix}.
$$

The difference of largest magnitude is $V_3 - V_{T3} = -0.504$, so this amount is added to β_{z3}. The vector of approximate weights is now

$$
\beta_z = \begin{bmatrix} 0.357 \\ 0.620 \\ 0.014 \end{bmatrix}.
$$

This process is repeated, adjusting one element of β_z at each iteration until V_T converges on V. In our example, this conversion occurs on the eighteenth iteration, as detailed in Table 13.3.

In multiple linear regression, the coefficient of determination R^2 is an estimate of the proportion of the variability of the dependent variable that can be attributed to or associated with variation of the predictors, the independent variables. Thus R^2 is a measure of the effectiveness of the multiple regression model.

R^2 can be found with a single matrix multiplication: $R^2 = \beta_z^{Tr}V$. In our example

$$
R^2 = \begin{bmatrix} 0.103 \\ 0.471 \\ 0.215 \end{bmatrix}^{Tr} \times \begin{bmatrix} 0.357 \\ 0.620 \\ 0.518 \end{bmatrix} = \begin{bmatrix} 0.103 & 0.471 & 0.215 \end{bmatrix} \times \begin{bmatrix} 0.357 \\ 0.620 \\ 0.518 \end{bmatrix} = 0.440.
$$

Forty-four percent of the observed variation of the dependent variable can be associated with the predictors X_1, X_2, and X_3.

THE RELATIVE IMPORTANCE OF PREDICTORS

Having found the multiple linear regression equation, we would like to evaluate the relative importances of the independent variables; which predictors have a large influence on the dependent variable, and which are less important?

TABLE 13.3. The iteration of the Kelley-Salisbury method.[a]

APPROXIMATIONS TO β_z			VECTOR OF TRIAL VALIDITIES		
β_1	β_2	β_3	V_1 0.357	V_2 0.620	V_3 0.518
0.357	0.620	0.518	0.803	1.014	1.022
0.357	0.620	0.014	0.563	0.742	0.518
0.151	0.620	0.014	0.357	0.676	0.420
0.151	0.620	0.112	0.403	0.729	0.518
0.151	0.511	0.112	0.368	0.620	0.459
0.151	0.511	0.171	0.397	0.652	0.518
0.111	0.511	0.171	0.357	0.639	0.499
0.111	0.492	0.171	0.350	0.620	0.489
0.111	0.492	0.200	0.364	0.635	0.518
0.111	0.477	0.200	0.360	0.620	0.510
0.111	0.477	0.208	0.363	0.625	0.518
0.105	0.477	0.208	0.357	0.623	0.515
0.105	0.474	0.208	0.356	0.620	0.514
0.105	0.474	0.212	0.358	0.622	0.518
0.105	0.474	0.212	0.358	0.620	0.516
0.105	0.472	0.214	0.359	0.621	0.518
0.103	0.422	0.214	0.3566	0.6204	0.5175
0.103	0.472	0.215	0.3571	0.6209	0.5185
0.103	0.471	0.215	0.3567	0.6199	0.5180

[a]The final values of the weights are $\beta_{z1} = 0.103$, $\beta_{z2} = 0.471$, and $\beta_{z3} = 0.215$.

Bottenberg and Christal have developed a straightforward technique for such evaluations using the multiple coefficient of determination R^2. To discover the importance of one or a group of predictors, the value of R^2 is calculated for the prediction model (called the *full model*) which includes all the predictor variables. Then a second model is constructed (the *reduced model*) from which the predictor or predictors of interest are *excluded*, and the value of R^2 for the reduced model is found.

The decrease in the value of R^2 from the full to the reduced model indicates the importance of the omitted predictors. If the drop in R^2 is large, the reduced model is not as effective as the full model in predicting the value of the dependent variable, and the omitted predictors must be important. If the drop in R^2 is small, the predictive ability of the model is not impaired, and the predictors being examined are unimportant.

A hypothesis test to determine if a drop in R^2 is significant can be performed with this F statistic:

$$F = \frac{(R^2_{FM} - R^2_{RM})/(df_{FM} - df_{RM})}{(1 - R^2_{FM})/(n - df_{FM})},$$

where $R^2_{FM} = R^2$ for the full model, $R^2_{RM} = R^2$ for the reduced model, $df_{FM} = $ (the number of linearly independent * predictors in the full model) -1, $df_{RM} = $ (the number of linearly independent predictors in the reduced model) -1, and $n = $ number of cases.

---

* A group of predictors is linearly dependent if, when one is predicted by the others, the multiple coefficient of determination of the model is 1.

The number of degrees of freedom associated with the numerator of the F statistic is $df_{FM} - df_{RM}$, and with the denominator, $n - df_{FM}$.

If the coefficient of determination of the reduced model is significantly less than that of the full model, the quantity $R^2_{FM} - R^2_{RM}$, and therefore the F statistic itself, will be large; values of F near zero occur when there is little difference between R^2_{FM} and R^2_{RM}. This is an *upper-tail* test of these hypotheses:

$$H_o: R^2_{FM} = R^2_{RM}$$

$$H_a: R^2_{FM} > R^2_{RM} .$$

Consider the reduced model created from the previous example by removing the predictor X_2. To compute R^2 for the reduced model, we perform another multiple linear regression. The intercorrelation matrix without X_2 is

	Y	X_1	X_3
Y	1.000	0.357	0.518
X_1	0.357	1.000	0.477
X_3	0.518	0.477	1.000

and the equation $R \times \beta_z = V$ becomes

$$\begin{bmatrix} 1.000 & 0.477 \\ 0.477 & 1.000 \end{bmatrix} \times \begin{bmatrix} \beta_{z_1} \\ \beta_{z_3} \end{bmatrix} = \begin{bmatrix} 0.357 \\ 0.518 \end{bmatrix} .$$

Computation of the vector β_z is left as an exercise; the values are $\beta_{z_1} = 0.142$ and $\beta_{z_3} = 0.450$, and the coefficient of determination for the reduced model is

$$R^2_{RM} = \beta_z^{Tr} V = [0.142 \quad 0.450] \times \begin{bmatrix} 0.357 \\ 0.618 \end{bmatrix} = 0.284.$$

For the full model, $R^2 = 0.440$. To test the drop in R^2 from the full model to the reduced model at the 5% significance level, we compute the F statistic

$$F = \frac{(R^2_{FM} - R^2_{RM})/(df^2_{FM} - df^2_{RM})}{(1 - R^2_{FM})/(n - df_{FM})} = \frac{(0.440 - 0.284)/(3 - 2)}{(1 - 0.440)/(36 - 3)}$$

$$= \frac{0.156/(3 - 2)}{0.560/(36 - 3)} = 9.47.$$

Corresponding to $(3-2)$ degrees of freedom in the numerator and $(36-3)$ in the denominator, the critical value of the F statistic is $F_{0.05} \doteq 4.17$. The F statistic exceeds the critical value, so the drop in the value of R^2 from the full model to the reduced model is significant at $\alpha = 5\%$; predictor X_2 is important in the regression model.

The Significance of R^2

We can also investigate the significance of the multiple coefficient of determination R^2 with an F test.

As with simple linear regression, the total sum of the squared deviations of the observed values of the dependent variable from the mean value of the dependent variable,

$$SS_{total} = \sum_{i=1}^{n} (Y_i - \bar{Y})^2,$$

is the sum of the regression sum of squares,

$$SS_{Reg} = \sum_{i=1}^{n} \left[\left(b_o + \sum_{j=1}^{p} b_j X_{ji} \right) - \bar{Y} \right]^2,$$

a measure of the variability of the dependent variable which can be related to the predictors, and the residual sum of squares,

$$SS_{Res} = \sum_{i=1}^{n} (Y_i - b_o - \sum_{j=1}^{p} b_j \bar{X}_{ji})^2,$$

a measure of the remaining variability of the dependent variable. Associated with the SS_{Reg} is a number of degrees of freedom equal to the number of linearly independent predictors in the model minus 1, while the SS_{Res} has an associated number of degrees of freedom equal to the number of cases in the regression minus the number of linearly independent predictors.

Dividing each sum of squares by its respective number of degrees of freedom, we obtain an estimate of the variance of the dependent variable associated with the predictors (called the *mean square for regression,* MS_{Reg}) and an estimate of the variance of the dependent variable *not* associated with the predictors (the *mean square for residuals,* MS_{Res}). That is,

$$MS_{Reg} = \frac{SS_{Reg}}{df_{SS_{Reg}}} \text{ and } MS_{Res} = \frac{SS_{Res}}{df_{SS_{Res}}} .$$

The upper-tail test of the hypotheses

$$H_o: \sigma^2_{Reg} = \sigma^2_{Res}$$

$$H_a: \sigma^2_{Reg} > \sigma^2_{Res}$$

is then carried out with the F statistic

$$F = \frac{MS_{Reg}}{MS_{Res}} .$$

This test is equivalent to testing $H_o: \varrho^2 = 0$ against $H_a: \varrho^2 > 0$, where ϱ^2 is the population coefficient of determination.

When a multiple linear regression is performed using a statistical program package (the SPSS subprogram REGRESSION, for example), the printout will generally include the ANOVA (Analysis of Variance) table (Table 13.4). Such a table assumes that all the predictors are linearly independent, and the significance of F corresponds to the test described above.

TABLE 13.4.   ANOVA table.

ANOVA	DF	SUM OF SQUARES	MEAN SQUARE	F
Regression	$p - 1$	SS_{Reg}	SS_{Reg}/df_{Reg}	MS_{Reg}/MS_{Res}
Residual	$n - p$	SS_{Res}	SS_{Res}/df_{Res}	Significance of F

Inferences about the Regression Coefficients

Such printouts often contain the variance-covariance matrix $s^2(\mathbf{b})$ of the variances and covariances of the sampling distributions of the b_i.

$$s^2(\mathbf{b}) = \begin{bmatrix} s^2(b_o) & s(b_o, b_1) & . & . & . & s(b_o, b_p) \\ s(b_1, b_o) & s^2(b_1) & . & . & . & s(b_1, b_p) \\ . & & & & & \\ . & & & & & \\ . & & & & & \\ s(b_p, b_o) & s(b_p, b_1) & . & . & s^2(b_p) \end{bmatrix},$$

where $s^2(b_i)$ is the variance of the sampling distribution of b_i, and $s(b_i, b_j)$ is the covariance between b_i and b_j. The variance-covariance matrix can be found in this way:

$$s^2(\mathbf{b}) = \mathbf{Y}^{Tr}\mathbf{Y} - \mathbf{b}^{TR}X^{TR}\mathbf{Y} \times (X^{TR}X)^{-1}.$$

Since we also know that $E(\mathbf{b}) = \boldsymbol{\beta}$, we have here important information about the sampling distributions of the elements of $\mathbf{b}$.

In particular, tests of the hypothesis $H_o: \beta_k = \beta_{k_o}$ can be performed using the statistic $t = (b_k - \beta_{k_o})/s(b_k)$, which has associated with it $n - p$ degrees of freedom. Also $b_k \pm t_{\alpha/2} s(b_k)$ are the endpoints of a confidence interval for β_k with confidence level $(1 - \alpha) \times 100\%$.

An Extended Example. A statistician at Arcane College has been asked to analyze the hiring policies of the college's administrators. From 330 recent applications, a 10 percent random sample is selected, and for each sample applicant, the statistician constructs the following profile:

APPLICANT PROFILE

Profile Items

X1—Age

X2—Sex (1 = male, 2 = female)

X3—Degree Held (1 − M.A., 2 = M.A. + 30 sem hr, 3 = Ph.D.)

X4—Cumulative GPA for graduate study

X5—Years of Teaching and/or Professional Experience

X6—Number of Publications

X7—Rank Requested (1 = instructor, 2 = asst. prof., 3 = assoc. prof., 4 = prof.)

	Low		Average		High		
X8—Personal Characteristics (enthusiastic, friendly, mature)	1	2	3	4	5	6	7
X9—Professional Qualities and Attitudes (knowledge, preparation)	1	2	3	4	5	6	7
X10—Interpersonal Relationships (approachable, fair, honest)	1	2	3	4	5	6	7
X11—Technical Skills of Teaching (testing, grading, methods, organization)	1	2	3	4	5	6	7
X12—Communication Skills (open, rapport, relevant)	1	2	3	4	5	6	7
X13—Recommendation of Major Professor	1	2	3	4	5	6	7
X14—Recommendation of Previous Employer	1	2	3	4	5	6	7
X15—Interview Outcome	1	2	3	4	5	6	7

Then ten judges (in this case, administrators responsible for hiring new personnel) are asked to sort the 33 applicants into seven categories as shown on this form:

JUDGMENTAL RATING SHEET

Rating Sheet   ID Number_____

1. Two Highest Profiles _____ _____
2. Four Higher Profiles _____ _____ _____ _____
3. Six High Profiles _____ _____ _____ _____ _____ _____
4. Nine Middle Profiles _____ _____ _____ _____ _____ _____ _____ _____ _____
5. Six Low Profiles _____ _____ _____ _____ _____ _____
6. Four Lower Profiles _____ _____ _____ _____
7. Two Lowest Profiles _____ _____

Note: Within a group (e.g., the Six High Profiles) the order of preference is not noted.

Data from the two sheets—the applicants' profiles and ranks as assigned by each of the ten judges—are coded and submitted to the college computer center (the entire data file is given in Appendix C) and an SPSS program uses the procedure CONDESCRIPTIVE to generate the following statistics. The variables J1 to J10 are the judges' ratings for each candidate.

VARIABLE	MEAN	STANDARD DEVIATION	STANDARD ERROR
J1	4.000	1.581	0.275
J2	4.000	1.581	0.275
J3	4.000	1.581	0.275
J4	4.000	1.581	0.275
J5	4.000	1.581	0.275
J6	4.000	1.581	0.275
J7	4.000	1.581	0.275
J8	4.000	1.581	0.275
J9	4.000	1.581	0.275
J10	4.000	1.581	0.275
X1	30.242	5.602	0.975
X2	1.576	0.502	0.087
X3	3.091	0.879	0.153
X4	3.582	0.248	0.043
X5	3.879	2.713	0.472
X6	2.000	2.179	0.379
X7	1.758	0.867	0.151
X8	5.636	1.055	0.184
X9	5.303	1.104	0.192
X10	5.455	1.063	0.185
X11	5.424	1.091	0.190
X12	5.273	0.944	0.164
X13	5.545	0.666	0.116
X14	5.364	0.653	0.114
X15	5.394	0.864	0.150

We see that on the average applicants tended to be near 30 years old (X1), equally likely to be male or female (X2), with either a master's degree plus 30 hours or a Ph.D., an A− average in the graduate minor, less than five years experience, and less than three publications. Most applicants sought assistant professorships, and received ratings of between 5 and 6 on all the variables X8 through X15. The judges performed their tasks correctly; the judges' ratings, J1 through J10, are identically distributed.

Restricting our attention to the ratings given by judge 1, we construct a multiple linear regression equation relating the judge's rating as the dependent variable to the elements of the applicant's profiles as predictors. In addition, we would like to make our prediction model as economical as possible, identifying and including the predictors which judge 1 considers important, but excluding those which have no significant impact on the judge's ratings.

We will construct a full model that relates the value of J1 to all fifteen predictors, then a series of reduced models from which various predictors or groups of predictors are excluded. If the drop in R^2 from the full model to a reduced model is significant, then the omitted predictors are making a contribution to the value of J1; if the drop in R^2 is not significant, then the omitted predictors are not important in predicting the value of J1.

Groups of predictors to be considered together when constructing reduced models can be identified by examining the intercorrelation matrix of all the variables involved in the regression. In SPSS, this table is produced by the REGRESSION subprogram; the command is

REGRESSION; VARIABLES = J1, X1 TO X15/REGRESSION = J1 WITH X1 TO X15

The output begins with these cross-correlations:

CORRELATION COEFFICIENTS.

A VALUE OF 99.00000 IS PRINTED
IF A COEFFICIENT CANNOT BE COMPUTED.

	J1	X1	X2	X3	X4	X5
X1	− .21170					
X2	− .27566	.21558				
X3	− .33725	.52849	.09015			
X4	− .43798	.19368	.18949	.12093		
X5	− .27684	.61481	.02991	.22753	.26176	
X6	− .29926	.58874	.19998	.65245	.35710	.36468
X7	− .27352	.69446	.11533	.68578	.27485	.53177
X8	.22474	− .17494	.05364	.03675	− .23319	− .20144
X9	.17910	.15963	.07010	− .12593	.06238	.14835
X10	− .72489	.13307	.13840	.08814	.12262	.10636
X11	.39869	− .18106	− .46020	− .17187	− .15313	− .23557
X12	.02093	.14660	.25171	.15740	.23348	.07429
X13	− .08907	.18132	.24658	.01942	.21721	.10696
X14	− .21194	− .22997	.00867	− .05941	− .07905	− .25667
X15	.64063	− .26577	− .32326	− .41902	− .50254	− .16567

	X6	X7	X8	X9	X10	X11
X7	.82679					
X8	.01359	− .20180				
X9	− .20789	.01385	− .35860			
X10	.12136	.22492	− .23798	− .20094		
X11	− .19721	− .18525	− .10614	− .26595	− .33317	
X12	.28845	.27404	− .43040	.12811	− .12729	− .35855
X13	− .04308	.07382	− .28711	.06574	− .00803	− .19956
X14	− .06589	− .11543	.19795	− .07099	.38472	− .04789
X15	− .28218	− .28572	.19633	− .26027	− .20103	.28145

	X12	X13	X14
X13	.50153		
X14	− .31794	− .25495	
X15	− .21240	− .27665	− .09572

Since X5, X6, and X7 have high positive correlations with respect to one another, it seems reasonable to consider them together when forming reduced models. Likewise, the mutual negative correlations among X13, X14, and X15 suggest grouping them together.

Considering the intercorrelations and a common-sense look at which predictors might reasonably be linked, the hierarchical chart of reduced models is constructed (Figure 13.6). Note that each model is reduced from the full model, not from those above it.

Significant drops in R^2, indicating significant contributions to the dependent variable J1 by the omitted predictors, are identified by asterisks. Only reduced models 4, 5, and 6 register significant drops in R^2 relative to the full model. This indicates that independent variables X8 to X15, dealing with personal characteristics, skills, and recommendations, should be included when predicting ratings given by judge 1, while predictors X1 through X8, demographic data, educational and professional history, can safely be ignored.

Note that if RM4 had shown a significant drop in R^2 from RM14, while RM15 and RM16 had not shown such a difference, then only X8 would be considered to be important among X8, X9, and X10, and X9 and X10 could be disregarded in predicting J1.

Having identified the independent variables which are important in the ratings given by judge 1, we complete this analysis by constructing a linear prediction equation for ratings by judge 1 based on the relevant predictors. Using SPSS, this process is initiated by the following commands:

REGRESSION; VARIABLES = J1,X1 TO X15/

REGRESSION = J1 WITH X8 TO X15/

STATISTICS; ALL

Note that in the REGRESSION statement we specify that J1 be related to X8 through X15, the predictors that our earlier work has shown are important.

The output produced by these commands begins with the mean and standard deviation of all the variables included in the regression, and continues with their intercorrelation matrix. The results of the regression conclude the printout.

FIGURE 13.6. Hierarchical chart of reduced models.

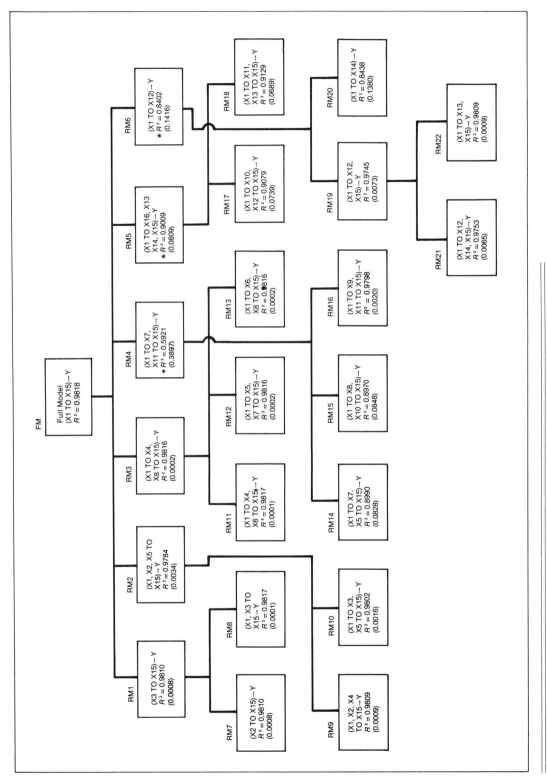

```
DEP. VAR...J1      JUDGE 1
MEAN RESPONSE      4.00000        STD DEV              1.58114

FINAL STEP
```

		ANOVA	DF	SQUARES	MEAN SQ.	F
MULTIPLE R	.9888					
R SQUARE	.9778	REGRESSION	8.	78.221	9.778	131.917
STD DEV	.2722	RESIDUAL	24.	1.779	.074	SIG. .000
ADJ R SQUARE	.9704	COEFF OF VARIABILITY		6.8PCT		

VARIABLE	B	S.E. B	F	SIG.	BETA	ELASTICITY
X8	1.175	.105	125.745	.000	.78405	1.65531
X11	1.133	.094	144.845	.000	.78157	1.53657
X14	− .060	.092	.428	.519	− .02477	− .08044
X15	1.179	.063	345.399	.000	.64438	1.59047
X13	.230	.087	6.988	.014	.09682	.31880
X9	1.097	.080	187.982	.000	.76597	1.45498
X12	1.060	.100	111.953	.000	.63319	1.39729
X10	.143	.099	2.089	.161	.09645	.19558
CONSTANT	− 28.274	2.374	141.852	.000		

ALL VARIABLES ARE IN THE EQUATION.

COEFFICIENTS AND CONFIDENCE INTERVALS.

VARIABLE	B	95P	C.I.
X8	1.1747	.9585	1.3910
X11	1.1331	.9388	1.3274
X14	− .0600	− .2493	.1293
X15	1.1794	1.0485	1.3104
X13	.2300	.0504	.4095
X9	1.0975	.9323	1.2627
X12	1.0600	.8532	1.2668
X10	.1434	− .0614	.3483
CONSTANT	− 28.2742	− 33.1738	− 23.3746

VARIANCE/COVARIANCE MATRIX OF THE UNNORMALIZED REGRESSION
COEFFICIENTS.

	X8	X9	X10	X11	X12	X13
X8	.01097					
X9	.00658	.00641				
X10	.00851	.00608	.00985			
X11	.00797	.00567	.00752	.00886		
X12	.00790	.00499	.00693	.00708	.01004	
X13	.00069	.00057	.00022	.00043	− .00175	.00757
X14	− .00353	− .00210	− .00418	− .00247	− .00144	.00106
X15	.00055	.00117	.00110	.00024	.00081	.00121

	X14	X15
X14	.00841	
X15	.00049	.00403

From the ANOVA Table 13.4, we see that the value of the F statistic (131.917) is significant; we can conclude, then, that $\sigma^2_{\text{Reg}} > \sigma^2_{\text{Res}}$, or equivalently, that $\varrho^2 > 0$. The sample estimate of ϱ^2 is $R^2 = 0.9778$, and the predictor equation for ratings given by judge 1 is:

$$J1' = b_o + \sum_{i=8}^{15} b_i X_i$$
$$= -28.2742 + 1.175X_8 + 1.097X_9 + 0.143X_{10}$$
$$+ 1.133X_{11} + 1.060X_{12} + 0.230X_{13} - 0.060X_{14}$$
$$+ 1.179X_{15},$$

with predictive efficiency 97.78%.

The coefficients b_8, b_9, b_{11}, b_{12}, and b_{15} can all be considered not equal to zero at the 1% significance level, but we are unable to reject $H_o\colon \beta_i = 0$ at $\alpha = 1\%$ for b_{10}, b_{13}, and b_{14}. Even at $\alpha = 5\%$, we cannot conclude that b_{10} and b_{14} are not zero; their 95% confidence intervals include zero.

There are other methods used to determine which independent variables should be retained in a regression model. The most popular of these is *stepwise regression*, in which a sequence of regression equations is constructed. At each step, an independent variable is added to or deleted from the model.

Forward stepwise regression begins by including in the regression equation only the single independent variable which, alone, produces the largest coefficient of determination R^2 as measured by an F test. In the second step, another predictor is added to the developing equation, the one which, with the predictor of the first step, produces the largest value of R^2. This second step is then iterated, including one new predictor at each repetition until all are included in the final full model. The predictors entered first are the most important, and we can gauge the relative importance of predictors by watching R^2 increase at each step of the development.

The stepwise regression process also allows predictors to be removed from the equation, if subsequent inclusions have made the earlier additions unimportant; this assumes that there is a single "best" set of predictors and seeks to identify them in the full model.

Stepwise regression can also be extended to FORWARD and BACKWARD SET SELECTION, in which the researcher selects sets of predictors to be included and deleted at each step. For more details, see Draper and Smith.

Curvilinear Regression

As illustrated in the scatter diagram examples early in Chapter 12, many relationships between a dependent variable and one or more independent predictor variables are not linear, but curvilinear. In business, for example, such equations often occur in the analysis of cost-estimation relationships (CER), in which categories of cost are related to cost-generating or explanatory variables. We can extend linear regression analysis to cover many nonlinear relationships by employing simple transformations of the nonlinear equations to produce the familiar linear forms. Solutions to these linear cases can then be re-transformed to describe the curvilinear relationship.

Table 13.5 illustrates six such transformations involving one predictor. In each case, X is the predictor; Y the dependent variable; and a, b, and c are constants.

TABLE 13.5.   Six curvilinear transformations involving one predictor.

EQUATION REPRESENTING THE CURVILINEAR RELATIONSHIP	*CORRESPONDING SIMPLE LINEAR REGRESSION EQUATION*
$Y = aX^b$	$\ln(Y) = \ln(a) + b\ln(X)$
$Y = aX^b + c$	$\ln(Y - c) = \ln(a) + b\ln(X)$
$Y = 10^{a + bX}$	$\log(Y) = a + bX$
$Y = 10^{a + bX} + c$	$\log(Y - c) = a + bX$
$Y = \dfrac{X}{a + bX}$	$\dfrac{1}{Y} = a\left(\dfrac{1}{X}\right) + b$
$Y = (ab)^{cX}$	$\ln(\ln(Y)/\ln(b)) = \ln(X) + \ln(c)$

   In analyzing a relationship of the first type given, for example, we would begin by taking the natural logarithm of each value in our data file to generate $(\ln(X_i), \ln(Y_i))$. From these new pairs of values, we find the values of b and $\ln(a)$ using simple linear regression. The value of b can be inserted directly into the equation $Y = aX^b$, and the value of a is the antilogarithm of $\ln(a)$. Note that we have not described the models on which these equations are based.

   Multiple linear regression can also be applied to nonlinear situations, particularly polynomial relationships, as shown in Table 13.6.

TABLE 13.6.   Multiple linear regression applied to polynomial relationships.[a]

EQUATION REPRESENTING THE CURVILINEAR RELATIONSHIP	*CORRESPONDING MULTIPLE LINEAR REGRESSION EQUATION*
$Y = a_0 + \displaystyle\sum_{i=1}^{n} a_i X^i$	Let $Z_i = X^i$, then find the coefficients in $$Y = a_0 + \sum_{i=1}^{n} a_i Z_i.$$
$Y = \sinh(X)$	Since $\sinh(X) = \dfrac{e^X - e^{-X}}{2}$, let $Z = \dfrac{e^X}{2}$ and $W = \dfrac{e^{-X}}{2}$, and find a_1 and a_2 in $Y = a_1 Z + a_2 W$.

[a] The latter technique above can be employed for any function f where $f(X) = f_1(X) + f_2(X)$.

   Developing representations of curvilinear relationships is something of an art, involving a certain amount of intuition. Given a data file that represents a nonlinear relationship, linear regression may be used if a transformation of the data can be found which produces a new file possessing linearity, homoscedasticity, and the other properties prerequisite to linear regression. This transformation, and the nature of the relationship between the predictors and the dependent variable may be found by examining scatter diagrams of the original data, or of new values obtained by applying a trial transformation to the original data. If nonlinearity persists, then other techniques must be used.

EXERCISES

1. From the general forms of the normal equations for simple linear regression, obtain the formulas for the regression coefficients A and B given on p. 275.

2. Let $\bar{X}$ be the mean of the values $X_1, X_2, \ldots X_n$.

   a. Show that $n \sum_{i=1}^{n} (X_i - \bar{X})^2 = n \sum_{i=1}^{n} X_i^2 - (\sum_{i=1}^{n} X_i)^2$.

   b. Show that $\sum_{i=1}^{n} (X_i - \bar{X})X_i = \sum_{i=1}^{n} (X_i - \bar{X})^2$.

3. Show that the regression coefficient B can be written

$$B = \frac{\left[\sum_{i=1}^{n} (X_i - \bar{X}) Y_i\right]}{\left[\sum_{i=1}^{n} (X_i - \bar{X})^2\right]}.$$

4. Consider the data given with Exercise 1 of Chapter 12 (repeated here for your convenience).

WEIGHT	HORSEPOWER	MPG
3205	125	14
2945	80	20
4380	125	21
2040	74	29
3295	165	16
3520	195	16
2195	77	25
3205	195	16
1760	66	28
2515	130	21
3205	97	17
2480	93	28

   a. Find the least squares regression equation relating mileage to automobile weight. Draw the line corresponding to this equation on the scatter diagram of mileage with weight.

   b. If an American car weighs 2700 pounds, what mileage is predicted by the regression equation?

   c. Find the standard error of estimate s_e. Of what is s_e an estimate?

   d. Find a 95% confidence interval for β_1, the slope of the true regression line. Based on this, can you reject the null hypothesis H_o: $\beta_1 = 0$ at the 5% significance level?

   e. Find a 95% confidence interval for β_o, the Y-intercept of the true regression line. What meaning, if any, does the value of A have in this problem?

   f. Corresponding to a weight of 2700 pounds, find a 95% confidence interval for the expected mileage.

5. Consider the data given in Exercise 4.

   a. Find the least squares regression equation to predict mileage from horsepower for American cars. Draw its graph on the scatter diagram of mileage with horsepower.

   b. If a car develops 145 horsepower, what mileage would we expect it to get?

c. Find s_e, the standard error of estimate. Of what is s_e an estimate?

d. At the 1% significance level, can we conclude that there is a significant relationship between horsepower and mileage? Examine B to find out.

e. Find a 95% confidence interval for the value of β_o, the Y-intercept of the regression line. What meaning, if any, does A have in this situation?

f. If the horsepower of a car is increased by 25, what effect would you expect this to have on mileage?

g. For a car that develops 145 horsepower, find a 95% confidence interval for the corresponding expected mileage.

6. Show that if two variables X and Y have a perfect positive linear relationship, then for any data point (X_i, Y_i),

$$\frac{X_i - \bar{X}}{s_x} = \frac{Y_i - \bar{Y}}{s_y}.$$

(*Hint:* "Perfect positive linear relationship" means $Y_i = A + BX_i$ for all i.)

7. If $t = (B - \beta_{1_o})/\sigma_B$, and s_e is an estimate of σ, show that

$$t = \frac{B - B_{1_o}}{s_e} \left(\sum_{i=1}^{n} X_i^2 - n\bar{X}^2 \right)^{1/2}.$$

8. In general, what is the meaning of the regression coefficient β_o? Does this always apply?

9. Use your computer center and statistical package to answer the following questions about the variables X3 and X5 from the USNEWS file:

a. Plot the scatter diagram of X3 with X5 and find the value of R. What information do these convey?

b. Is there a significant relationship between these two variables? Explain.

c. Determine the least squares regression equation for predicting X3 from X5.

d. What is the value of the coefficient of determination R^2, and what does it tell us?

e. Find a 95% confidence interval for β_1, the slope of the true regression line.

f. Why is the significance of B the same as the significance of R? What does it mean?

g. Find the expected value of X3 if X5 is $18,500.

10. This table gives the number of BASIC jobs submitted and the number of downtimes at a college computer center for each month of a 12-month period.

MONTH	BASIC JOBS	DOWNTIMES
1	800	3
2	500	0
3	1000	1
4	1600	2
5	1300	3
6	2100	3
7	1900	3
8	450	0
9	700	1
10	800	2
11	1200	4
12	750	2

    a. By hand, find the least squares regression line for predicting the number of downtimes from the number of BASIC jobs.

    b. Use the computer to repeat and verify the results of part (a).

    c. At the 5% significance level, test, by hand, the hypothesis $H_o:B_1 = 0$.

    d. Use the computer to verify the results of part (c).

11. For the regression equation found in Exercise 10,

    a. Construct a 95% confidence interval for β_1, the slope of the true regression line.

    b. What does the value of B tell us?

    c. Construct a 95% confidence interval for β_o, the Y-intercept of the true regression line.

    d. What does the value of A tell us?

12. Using the fact that any linear combination of normal random variables is normal, argue that the sampling distribution of the least squares regression coefficient A is normal.

13. Show that

    a. $B = \dfrac{S_Y}{S_X} R_{XY}$, and

    b. $z_{Y'} = R_{XY} z_X$,

    where R_{XY} is the Pearson correlation coefficient of the variables X and Y.

14. Consider the regression equation constructed for Exercise 10.

    a. Find a 95% confidence interval for the expected number of downtimes corresponding to the submission of 1000 BASIC jobs.

    b. The sampling distribution of a *single* observation of Y corresponding to a given value X_0 is identical to that of $\bar{Y}_{x_0}$ except that it has standard deviation

$$\sigma \left[1 + \frac{1}{n} + \frac{(X_o - \bar{X})^2}{\sum\limits_{i=1}^{n} X^2 - n\bar{X}^2} \right]^{1/2}.$$

    Use this information to find a 95% PREDICTION INTERVAL for the value of a single future number of downtimes in a month in which 1000 BASIC jobs are submitted.

15. Consider the regression equation developed in Exercise 5. Find a 95% prediction interval for the mileage of a car which developed 145 horsepower. [*Hint:* See Exercise 14(b).] What is the difference in meaning between this result and the result of Exercise 5(g)?

16. For the values given in Exercise 10,

    a. Test $H_o: \varrho = 0$ at the 5% significance level.

    b. Test $H_o: \beta_1 = 0$ at the 5% significance level.

    c. Show that, when $\varrho = \beta_1 = 0$, the statistics of these two tests are equal.

17. Suppose that in Exercise 5 we wish to predict the horsepower of a car from its mileage.

    a. Using the formulas for A and B, find the equation of the regression line for this situation.

    b. In the regression equation computed in Exercise 5, solve for horsepower in terms of mileage. Note that this is not the same equation as found in part (b).

    c. In general, to predict X from Y, which of these methods would you use? Why?

18. Compute the coefficient of determination R^2 for the data given in Exercise 10. What does this value tell us?

19. These are the statistics produced by the SCATTERGRAM subprogram from a file of 28 data pairs (X_i, Y_i).

CORRELATION (R)	—	.91745	R SQUARED	—	.84171
SIGNIFICANCE R	—	.00001	STD ERR OF EST	—	.96962
INTERCEPT (A)	—	.98862	STD ERR OF A	—	.49197
SIGNIFICANCE A	—	.02749	SLOPE (B)	—	.09611
STD ERR OF B	—	.00817	SIGNIFICANCE B	—	.00001

PLOTTED VALUES — 28

EXCLUDED VALUES — 0

MISSING VALUES — 0

a. Is there a significant linear relationship between X and Y here? Describe that relationship.

b. State the least squares regression equation. What is the expected value of Y corresponding to $X_o = 28$?

c. Find a 95% confidence interval for β_1. Is there a way to do this without using the value STD ERR OF EST?

d. Find a 95% confidence interval for β_o.

20. These are the statistics produced by the SPSS subprogram SCATTERGRAM from a file of 50 data pairs (X_i, Y_i):

CORRELATION (R)	—	.89733	R SQUARED	—	.80520
SIGNIFICANCE R	—	.00001	STD ERR OF EST	—	8.08456
INTERCEPT (A)	—	−10.59451	STD ERROR OF A	—	4.57615
SIGNIFICANCE A	—	.01246	SLOPE (B)	—	.78906
STD ERROR OF B	—	.05602	SIGNIFICANCE B	—	.00001

PLOTTED VALUES — 50

EXCLUDED VALUES — 0

MISSING VALUES — 0

a. State the regression equation, and find the value of Y expected to correspond to $X_o = 95$.

b. Is the value of B significant? What does this tell us?

c. Find a 95% confidence interval for β_1.

d. Find a 95% confidence interval for β_o.

e. What is the meaning of R SQUARED?

21. Compute the values of the elements of the vector $\boldsymbol{\beta}_z$ for the reduced model in the example on p. 290.

22. This intercorrelation matrix relates overhead cost in a factory (Y) to units purchased (X_1), direct labor costs (X_2), weight of output (X_3), and research and development costs (X_4).

		Y	X_1	X_2	X_3	X_4
	Y	1.000	0.562	0.401	0.197	0.465
	X_1	0.562	1.000	0.396	0.215	0.583
$R =$	X_2	0.401	0.396	1.000	0.345	0.546
	X_3	0.197	0.215	0.345	1.000	0.365
	X_4	0.465	0.583	0.546	0.365	1.000

    a. Write out the equation $R\beta_z = \mathbf{V}$.

    b. Find the normalized regression equation $z_Y = \sum_{i=1}^{4} \beta_i z_{x_i}$ by the iterative Kelley-Salisbury technique.

    c. Find and interpret the value of R^2.

23. Consider the data provided in Exercise 4. We will predict mileage (Y) with the two predictors weight (X_1) and horsepower (X_2).

    a. Find the mean and standard deviation of X_1, X_2, and Y.

    b. Construct the intercorrelation matrix of X_1, X_2, and Y.

    c. Write the vector equation $R\beta_z = \mathbf{V}$.

    d. Use the Kelley-Salisbury technique to find the coefficients β_{z1} and β_{z2} in the normalized regression equation $z_Y = \beta_{z1}X_1 + \beta_{z2}X_2$.

    e. Using the normalized equation found in part (d), predict the mileage of a 2500-pound car which develops 105 horsepower. (*Hint*: In the equation, use

$$Z_{x_1} = \frac{2500 - \bar{X}_1}{s_{x_1}} \text{ and } Z_{x_2} = \frac{105 - \bar{X}_2}{s_{x_2}} ,$$

then find Y' so that $z_Y = (Y' - \bar{Y})/s_Y$.)

    f. Find the coefficients b_o, b_1, and b_2 in the regression equation $Y = b_o + b_1X_1 + b_2X_2$. *Hint*: Start here:

$$\frac{Y' - \bar{Y}}{s_Y} = \beta_{z2} \frac{X_1 - \bar{X}_1}{s_{x_1}} + \beta_{z2} \frac{X_2 - \bar{X}_2}{s_{x_2}} .)$$

    g. Use the equation found in part (f) to predict the mileage of a 2500-pound car which develops 105 horsepower. Does this result agree with the answer in part (e)?

24. Using the data file in Appendix C from the Arcane College Study,

    a. Determine which predictors are significant in the ratings given by judge 3.

    b . Develop a prediction equation for judge 3's ratings which involves only the predictors found to be important in part (a). (*Hint:* This problem will be hopelessly large unless you use the computer to find the R^2 values.)

25. The value of R^2 computed from a least squares solution to a multiple linear regression problem is a biased estimator of the population coefficient of determination of ϱ^2. It tends to be inflated. For example:

    a. Use the first 16 cases of the Arcane College study to develop a prediction equation for J1 based on X13, X14, and X15. Find the values of R^2 associated with these variables.

    b. Use the regression equation to predict J1 from the values of X13, X14, and X15 for the last 17 cases in the data file.

    c. Compute the coefficient of determination between the predicted values of J1 found in part (b) and the observed values. Compare this value to that found in part (a). R^2 can be "shrunk" to a more probable population value by this correction:

$$R^2_{\text{Adjusted}} = 1 - (1 - R^2) \times \left(\frac{n - 1}{n - m}\right),$$

where m is the number of variables correlated, and n is the number of cases.

   d. Adjust the value of R^2 found in part (a) with this formula. Does it now agree more closely with the observed value of part (c)?

26. In a study of subscribers to two computer science journals, a regression analysis was based on the data file COMPSCI to predict the value of the dependent variable TYPE (1 = *Communications of the ACM,* 0 = *SIGCSE Bulletin*) from the predictors ORGAN (professional organization affiliation), EDUC (degree held), INCOME (total household income), and INTEREST (research interests). The data file contained the values of these five variables for 225 individuals, and this partial chart summarizes the results:

FIGURE 13.26

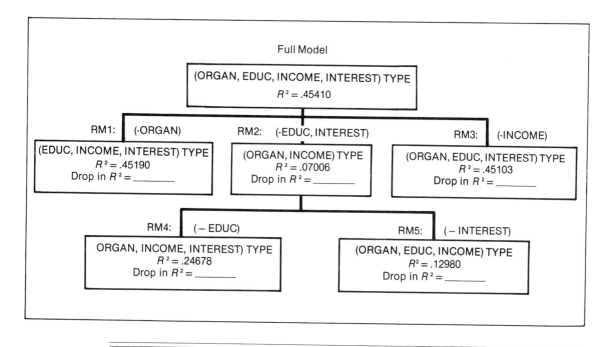

This partial printout corresponds to the full regression model:

MULT R	0.67387	ANOVA	DF	F
RSQUARE	0.45410	REGRESSION	2	23.09671
ADJ R SQ	0.45308	RESIDUAL	222	SIGN
ST ERR	0.79004			.002

   a. Is the full model a "good" model from which to develop the regression equation? Why?

   b. Which predictors are significant in predicting TYPE? Which should be included in the final regression equation? Why?

27. a. Have a group of at least 10 members of your class rate 20 famous paintings on these three scales:

FIGURE 13.27(a)

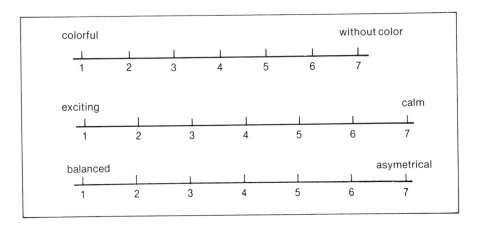

b. Provide the remaining members of the class with the ratings (but not the paintings), and have them rate the paintings on this scale:

FIGURE 13.27(b)

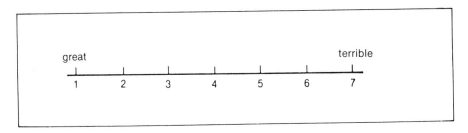

c. Develop a regression equation for each participant in part (b) which relates the final judgment to the three predictors of part (a).

28. Consider this data file:

CASE	X_1	X_2	Y
1	2	6	23
2	8	9	83
3	6	8	63
4	10	10	103

a. Find a least squares regression equation to predict Y from X_1 and X_2, and its associated value of R^2.

b. Show that the equations $Y = -7 + 9X + 2X_2$ and $Y = -87 + X_1 + 18X_2$ are just as good as the one found in part (a) by computing their coefficients of determination.

c. Why are there several equally good regression equations? (*Hint:* Consider $R_{x_1 x_2}$.)

29. Consider the following data file:

X	1	1	2	2	2	3	3	3	4	4
Y	1	1	3	4	4	8	9	10	15	16

    a. Plot these data pairs on a scatter diagram with Y as the dependent variable.

    b. What type of curve might fit these points?

    c. Use simple linear regression to find the equation relating Y to X. [*Hint:* If $Y = AX^B$, then $\ln(Y) = \ln(A) + B\ln(X)$.]

    d. What value of Y would be expected to correspond to $X = 3.5$?

30. A company needs to determine the relationship between cost per unit produced (Y) and number of units produced (X). They collect the following data:

NUMBER OF UNITS (X)	COST PER UNIT (Y)
10	509.2
10	502.1
20	573.2
20	569.4
30	652.7
30	658.0
50	755.2
50	760.0
60	789.2
60	791.2
70	841.2
70	839.7
80	855.1
80	872.4

    a. Plot the scatter diagram of these data with Y the dependent variable.

    b . Use multiple linear regression to fit an equation of the form $Y = b_o + b_1X + b_2X^2$ to the data.

    c. What cost per unit would be expected from a production of 45 units?

31. For the following data:

X	Y	X	Y	X	Y
− 3.0	1.8	− 0.6	0.0	2.0	5.0
− 3.0	1.5	0.0	0.0	2.0	4.6
− 2.6	1.2	0.4	0.2	2.0	5.4
− 2.6	0.4	0.4	0.2	2.6	7.4
− 1.8	0.4	0.4	1.0	2.6	7.0
− 1.8	0.2	1.0	1.5	2.6	6.6
− 1.0	0.0	1.0	0.8	3.0	8.0
− 1.0	0.5	1.0	2.0	3.0	7.8
− 1.0	0.6	1.6	4.6	3.0	7.5
− 0.6	0.2	1.6	3.6	3.0	6.6

    a. Plot a scatter diagram of these data with Y as the dependent variable.

    b. State a type of curve that would fit these data.

    c. Use multiple linear regression to identify the appropriate curve and state its equation.

    d. What value of Y would be expected to correspond to $X = 2.3$?

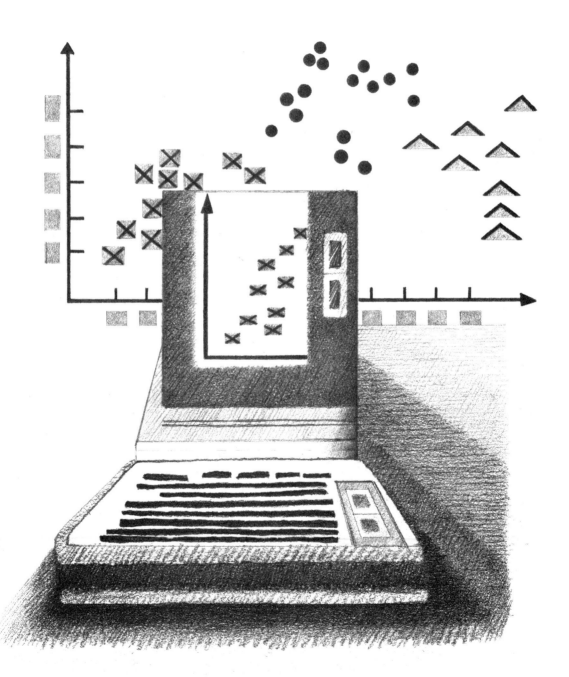

Analysis of Variance

INTRODUCTION

In Chapter 10, we developed the t-test to decide, using independent samples, whether the means of two populations are equal. To compare the means of three or more populations, we employ a technique which compares the variabilities of the values of the dependent variable in the several samples.

In linear regression, the total variability of the values of the dependent variable was shown to be the sum of the variation associated with the independent variables and the unexplained variation not associated with the predictors. This idea —examining the sources of the variation of the dependent variable—is at the heart of a technique which can be thought of as comparing the means of several populations, or as regression when the predictors are nominal rather than at least ratio level. Variation in the values of the dependent variable is broken into components attributed to the independent variables and other, unexplained, variation. Ratios of these variations (sums of squared deviations) are used to make inferences about the population means. For this reason, the technique is called analysis of variance, or ANOVA.

One-Way Analysis of Variance

We begin with the case, corresponding to simple linear regression, in which the value of the dependent variable is influenced by a single *qualitative* predictor. This is called *one-way* analysis of variance. The predictor is called a *factor*, and its values, called *levels* or *treatments*, divide the observed cases into several groups. The corresponding groups of values of the dependent variable are then samples from several populations. The levels of the factor are *fixed* if all possible levels are included in the collected data or if inferences are to be drawn about only the observed levels. If the observed levels of the factor are a sample from the population of all possible levels and inferences are to be drawn over all levels, whether represented in the data or not, the levels are *random*. Consider, for example, drawing conclusions about all possible fertilizing patterns for wheat based on a study involving four such patterns.

In our first example, the levels of the factor are fixed, since conclusions are to be drawn only over the levels mentioned in the data.

Example 1 The Acme Publishing Company has released a paperback book in three differently colored covers. Each bookstore receives copies in only one color, and from twenty bookstores in similar markets they receive the sales totals, as shown in Table 14.1, for the first month after the book's publication.

TABLE 14.1. Sales totals for one month for a paperback book.

	COVER COLOR		
	RED	*BLUE*	*GREEN*
Number	145	255	226
of copies	210	190	240
sold	158	276	185
	205	222	175
	182	248	206
		206	221
		269	231
		230	

The publishers wish to know whether the mean sales for the book are the same for each cover color. *Sales* is the dependent variable, and *cover color* the factor, with treatments (levels) red, blue, and green. The three groups of sales values are independent samples from three populations, and we must decide if the means of these populations are not equal. These hypotheses are being tested:

$$H_o: \mu_{red} = \mu_{blue} = \mu_{green}$$

H_a: At least one mean differs from the other two.

The test statistic will be based on a comparison of sums of squared deviations, and will have an F distribution when the null hypothesis is true. We now consider this analysis in general, before completing the example.

The one-way ANOVA problem is a test of these hypotheses:

$$H_o: \mu_1 = \mu_2 = \ldots = \mu_k$$

H_a: At least one population mean differs from the others.

The data for the test are organized as shown in Table 14.2.

TABLE 14.2. Data for one way analysis of varients.

	TREATMENT 1	TREATMENT 2 ...	TREATMENT k
Values of the dependent variable	X_{11} X_{21} . . . X_{n1}	X_{12} X_{22} . . . X_{n2}	X_{1k} X_{2k} . . . X_{nk}

The observations of the values of the dependent variable X_{ij} form samples of sizes n_j from k populations, as determined by the treatments. The populations are assumed to have common variance σ^2. Let $\bar{X}_{.j}$ be the mean of the sample of treatment j, and let $\bar{X}..$ be the grand mean of all the values X_{ij}. Then if $N = \sum_{j=1}^{k} n_j$ is the total number of observations,

$$\bar{X}_{.j} = \frac{1}{n_j} \sum_{i=1}^{n_j} X_{ij} \text{ and } \bar{X}.. = \frac{1}{N} \sum_{j=1}^{k} \sum_{i=1}^{n_j} X_{ij}.$$

The total variation of the values of the dependent variable is measured, as in linear regression, by the *total sum of squares*, SS_{tot}:

$$SS_{tot} = \sum_{j=1}^{k} \sum_{i=1}^{n_j} (X_{ij} - \bar{X}..)^2.$$

Each value of the dependent variable contributes one term to this sum.

It can be shown with some tedious algebra that

$$SS_{tot} = \sum_{j=1}^{k} \sum_{i=1}^{n_j} (X_{ij} - \bar{X}..)^2$$

$$= \sum_{j=1}^{k} \sum_{i=1}^{n_j} [(X_{ij} - \bar{X}_{.j}) + (\bar{X}_{.i} - \bar{X}..)]^2$$

$$= \sum_{j=1}^{k} \sum_{i=1}^{n_j} (X_{ij} - \bar{X}_{.j})^2 + \sum_{j=1}^{k} n_j (\bar{X}_{.j} - \bar{X}..)^2 +$$

$$(X_{ij} \sum_{j=1}^{k} [(\bar{X}_{.j} - \bar{X}..) \sum_{i=1}^{n_j} (x_{ij} - \bar{X}_{.j})].$$

But $\bar{X}._j$ is the mean of the values of the jth treatment group, so

$$\sum_{i=1}^{n_j} (X_{ij} - \bar{X}._j) = 0$$

for all j. Therefore,

$$SS_{tot} = \sum_{j=1}^{k} \sum_{i=1}^{n_j} (X_{ij} - \bar{X}._j)^2 + \sum_{j=1}^{k} n_j (\bar{X}._j - \bar{X}..)^2.$$

The double summation is the sum of the squared deviations of each value from its treatment group mean, and is a measure of the variation observed *within* the groups. It is called the *sum of squares within*, SS_{within}. The second sum is the total of the squared deviations of the group means from the grand mean, a measure of the variation *between* groups. It is called the *sum of squares between*, $SS_{between}$, so that

$$SS_{tot} = SS_{within} + SS_{between}.$$

We have partitioned the total variability observed in all the values of the dependent variable into two quantities, a measure of the variability within each group and a measure of the variability between the groups. From these we derive a test statistic.

If the null hypothesis of the test is true and the population means are equal across the treatments, then $\bar{X}..$ and all the $\bar{X}._j$ are unbiased estimates of the common population mean μ. Further, estimators of the common population variance σ^2 can be developed from the sums of squared deviations from these sample means.

For example, let

$$(S_j^2) = \frac{1}{n_j - 1} \sum_{i=1}^{n_j} (X_{ij} - \bar{X}._j)^2$$

be the variance of treatment group j, so that $E(S_j^2) = \sigma^2$. Then:

$$E(SS_{within}) = E\left(\sum_{j=1}^{k} \sum_{i=1}^{n_j} (X_{ij} - \bar{X}._j)^2\right)$$

$$= E\left(\sum_{j=1}^{k} (n_j - 1) S_j^2\right)$$

$$= \sum_{j=1}^{K} (n_j - 1)E(S_j^2)$$

$$= \sum_{j=1}^{k} (n_j - 1) \sigma^2$$

$$= \sigma^2 \left[\sum_{j=1}^{k} n_j - \sum_{j=1}^{k} 1\right]$$

$$= \sigma^2 (N - k).$$

Therefore, $E(\text{SS}_{\text{within}}/(N - k)) = (N - k)\sigma^2/(N - k) = \sigma^2$. The quantity $\text{SS}_{\text{within}}/(N - k)$ is called the *mean square error within groups*, $\text{MS}_{\text{within}}$, and is an unbiased estimator of the common population variance σ^2.

Similarly, it can be shown that

$$E(\text{SS}_{\text{between}}) = (k - 1)\,\sigma^2 + \sum_{j=1}^{k} n_j(\mu_j - \mu)^2,$$

where μ_j is the mean of the population corresponding to treatment j and $\mu = N^{-1}\sum_{j=1}^{k} n_j\mu_j$. We define the mean square error between groups to be

$$\text{MS}_{\text{between}} = \text{SS}_{\text{between}}/(k - 1),$$

so that, in general,

$$E(\text{MS}_{\text{between}}) = \sigma^2 + \frac{1}{k-1} \sum_{j=1}^{k} n_j(\mu_j - \mu)^2.$$

When the null hypothesis is true, $\mu_j = \mu$ for all j, and the second term of this sum is 0, so that $E(\text{MS}_{\text{between}}) = \sigma^2$. In this case, the observed values of $\text{MS}_{\text{within}}$ and $\text{MS}_{\text{between}}$ should be approximately equal, and their *ratio* should be near 1. If H_o is false, some of the population means are unequal and $\sum_{j=1}^{k} n_j(\mu_j - \mu)^2 > 0$, so that $E(\text{SS}_{\text{between}}) > \sigma^2$. We would expect $\text{MS}_{\text{between}}$ to be larger than $\text{MS}_{\text{within}}$.

The test statistic, then, is $F = \text{MS}_{\text{between}}/\text{MS}_{\text{within}}$, whose sampling distribution, if H_o is true, is an F distribution with $k - 1$ degrees of freedom in the numerator and $N - k$ degrees of freedom in the denominator. The critical value for this upper-tail test can be found in Table A-7. If F exceeds the critical value, we reject the null hypothesis and accept the alternative, concluding that some of the population means differ and that the treatments do affect the dependent variable.

The information used in a one-way analysis of variance is often summarized in an ANOVA table. (See Table 14.3.)

TABLE 14.3.   **One-way ANOVA table.**

SOURCE	DEGREES OF FREEDOM	SUM OF SQUARES	MEAN SUM OF SQUARES	F
Between groups	$k - 1$	$\text{SS}_{\text{between}}$	$\text{MS}_{\text{between}}$	$\dfrac{\text{MS}_{\text{between}}}{\text{MS}_{\text{within}}}$
Within groups	$N - k$	$\text{SS}_{\text{within}}$	$\text{MS}_{\text{within}}$	
Total	$N - 1$	SS_{tot}		

The following assumptions necessary for a one-way analysis of variance should also be noted:

1. The variances of the populations from which the treatment groups are drawn must be equal. This condition is called *homogeneity of variance*.

2. The effect of the factor on the dependent variable is additive:

$$X_{ij} = \mu + \alpha_j + \epsilon_{ij},$$

where α_j is the effect of the jth treatment, and ϵ_{ij} is the random variation of observation X_{ij}.

It is also generally assumed that the ϵ_{ij} are $N(0, \sigma^2)$, and that the α_j sum to 0, since this simplifies the model.

To complete Example 1, we perform the calculations shown in Table 14.4.

TABLE 14.4.   Completion of calculations of Table 14.1.

		COVER COLOR	
	RED	BLUE	GREEN
Number	145	255	226
of copies	210	190	240
sold	158	276	185
	205	222	175
	182	248	206
		206	221
		269	231
		230	
	900	1896	1484

Calculations.

$$\bar{X}_{.1} = 180 \qquad \bar{X}_{.2} = 237 \qquad \bar{X}_{.3} = 212$$

$$\bar{X}_{..} = \frac{900 + 1896 + 1484}{20} = 214$$

$$SS_{between} = \sum_{j=1}^{k} n_j (\bar{X}_{.j} - \bar{X}_{..})^2 = 5(180 - 214)^2 + 8(237 - 214)^2$$

$$+ 7(212 - 214)^2 = 10040$$

$$MS_{between} = \frac{10040}{3 - 1} = 5020$$

$$SS_{within} = \sum_{j=1}^{k} \sum_{i=1}^{n_j} (X_{ij} - \bar{X}_{.j})^2 = (145 - 180)^2 + (210 - 180)^2$$

$$+ \ldots + (231 - 212)^2 = 13228$$

$$MS_{within} = \frac{13228}{20 - 3} = 778.12.$$

$$F = \frac{MS_{between}}{MS_{within}} = \frac{5020}{778.12} = 6.45.$$

The ANOVA Table 14.5 summarizes the calculations of Table 14.4.

TABLE 14.5.   ANOVA summary for Example 1.

SOURCE	DEGREES OF FREEDOM	SUM OF SQUARES	MEAN SUM OF SQUARES	F
Between groups	$3 - 1 = 2$	10040	5020	
				6.45
Within groups	$20 - 3 = 17$	13228	778.12	
Total	$20 - 1 = 19$	23268		

At the 5% significance level, the critical value of the F statistic, corresponding to 2 and 17 degrees of freedom, is 3.59. The observed value of the statistic exceeds the critical value, so we reject H_o and conclude that at least one of the population means differs from the others. The color of the book cover does affect sales.

The SPSS procedure ONEWAY performs the calculations associated with one-way analysis of variance and produces the ANOVA table. In the USNEWS file, the variable X1 (region) is nominal, and divides the cases of the file into three groups. The following program tests the null hypothesis that mean changes in department store sales (X2) are equal across the three regions against the alternative hypothesis that at least one regional mean differs from the other two.

```
10.    GET FILE
10.005 USNEWS
20.    ONEWAY
20.005 X2 BY X1(1,3)
```

Note that the levels (treatments) of the factor X1 are encoded numerically. In the specification field of the ONEWAY command, we must give the lowest and highest values of the factor which we want used in the analysis. The output of the program is this:

```
VARIABLE X2          CHANGE IN DEPARTMENT STORE SALES
    BY   X1              CITY REGION

                   ANALYSIS OF VARIANCE
          SOURCE      D.F.    SUM OF SQ.  MEAN SQ.   F RATIO  F PROB
BETWEEN GROUPS         2       329.904    164.952    4.014    .023
WITHIN GROUPS         67      2753.188     41.092
TOTAL                 69      3083.092
```

If the null hypothesis is true, the probability that the F statistic will be at least 4.014 is 0.023. This is less than 5%, a reasonable significance level, so we reject the null hypothesis. We conclude that mean change in department store sales varies across the three regions.

The inclusion of the command STATISTICS; ALL following the ONEWAY statement will cause the printing not only of the ANOVA table but also of the means and standard deviations of the treatment groups (the statistics associated with the BREAKDOWN procedure) and of statistics and probabilities to test the null hypothesis that variance is equal across the treatments (homogeneity of variance). Appending this command to the preceeding program, we obtain:

```
10.     GET FILE
10.005 USNEWS
20.     ONEWAY
20.005 X2 BY X1(1,3)
30.     STATISTICS
30.005 ALL
```

The output of the program is this:

VARIABLE X2   CHANGE IN DEPARTMENT STORE SALES
     BY   X1   CITY REGION

ANALYSIS OF VARIANCE

SOURCE	D.F.	SUM OF SQ.	MEAN SQ.	F RATIO	F PROB
BETWEEN GROUPS	2	329.904	164.952	4.014	.023
WITHIN GROUPS	67	2753.188	41.092		
TOTAL	69	3083.092			

GROUP	COUNT	MEAN	STAND. DEV.	STAND. ERROR	MIN.	MAX	95 PERCENT CONF INT FOR MEAN	
GRP 1	25	8.32	5.78	1.16	− 7.30	19.00	5.94 TO	10.70
GRP 2	30	12.69	7.08	1.29	− 1.70	24.30	10.04 TO	15.33
GRP 3	15	8.27	5.96	1.54	− 3.00	22.10	4.96 TO	11.57
TOTAL	70	10.18			− 7.30	24.30		
UNGROUPED DATA			6.68	.80			8.59 TO	11.77
FIXED EFFECTS MODEL			6.41	.77			8.65 TO	11.57
RANDOM EFFECTS MODEL			2.77	1.60			3.31 TO	17.05

RANDOM EFFECTS MODEL — ESTIM. OF BETWEEN COMPONENT VARIANCE   5.5049

TESTS FOR HOMOGENEITY OF VARIANCES

COCHRANS C = MAX. VARIANCE/SUM(VARIANCES) =   .4212, P =   .428 (APPROX)
BARTLETT-BOX F =   .608, P =   .545
MAXIMUM VARIANCE / MINIMUM VARIANCE =   1.504

A number of statistics of the data values are produced here, with the ANOVA table. Of particular interest to us are Cochran's C and the Bartlett-Box F. Both are used to test these hypotheses:

$$H_o: X_1^2 = \sigma_2^2 = \ldots = \sigma_k^2$$

H_a: At least one variance differs from the others.

The associated P value of each of these statistics—the probability of obtaining a value at least as extreme as that observed under the null hypothesis—is also given, and from them, we can determine whether to reject the null hypothesis of homogeneity of variance. In this case, neither P is less than any reasonable significance level, so we do not reject H_o. The several variances appear to be homogeneous, and we may apply the analysis of variance given above. (Recall that analysis of variance is based on the assumption of homogeneity.)

Tests of Homogeneity of Variance

To test the uniformity of several population means by an analysis of variance, it is necessary that the variances of the populations be the same. For samples from normally distributed populations, several tests of homogeneity of variance have been developed. We describe here those performed by the SPSS procedure ONEWAY with the command STATISTICS; 3. For a study involving k samples, each generates a statistic to test these hypotheses:

$$H_o: \sigma_1^2 = \sigma_2^2 = \ldots = \sigma_k^2$$

H_a: At least one variance differs from the others.

The simplest test of homogeneity of variance is the **Hartley test**, which may be used when all the samples are of the same size n. The test statistic F_{max} is the ratio of the largest sample variance to the smallest:

$$F_{max} = \frac{s_{max}^2}{s_{min}^2}$$

If the population variances are uniform, the statistic will tend to be near one; if F_{max} is large enough, we have evidence that the variances of the several populations are not equal and we reject the null hypothesis of the test. Critical values of this test correspond to $n - 1$ degrees of freedom and to the number k of sample variances. For the 5% and 1% significance levels, critical values of the Hartley test are given in Table A-13.

The Hartley test loses power when the number of samples (treatment groups) grows beyond five. Another test of homogeneity of variance which requires uniform sample sizes and is nearly as simple as Hartley's is the **Cochran test**. The variance for each sample is found, and the test statistic is

$$C = \frac{s^2{}_{max}}{\sum\limits_{i=1}^{k} s_i{}^2}$$

Critical values of C correspond to n and k, and are given for the 5% significance level in Table A-14.

Suppose that in a one-way analysis of variance involving five treatment groups, each of size 10, and selected from normally distributed populations, the following variances are computed:

$$s_1^2 = 16.5; \; s_2^2 = 37.8; \; s_3^2 = 14.1; \; s_4^2 = 22.6; \; s_5^2 = 42.8.$$

The value of Hartley's statistic for these values is

$$F_{max} = \frac{s_{max}^2}{s_{max}^2} = \frac{42.8}{14.1} = 3.04.$$

Corresponding to $n-1=9$ and $k=5$, the critical value of F_{max} at the 5% significance level is 7.11. The value of the statistic is less than 7.11, so the null hypothesis of the test is not rejected; it cannot be concluded that the population variances are unequal.

For the same sample variances, the value of the Cochran statistic is

$$C = \frac{s_{max}^2}{\sum\limits_{i=1}^{5} s_i^2} = \frac{42.8}{16.5 + 37.8 + 14.1 + 22.6 + 42.8} = 0.320.$$

At the 5% significance level, the critical value of C corresponding to $n=10$ and $k=5$ is 0.424. The observed value of the statistic does not exceed the critical value, and again the null hypothesis of equal variances is not rejected.

A test of homogeneity of variance which does not require equal sample sizes, though its calculations are relatively involved, is the **Bartlett test**. For k samples of size $n_j (j = 1, 2, \ldots, k)$ from normally distributed populations, the value of the Bartlett statistic is

$$B = \frac{1}{C} \left[(N-k) \ln \left(\frac{SS_{within}}{N-k} \right) - \left[\sum_{j=1}^{k} [(n_j - 1) \ln s_j^2] \right] \right],$$

$$\text{where } N = \sum_{j=1}^{k} n_j \text{ and } C = 1 + \frac{1}{3(k-1)} \left[\sum_{j=1}^{k} \left(\frac{1}{n_j - 1} \right) - \frac{1}{N-k} \right].$$

For sufficiently large sample sizes ($n_j \geq 5$), the sampling distribution of B is approximately a chi-square distribution with $k-1$ degrees of freedom. Large values of B tend to support the alternative hypothesis of unequal variances, and critical values of B may be found in Table A-6.

Consider the data given in Table 14.1, the first example of this chapter. We perform the Bartlett test of homogeneity of variance for the populations from which the three treatment groups were selected with these calculations:

$$s_1^2 = 809.5 \qquad s_2^2 = 919.1 \qquad s_3^2 = 592.7$$

$$N = 20 \qquad SS_{within} = 13228$$

$$C = 1 + \frac{1}{3(3-1)} \left[\frac{1}{5-1} + \frac{1}{8-1} + \frac{1}{7-1} - \frac{1}{20-3} \right] = 1.083$$

$$B = \frac{1}{1.083} \left[(20 - 3) \ln \left(\frac{13228}{17-3} \right) - \sum_{j=1}^{k} [(n_j - 1) \ln s_j^2] \right]$$

$$= \frac{1}{1.083} [17 \times 6.657 - (5 - 1) 6.696 - (8 - 1) 6.823 - (7 - 1) 6.385]$$

$$= 0.285$$

Note that the C used to find B is not the C of the Cochran test.

Corresponding to $3-1=2$ degrees of freedom, the critical value of B at the 5% significance level is 5.99. The value of the statistic is less than this, so we do not reject the null hypothesis of the test. We do not conclude that the variances of the respective populations of book sales are unequal.

The tests of homogeneity of variance just described are all very sensitive to violations of the assumption of normally distributed populations. Box* examined the effects of departures from normality on the Bartlett test, and developed from it a more robust one. Analysis of variance is less sensitive to departures from normality than are these tests of homogeneity of variance, so when nonnormality is suspected in the underlying populations, a small significance level may be used with the tests of homogeneity of variance.

Multiple Comparisons

When the ANOVA F test results in the rejection of the null hypothesis that all treatment means are equal, this does not indicate which mean differs from another. Some pairs of means may exhibit significant differences while others may not. One category of procedures for making *multiple comparisons* among pairs of treatment means is called *post hoc* or *a posteriori*. Such procedures make all possible comparisons between treatment means; if there are k treatments, there will be $k(k-1)/2$ comparisons. We briefly describe two such tests.

The Scheffé test calculates an F statistic for each comparison being performed, and compares each to a single critical value. If $\bar{X}_i$ and $\bar{X}_j$ are the means of treatment groups i and j containing n_i and n_j values, respectively, and if there are k treatment groups in the experiment, the corresponding F statistic is

$$F = \frac{(\bar{X}_i - \bar{X}_j)^2}{MS_{within}\left(\dfrac{1}{n_i} + \dfrac{1}{n_j}\right)(k - 1)} \quad,$$

where MS_{within} is the mean sum of squares within groups from the ANOVA Table 14.3. If there are N values of the dependent variable in the experiment, then the numbers of degrees of freedom associated with the F statistic are $k-1$ and $N-k$, as in the original analysis of variance.

Consider Example 14.1 in which cover color influenced the number of copies of a book sold. The Scheffé F statistic for the first two treatments is:

$$F = \frac{(180 - 237)^2}{778.12\left(\dfrac{1}{5} + \dfrac{1}{8}\right)(3 - 1)} = \frac{3249}{505.778} = 6.42.$$

Similar computations complete this table of Scheffé F statistics for the three comparisons possible in the example:

GROUPS	F
1, 2	6.42
2, 3	1.50
1, 3	1.92

---

*Box, G.E.P., "Nonnormality and Tests on Variances," *Biometrika*, 1953, 40, 318-335.

Corresponding to $3-1=2$ and $20-3=17$ degrees of freedom, $F_{0.05} = 3.59$. At the 5% significance level, only the difference between $\bar{X}_1$ and $\bar{X}_2$ is significant.

The Scheffé test is very rigorous with respect to Type I errors and will show fewer significant differences than other multiple comparison techniques. The use of a larger significance level, perhaps 10% instead of 5%, is often recommended.

Tukey's test may be applied when the treatment groups are of equal size, and is based on the statistic

$$q = \frac{\bar{X}_i - \bar{X}_j}{(\text{MS}_{within}/n)^{1/2}} ,$$

where n is the size of the treatment groups. Critical values for this test correspond to $N-k$ degrees of freedom, where k is the number of treatment groups and $N=nk$, and are found in Table A-12. A difference between two means is significant if the value of q for that pair is more extreme than the tabled critical value for the chosen significance level.

Note that comparisons made with the Scheffé technique are necessarily two-tailed, while the Tukey test may be one- or two-tailed.

Two-Way Analysis of Variance

In Chapter 13, we extended regression from the simple linear case, in which the dependent variable is related to a single predictor, to multiple linear regression, which involves two or more predictors. Similarly, we perform analysis of variance when a dependent variable is affected by two or more factors, as in this example.

Example 2 A manufacturer of soap products wants to find both the most attractive color for the container of a new shampoo and the most effective advertising campaign for its introduction. The package is produced in four colors, and three advertising campaigns are constructed; these are used in all possible combinations in thirty-six similar market areas. The following sales are recorded in those areas for the first month of the test period:

		PACKAGE COLOR (Sales in $1000s)			
		YELLOW	RED	BLUE	GREEN
	A_1	2.6	1.9	3.9	3.7
		3.0	1.6	3.5	2.6
		2.0	1.3	3.0	3.3
Advertising	A_2	3.6	3.4	4.6	3.4
campaign		3.0	2.6	4.1	3.7
		3.9	2.3	3.7	4.3
	A_3	4.9	3.8	4.9	4.7
		3.9	3.1	4.3	5.0
		4.2	2.8	5.4	4.1

From these data, the manufacturer can derive answers to three separate but related questions. Does the package color affect sales? Does the choice of advertising campaign affect sales? Do the two factors act *together*, interact, to affect sales? (A highly visual television campaign may more successfully promote a brightly colored package.) That is, these three null hypotheses are being tested simultaneously:

H_o: Mean sales are equal across package colors.

H_o: Mean sales are equal across advertising campaigns.

H_o: The factors do not interact to affect sales.

Interaction effects occur when the *response* of the dependent variable X to the levels of factor A changes with the levels of factor B. This is illustrated in Figure 14.1. In graph (a), the mean value of X changes over the levels of factor A in the same way for each level of factor B; that is, factors A and B do not interact. In graphs (b) and (c), the nature of the change of $\bar{X}$ with A depends on the level of factor B; A and B interact.

In general, with n observations of the dependent variable for each *pair* of factor levels, the data may be organized in this way:

		COLUMNS; FACTOR B			
		B_1	B_2	$\ldots$	B_c	ROW MEANS
	A_1	X_{111} X_{112} $\cdot$ $\cdot$ $\cdot$	X_{121} X_{122} $\cdot$ $\cdot$ $\cdot$	$\ldots$	X_{1c1} X_{1c2} $\cdot$ $\cdot$ $\cdot$	$\bar{X}_{1..}$
ROWS;		X_{11n}	X_{12n}		X_{1cn}	
FACTOR A	A_2	X_{211} X_{212} $\cdot$ $\cdot$ $\cdot$	x_{211} X_{222} $\cdot$ $\cdot$ $\cdot$	$\ldots$	X_{2c1} X_{2c2} $\cdot$ $\cdot$ $\cdot$	$\bar{X}_{2..}$
		X_{21n}	X_{22n}		X_{2cn}	
	$\cdot$ $\cdot$ $\cdot$	$\cdot$ $\cdot$ $\cdot$	$\cdot$ $\cdot$ $\cdot$		$\cdot$ $\cdot$ $\cdot$	$\cdot$ $\cdot$ $\cdot$
	A_r	X_{r11} X_{r12} $\cdot$ $\cdot$ $\cdot$	X_{r21} X_{r22} $\cdot$ $\cdot$ $\cdot$	$\ldots$	X_{rc1} X_{rc2} $\cdot$ $\cdot$ $\cdot$	$\bar{X}_{r..}$
		X_{r1n}	X_{r1n}		X_{rcn}	
COLUMN MEANS		$\bar{X}.1.$	$\bar{X}.2.$	$\ldots$	$\bar{X}._c.$	

The mean of the values in cell ij. is $\bar{X}_{ij.}$, the grand mean of all the observations is $\bar{X}...$, and there are a total of $N = nrc$ observations. We assume again that the populations of values corresponding to the samples listed in the cells have common variance σ^2.

FIGURE 14.1. Illustrations of interaction effects in two-way analysis of variance.

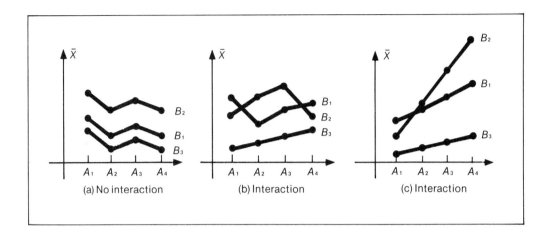

(a) No interaction   (b) Interaction   (c) Interaction

It can be shown that the total sum of squares for such a two-way analysis of variance can be partitioned in this way:

$$SS_{tot} = \sum_{i=1}^{r} \sum_{j=1}^{c} \sum_{k=1}^{n} (X_{ijk} - \bar{X}...)^2$$

$$= cn \sum_{i=1}^{r} (\bar{X}_{i..} - \bar{X}...)^2 + rn \sum_{j=1}^{c} (\bar{X}_{.j.} - \bar{X}...)^2$$

$$+ n \sum_{i=1}^{r} \sum_{j=1}^{c} (\bar{X}_{ij.} - \bar{X}_{i..} - \bar{X}_{.j.} + \bar{X}...)^2$$

$$+ \sum_{i=1}^{r} \sum_{j=1}^{c} \sum_{k=1}^{n} (X_{ijk} - \bar{X}_{ij.})^2 .$$

On the right-hand side of this equation, the first sum of squares, SS_{row}, measures the effect of the row factor (A) on the dependent variable. The second term, SS_{col}, measures the effect of the column factor (B); the third, SS_{int}, measures the effect of interaction of the two factors; the last, SS_{within}, is due to variation within the cells of the table.

There are $(r-1)$ and $(c-1)$ degrees of freedom, respectively, associated with the row and column sums of squares, $(r-1)(c-1)$ degrees of freedom associated with the interaction sum of squares, and $rc(n-1)$ degrees of freedom associated with the sum of squares within cells. Note that there are $N-1 = nrc-1$ degrees of freedom associated with SS_{tot}, and that $(r-1) + (c-1) + (r-1)(c-1) + rc(n-1) = nrc - 1$.

If all the null hypotheses stated above are true, each sum of squares divided by its number of degrees of freedom—each mean sum of squares—is an unbiased estimate of the common variance σ^2 of the X_{ijk}. Ratios of mean sums of squares form F statistics with which to test the hypotheses, as shown in the two-way ANOVA Table 14.6. Computation of the F statistics for row and column effects depends on whether the row and column factor levels are random or fixed.

Note that the caluclations to construct this ANOVA table require equal cell frequencies in the data. For unequal cell frequencies, the calculations become more complicated, but the approach and interpretation remain the same (see Winer, 1962).

Example 14.4 produces the ANOVA Table 14.7. Since no other package colors or ad campaigns are being considered, the effects of both factors are fixed.

If there is significant interaction between the factors, we cannot draw reliable conclusions about their individual effects on the dependent variable. Therefore, we look first at the F statistic associated with the interaction sum of squares. At the 5% significance level, the critical F value corresponding to 6 degrees of freedom in the numerator and 24 in the denominator is 2.51. The F statistic for interaction is 0.33, less than the critical value, so at α = 5% we cannot conclude that the two factors interact to affect sales. We examine the other F statistics to test the row and column effects.

Corresponding to 2 and 24 degrees of freedom $F_{0.05} = 3.40$. $MS_{row}/MS_{within} = 30.50 >$ 3.40, so we conclude that the choice of advertising campaign affects sales. Similarly, for 3 and 24 degrees of freedom, $F_{0.05} = 3.01$. $MS_{col}/MS_{within} = 39.53 > 3.01$; package color also affects sales.

TABLE 14.6.   Two-way ANOVA

SOURCE OF VARIANCE	SUM OF SQUARES	DEGREES OF FREEDOM	MEAN SUM OF SQUARES (MS)	F (fixed effects)	F (random effects)
Rows	$SS_{row} = cn \sum_{i=1}^{r} (\bar{X}_{i..} - \bar{X}...)^2$	$r - 1$	$\dfrac{SS_{row}}{(r-1)}$	$\dfrac{MS_{row}}{MS_{within}}$	$\dfrac{MS_{row}}{MS_{int}}$
Columns	$SS_{col} = rn \sum_{j=1}^{c} (\bar{X}_{.j.} - \bar{X}...)^2$	$c - 1$	$\dfrac{SS_{col}}{(c-1)}$	$\dfrac{MS_{col}}{MS_{within}}$	$\dfrac{MS_{col}}{MS_{int}}$
Interaction	$SS_{int} = n \sum_{i=1}^{r} \sum_{j=1}^{c} (\bar{X}_{ij.} - \bar{X}_{i..}$ $- \bar{X}_{.j.} + \bar{X}...)^2$	$(r-1)(c-1)$	$\dfrac{SS_{int}}{(r-1)(c-1)}$	$\dfrac{MS_{int}}{MS_{within}}$	$\dfrac{MS_{int}}{MS_{within}}$
Within Groups)	$SS_{within} = \sum_{i=1}^{r} \sum_{j=1}^{c} \sum_{k=1}^{n} (X_{ijk} - \bar{X}_{ij.})^2$	$rc(n-1)$	$\dfrac{SS_{within}}{rc(n-1)}$		
Total	$SS_{tot} = \sum_{i=1}^{r} \sum_{j=1}^{c} \sum_{k=1}^{n} (X_{ijk} - \bar{X}...)^2$	$n - 1$			

TABLE 14.7.   ANOVA Table 4, Example 14.2.

SOURCE OF VARIATION	SUM OF SQUARES	DEGREES OF FREEDOM	MEAN SUM OF SQUARES	VALUE OF F
Rows (ad campaign)	14.64	$3 - 1 = 2$	$\frac{14.64}{2} = 7.32$	$\frac{7.32}{0.24} = 30.50$
Columns (color)	28.46	$4 - 1 = 3$	$\frac{28.46}{3} = 9.49$	$\frac{9.49}{0.24} = 39.53$
Interaction	0.45	$(3-1)(4-1)=6$	$\frac{0.45}{6} = 0.08$	$\frac{0.08}{0.24} = 0.33$
Within cells	5.76	$3 \times 4 \times (3-1) = 24$	$\frac{5.76}{24} = 0.24$	
Total	49.31	$36 - 1 = 35$		

Analyses of variance involving two or more factors can be performed in SPSS with the procedure ANOVA (which also performs one-way tests). This example program using the USNEWS file groups the values of X7 in order to use it as a factor, then performs a two-way analysis of variance to determine if X3 (unemployment) is affected by X1 and X7.

```
10.     GET FILE
10.005  USNEWS
20.     RECODE
20.005  X7 ( - 20 THRU 0  =  1)(1.01 THRU 20  =  2)
20.010  (20.01 THRU HI  =  3)
30.     ANOVA
30.005  X3 BY X1(1,3),X7(1,3)
```

The output of the program is this:

ANOVA TABLE

```
    X3      UNEMPLOYMENT
BY X1      CITY REGION
    X7      CHANGE IN CONSTRUCTION ACTIVITY
```
****

SOURCE OF VARIATION	SUM OF SQUARES	DF	MEAN SQUARE	F	SIGNIF OF F
MAIN EFFECTS	2.933	4	.733	.255	.905
X1	2.595	2	1.297	.452	.639
X7	.110	2	.055	.019	.981
2-WAY INTERACTIONS	9.006	4	2.252	.784	.540
X1 X7	9.006	4	2.252	.784	.540
EXPLAINED	11.939	8	1.492	.520	.837
RESIDUAL	175.195	61	2.872		
TOTAL	187.135	69	2.712		

74 CASES WERE PROCESSED.
4 CASES (5.4 PCT) WERE MISSING.

None of the computed F statistics is significant, so we do not reject any of the null hypotheses in the three tests that this analysis performs. We cannot conclude that the mean values of X3 change across X1 or X7, or that X3 is affected by an interaction of X1 and X7. Note that the main effects are the row and column effects.

TWO—WAY ANOVA WITH ONE ENTRY PER CELL

In the two-way ANOVA (Table 14.6) of the previous section, in which each cell of the data table contained several entries, we tested for an interaction effect by computing $F = \mathrm{MS}_{int}/\mathrm{MS}_{within}$, where the denominator of the fraction was computed from the squared deviations of the observed values from their means *within* each cell. When there is only one entry per cell, however, this value is always zero; we cannot test for an interaction effect, nor can we use MS_{within} to form statistics to test for treatment effects. To test for treatment effects in this case, we *assume* that there is no treatment interaction and use MS_{int} in the denominators of the row and column F statistics. Table 14.8 indicates the appropriate calculations for two-way ANOVA with one entry per cell.

Consider this example, simplified from the previous section by using only the first entry in each cell.

		PACKAGE COLOR				
		YELLOW	RED	BLUE	GREEN	
	A_1	2.6	1.9	3.9	3.7	3.03
Advertising campaign	A_2	3.6	3.4	4.6	3.4	3.75
	A_3	4.9	3.8	4.9	4.7	4.58
		3.70	3.03	4.47	3.93	$3.78 = \bar{X}..$

TABLE 14.8. **Calculations for two-way ANOVA with one entry per cell.**

SOURCE OF VARIATION	SUM OF SQUARES	DEGREES OF FREEDOM	MEAN SUM OF SQUARES	F
Rows	$c \sum_{i=1}^{r} (\bar{X}_{i.} - \bar{X}..)^2$	$(r - 1)$	$\mathrm{SS}_{row}/(r - 1)$	$\mathrm{MS}_{row}/\mathrm{MS}_{int}$
Columns	$r \sum_{j=1}^{c} (\bar{X}_{.j} - \bar{X}..)^2$	$(c - 1)$	$\mathrm{SS}_{col}/(c - 1)$	$\mathrm{MS}_{col}/\mathrm{MS}_{int}$
Interaction	$\sum_{i=1}^{r} \sum_{j=1}^{c} (X_{ij} - \bar{X}_{i.} - \bar{X}_{.j} + \bar{X}..)^2$	$(r-)(c-1)$	$\mathrm{SS}_{int}/(r-1)(c-1)$	
Total	$\sum_{i=1}^{r} \sum_{j=1}^{c} (X_{ij} - \bar{X}..)^2$	$(rc - 1)$		

Note the row, column, and grand means in the margins. The ANOVA table derived from these data is Table 14.9.

Corresponding to 2 and 6 degrees of freedom, $F_{0.05} = 5.14$; for 3 and 6 degrees of freedom, $F_{0.05} = 4.76$. In both tests, the F statistic is greater than the critical value, and we reject the null hypothesis. Both package color and advertising campaign affect sales of the new shampoo.

TABLE 14.9. ANOVA derived from simplified table on p.325.

SOURCE OF VARIATION	SUM OF SQUARES	DEGREES OF FREEDOM	MEAN SUM OF SQUARES	VALUE OF F
Rows (ad campaign)	4.81	$3 - 1 = 2$	$\frac{4.81}{2} = 2.41$	$\frac{2.41}{0.22} = 10.95$
Columns (color)	3.29	$4 - 1 = 3$	$\frac{3.29}{3} = 1.10$	$\frac{1.10}{0.22} = 5.00$
Interaction	1.31	$(3 - 1)(4 - 1) = 6$	$\frac{1.31}{6} = 0.22$	
Total	9.30	$12 - 1 = 11$		

Randomized-Block Designs

In the ANOVA designs considered so far, the experimental units have been randomly assigned to each factor level, and we have been interested in the effects on the dependent variable of all the factors in the experiment. Such experimental designs are called *completely randomized*. If an experiment includes two factors of which only one is to be investigated, the effect of the "uninteresting" factor can be controlled through the use of a *randomized-block* design. In the following example, a researcher investigates the mean lifetimes of spark plugs of several brands (the experimental treatments, the factor of interest), while controlling for the types of cars used in the tests.

		SPARK PLUG TYPE (treatments)				
		A	B	C	D	
Auto Type (control)	Y	14.1	15.7	13.7	19.4	Thousands of miles before falure
		17.5	12.5	12.0	19.3	
		18.0	17.8	12.6	16.1	
		16.2	11.2	16.4	21.6	
	Z	15.2	15.2	12.4	21.5	
		18.7	11.8	11.7	20.1	
		18.5	18.2	11.6	15.0	
		15.9	11.2	18.1	21.0	

The model is the same as that for two-way ANOVA with multiple entries per cell (Table 14.6), except for the computation of certain F statistics, as shown in Table 14.10. Also, in the

TABLE 14.10. ANOVA table for a randomized-block design.

SOURCE OF VARIATION	SUM OF SQUARES	DEGREES OF FREEDOM	MEAN SUM OF SQUARES	F
Treatments	$SS_{Tr} = Bn \sum\limits_{j=1}^{B} (\bar{X}_{\cdot j\cdot} - \bar{X}...)^2$	$4 - 1 = 3$	$\dfrac{SS_{Tr}}{(T-1)}$	$\dfrac{MS_{Tr}}{MS_{int}}$
	$= 162.5640$		$= 54.1880$	$= 627.900$
Blocks	$SS_{B1} = Tn \sum\limits_{i=1}^{T} (\bar{X}_{i\cdot\cdot} - \bar{X}...)^2$	$2 - 1 = 1$	$\dfrac{SS_{B1}}{(B-1)}$	$\dfrac{MS_{B1}}{MS_{int}}$
	$= 0.1360$		$= 0.1360$	$= 1.576$
Interaction	$SS_{int} = \sum\limits_{i=1}^{T} \sum\limits_{j=1}^{B} (\bar{X}_{ij\cdot} - \bar{X}_{i\cdot\cdot}$ $- \bar{X}_{\cdot j\cdot} + \bar{X}...)^2$	$(4-1)(2-1) = 3$	$\dfrac{SS_{int}}{(T-1)(B-1)}$ $= 0.0863$	$\dfrac{MS_{int}}{MS_{res}}$ $= 0.012$
	$= 0.2588$			
Residual (res)	$SS_{res} = \sum\limits_{i=1}^{T} \sum\limits_{j=1}^{B} \sum\limits_{k=1}^{n} (X_{ijk} - \bar{X}_{ij\cdot})^2$	$4 \times 2 \times (3-1) = 16$	$\dfrac{SS_{res}}{(n-1)tB}$	
	$= 178.2352$		$= 7.4265$	
Total	$SS_{tot} = \sum\limits_{i=1}^{T} \sum\limits_{j=1}^{B} \sum\limits_{k=1}^{n} (X_{ijk} - \bar{X}...)^2$	$3 \times 2 \times 3 - 1 = 23$	$(nTB-1)$	
	$= 341.1944$			

table, we refer to treatments and blocks rather than columns and rows, with T the number of treatments and B the number of blocks.

Comparing the computed F statistics to the appropriate critical values from Table A-7, we find that we can reject the null hypothesis that the mean spark plug lifetimes are equal across the treatments; we conclude that there is a difference in mean lifetimes among the four brands. We do not reject the null hypotheses of no interaction between treatments and blocks and of no effect by the blocking factor.

Latin Square Design

We have seen that a factor whose effect on the dependent variable we wish to control can be treated as a blocking factor. This technique can be efficiently extended to two blocking factors by using an elegant design called a *Latin square*.

A Latin square is an $n \times n$ array whose elements each appear exactly once in each row and column of the array. By creating a rectangular arrangement of the levels of the two blocking factors, we can assign treatment levels in order to form a Latin square and thereby control the effects of both blocking factors (Figure 14.2).

FIGURE 14.2. **A 3 × 3 Latin square.**

$$
\begin{array}{ccc}
X & Y & Z \\
Z & X & Y \\
Y & Z & X
\end{array}
$$

Suppose we wish to find if differences exist among the mean lifetimes of four brands of tires, T_1, T_2, T_3, and T_4. We test the tires in several climates (A) and on several road surfaces (B). We are not directly concerned with the effect of climate or road surface on tire life in this experiment, so A and B are blocking factors.

If the blocking factors each have four levels, we could arrange them with the treatment levels in their $4 \times 4 \times 4 = 64$ possible combinations, and measure tire lifetime in each case. It is possible, however, to pair each climate with each road condition just once, and assign the treatments in a Latin square to obtain this design:

		ROAD CONDITION				
		B_1	B_2	B_3	B_4	
Climate	A_1	T_1	T_2	T_3	T_4	Treatment
	A_2	T_2	T_3	T_4	T_1	level (brand
	A_3	T_3	T_4	T_1	T_2	of tire)
	A_4	T_4	T_1	T_2	T_3	

Each treatment is paired with each level of both blocking factors, and only 16 experimental measurements are required (though multiple entries per cell are possible). Note that to construct an experiment of this design, the number of levels of the blocking factors must equal the number of treatments.

Let the measurements of the dependent variable (corresponding to the kth treatment in the ith row and jth column) be indicated by X_{ijk}, the treatment means by $\bar{X}_{..k}$, the row means by $\bar{X}_{i..}$, the column means by $\bar{X}_{.j.}$, and the grand mean by $\bar{X}_{...}$. The ANOVA table for a Latin square design with n levels for all the factors is Table 14.11. In the case of multiple entries per cell, X_{ijk} is the mean of the values corresponding to treatment k in position ij.

The statistic $F = MS_{tr}/MS_{res}$ is used to perform a test of these hypotheses:

H_o: The dependent variable means are equal across the treatments.

H_a: The dependent variable means are unequal across the treatments.

TABLE 14.11. ANOVA table for a Latin square design.

SOURCE OF VARIATION	SUM OF SQUARES	DEGREES OF FREEDOM	MEAN SUM OF SQUARES	F
Rows (A)	$n \sum_{i=1}^{n} (\bar{X}_{i..} - \bar{X}...)^2$	$(n-1)$	$\dfrac{SS_{row}}{(n-1)}$	$\dfrac{MS_{row}}{MS_{res}}$
Columns (B)	$n \sum_{j=1}^{n} (\bar{X}_{.j.} - \bar{X}...)^2$	$(n-1)$	$\dfrac{SS_{col}}{(n-1)}$	$\dfrac{MS_{col}}{MS_{res}}$
Interaction (T)	$n \sum_{k=1}^{n} (\bar{X}_{..k} - \bar{X}...)^2$	$(n-1)$	$\dfrac{SS_{Tr}}{(n-1)}$	$\dfrac{MS_{tr}}{MS_{res}}$
Residual	$\sum_{i=1}^{n}\sum_{j=1}^{n}\sum_{k=1}^{n} (X_{ijk} - \bar{X}_{i..}$ $- \bar{X}_{.j.} - \bar{X}_{..k} + 2\bar{X}...)^2$	$(n-1)(n-2)$	$\dfrac{SS_{res}}{(n-1)(n-2)}$	
Total	$\sum_{i=1}^{n}\sum_{j=1}^{n}\sum_{k=1}^{n} (X_{ijk} - \bar{X}...)^2$	$n^2 - 1$		

Analysis of Covariance

In many situations, the effect of an "uninteresting" factor cannot be controlled in the design of the experiment. In some such cases, linear regression can be used to *adjust* the values of the dependent variable to remove the unwanted effect. Analysis of variance is then applied to the adjusted values to determine the effects on the dependent variable of the treatments of interest. This procedure is called *analysis of covariance*, or ANCOVA. Consider this example.

Example 3 Annoying Novelties, Inc., investigates the relationship between sales of its products and three distinct package designs. Knowing that advertising also affects sales, they have recorded advertising expenditures as well as sales totals for several months for each package design. See Table 14.12.

A one-way analysis of variance of sales by package design to test these hypotheses:

H_o: Mean sales are equal across package design.

H_a: Package design affects mean sales.

TABLE 14.12.   **Advertising expenditures and sales of three package designs of Annoying Novelties, Inc.**

			PACKAGE DESIGN		
P_1		P_2		P_3	
Advertising (in $1000s)	Sales (in $1000s)	Advertising (in $1000s)	Sales (in $1000s)	Advertising (in $1000s)	Sales (in $1000s)
4.9	42.1	2.9	44.8	8.6	91.2
5.7	69.8	2.9	44.7	8.5	85.8
4.7	44.5	2.7	38.5	8.8	95.1
5.4	61.6	2.8	38.4	7.4	80.9
4.2	51.7	3.8	50.0	9.5	102.4
6.0	69.0	1.8	25.6	7.4	78.4
5.3	66.9	1.7	26.5	8.6	88.9
4.0	42.9	1.0	26.5	7.0	74.7
4.0	43.1	1.5	29.3	8.6	83.4
6.2	71.6	3.9	47.0		
6.5	65.4	1.9	29.5		
		1.0	18.2		

produces this output:

VARIABLE Y		SALES	
BY P		PACKAGE DESIGN	

ANALYSIS OF VARIANCE

SOURCE	D.F.	SUM OF SQ.	MEAN SQ.	F RATIO	F PROB
BETWEEN GROUPS	2	13820.249	6910.124	61.284	.000
WITHIN GROUPS	29	3269.906	112.775		
TOTAL	31	17090.155			

The negligible probability associated with the F statistic seems to indicate that mean sales do vary with package design. However, this contention ignores the possible influence of advertising on sales, which might be causing the observed variation.

Let advertising expenditures be X, and sales Y. Figure 14.3 is a scatter diagram of the data, with packaging technique also indicated.

There is a linear relationship between X and Y both within each of the treatment groups and across the groups. We must adjust the values of Y (sales) to remove the effects of these relationships. We begin by considering each treatment group separately.

For the cases in treatment group P_1, the least squares regression equation is:

$$\bar{Y}_X = A_1 + B_1 X = -4.26224 + 11.87144X,$$

and the coefficient of determination $R^2 = 0.73244$ is significant at $\alpha = 1\%$. The mean value of X for the first treatment group is $\bar{X}_1 = 5.173$, and we adjust the values of Y in this group by computing

$$Y_{1j}^* = Y_{1j} - B_1(X_{1j} - \bar{X}_1) = Y_{1j} - 11.871\ (X_{1j} - 5.173).$$

FIGURE 14.3. . = **Package design P_1**   x = **Package design P_2**   ∧ = **Package design P_3**

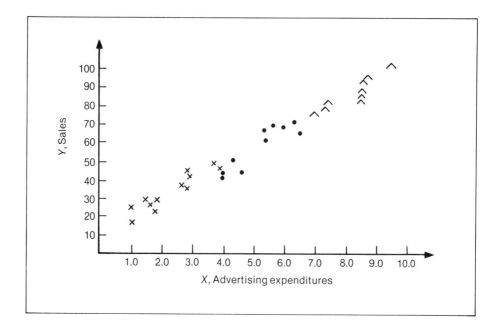

For the first case in treatment group P_1,

$$Y_{11}^* = 42.1 - 11.871\,(4.9 - 5.173) = 45.3.$$

For group P_2, $B_2 = 9.88342$ and $\bar{X}_2 = 2.235$, and for group P_3, $B_3 = 9.84645$ and $\bar{X}_3 = 8.267$. Calculations like that just given produce this table of adjusted values:

PACKAGE DESIGN					
P_1		P_2		P_3	
X	Y*	X	Y*	X	Y*
4.9	45.3	2.9	38.2	8.6	87.9
5.7	63.5	2.9	38.1	8.5	83.5
4.7	50.1	2.7	33.9	8.8	89.9
5.4	58.9	2.8	32.8	7.4	89.4
4.2	63.3	3.8	34.5	9.5	90.3
6.0	59.2	1.8	29.9	7.4	86.9
5.3	65.4	1.7	31.8	8.6	85.6
4.0	56.8	1.0	38.7	7.0	87.2
4.0	57.0	1.5	36.6	8.6	80.1
6.2	59.4	3.9	30.5		
6.5	49.6	1.9	32.8		
		1.0	30.4		

As you can see from the scatter diagram of Figure 14.4, we have adjusted for the linear relationship between sales and advertising *within* each of the three treatment groups, but a linear relationship between these quantities remains *across* the groups.

Assuming that the three regression slopes B_1, B_2, and B_3 are equal, we can find the least squares regression line for X and the adjusted values Y^*:

$$Y_X^* = A + BX = 18.896 + 7.687\ X.$$

The mean of all the values of X is $\bar{X} = 4.975$, and the final adjustment of the Y values takes this form:

$$\hat{Y}_{ij} = Y_{ij}^* - B(X_{ij} - \bar{X}) = Y_{ij}^* - 7.687\ (X_{ij} - 4.975).$$

For the first case in treatment group P_1, for example,

$$\hat{Y}_{11} = 45.3 - 7.687\ (4.9 - 4.975) = 45.9.$$

Similar calculations complete this table of fully adjusted values:

			PACKAGE DESIGN			
	P_1		P_2		P_3	
X	Y	X	$\hat{Y}$	$X,$	$\hat{Y}$	
4.9	45.9	2.9	54.2	8.6	60.0	
5.7	57.9	2.9	54.1	8.5	56.4	
4.7	52.2	2.7	51.4	8.8	60.5	
5.4	55.6	2.8	49.5	7.4	70.8	
4.2	69.2	3.8	43.5	9.5	55.5	
6.0	51.3	1.8	54.3	7.4	68.3	
5.3	62.9	1.7	57.0	8.6	57.7	
4.0	64.3	1.0	69.2	7.0	71.6	
4.0	64.5	1.5	63.3	8.6	52.2	
6.2	50.0	3.9	38.8			
6.5	37.9	1.9	56.4			
		1.0	61.0			

The coefficient of determination of X and the adjusted values $\hat{Y}$ is essentially 0, as illustrated by the scatter diagram of Figure 14.5. It also appears that there is no difference in sales, when adjusted for the effect of advertising, across the three treatment groups.

We test this last contention by performing the following analysis of variance, which uses the fact that variation in the values of the dependent variable, Y, is due to:

1. overall linear dependence between Y and X, the *covariate*,

2. the treatments, P,

3. differences in the linear relationships between X and Y across the treatment groups, and

4. random variation within the treatment groups.

This analysis is performed on the original data values, organized in this way:

FIGURE 14.4.   . = **Package design** P_1 x = **Package design** P_2 ∧ = **Package design** P_3.

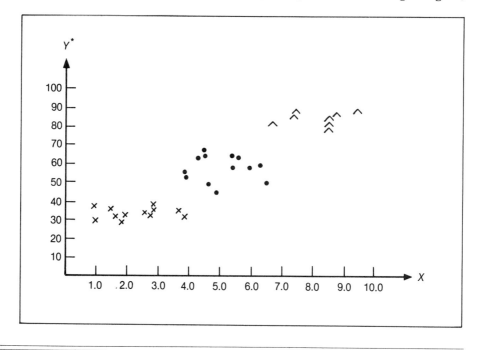

FIGURE 14.5.   . = **Package design 1**   x = **Package design 2**   ∧ = **Package design 3.**

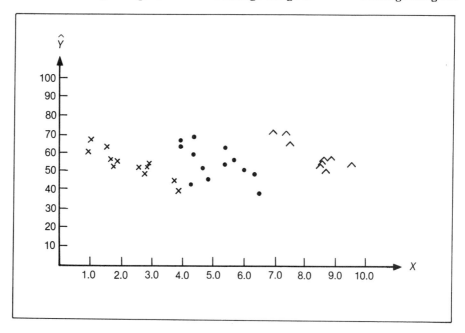

		TREATMENT GROUPS			
	1		2	...	k
X	Y	X	Y	X	Y
X_{11}	Y_{11}	X_{12}	Y_{12}	X_{1k}	Y_{1k}
X_{21}	Y_{21}	X_{22}	Y_{22}	X_{2k}	Y_{2k}
.	.	.	.	.	.
.	.	.	.	.	.
.	.	.	.	.	.
$X_{n_1 1}$	$Y_{n_1 1}$	$X_{n_2 2}$	$Y_{n_2 2}$	$X_{n_k k}$	$Y_{n_k k}$

Various sums of products and squares are computed in these ways:

$$SP_{(j)xy} = \sum_{i=1}^{n_j} (X_{ij} - \bar{X}._j)(Y_{ij} - \bar{Y}._j)$$

$$SP_{(j)xx} = \sum_{i=1}^{n_j} (X_{ij} - \bar{X}._j)^2$$

$$SP_{yy} = \sum_{j=1}^{k} \sum_{i=1}^{n_j} (Y_{ij} - \bar{Y}._j)^2$$

$$SP_{xy} = \sum_{j=1}^{k} SP_{(j)xy}$$

$$SP_{xx} = \sum_{i=1}^{k} SP_{(j)xx}$$

$$S_{xy} = \sum_{j=1}^{k} \sum_{i=1}^{n_j} (X_{ij} - \bar{X})(Y_{ij} - \bar{Y})$$

$$SS_{xx} = \sum_{j=1}^{k} \sum_{i=1}^{n_j} (X_{ij} - \bar{X})^2$$

$$S_{yy} = \sum_{j=1}^{k} \sum_{i=1}^{n_j} (Y_{ij} - \bar{Y})^2$$

Note that these values are related to the variances and covariances of the variables within and across the treatment groups. The ANOVA table for this analysis is shown in Table 14.13: 14.13):

The F statistic MS_{diff}/MS_{within}, with $k-1$ and $N-(k-1)$ degrees of freedom, is used to test the null hypothesis that the regression slopes are equal, while MS_{tr}/MS_{within}, also with $k-1$ and $N-(k-1)$ degrees of freedom, tests the null hypothesis that the mean value of the dependent variable does not vary across the treatment groups.

TABLE 14.13. ANOVA Table for analysis of covariants.

SOURCE OF VARIATION	DEGREES OF FREEDOM	SUM OF SQUARES	MEAN SUM OF SQUARES	VALUE OF F
Overall linear dependence	1	$SS_{lin\ dep} = S_{xy}^2/S_{xx}$	$MS_{lin\ dep} = \dfrac{SS_{lin\ dep}}{1}$	
Treatments	$k - 1$	$SS_{tr} =$	$MS_{Tr} = \dfrac{SS_{tr}}{(k-1)}$	$\dfrac{MS_{tr}}{MS_{within}}$

$$\left(S_{yy} - \frac{S_{xy}^2}{S_{xx}} \right)\left(- SP_{yy} - \frac{SP_{xy}^2}{SP_{xx}} \right)$$

Differences in slopes	$k - 1$	$SS_{diff} =$	$MS_{diff} = \dfrac{SS_{diff}}{(k-1)}$	$\dfrac{MS_{diff}}{MS_{within}}$

$$\sum_{j=1}^{k} \frac{SP_{(j)xy}^2}{SP_{(j)xx}} - \frac{SP_{xy}^2}{SP_{xx}}$$

Within groups	$N - (k-1)$	$SS_{within} =$	$MS_{within} = \dfrac{SS_{within}}{N-(k-1)}$	

$$SP_{yy} - \sum_{j=1}^{k} \frac{SP_{(j)xy}^2}{SP_{(j)xx}}$$

Total	$N - 1$	$SS_{tot} = S_{yy}$		

In our example, we compute the following sums:

$$SP_{(1)xy} = 93.14 \qquad SP_{(2)xy} = 105.94 \qquad SP_{(3)xy} = 52.19$$

$$SP_{(1)xx} = 7.84 \qquad SP_{(2)xx} = 10.72 \qquad SP_{(3)xx} = 5.30$$

$$SP_{yy} = 3269.84 \qquad SP_{xy} = 251.27 \qquad SP_{xx} = 23.86$$

$$S_{yy} = 17090.15 \qquad S_{xy} = 1835.43 \qquad S_{xx} = 206.09$$

From these, we construct the ANOVA Table 14.14.

Corresponding to 2 and 30 degrees of freedom, $F_{0.05} = 3.32$. The treatments F statistic is not greater than this value, so we cannot conclude that sales are influenced by the choice of package design. The observed variation in the values of Y (sales) was due to the influence of the covariate X (advertising). (Note the magnitude of $MS_{lin\ dep}$.) Finally, the F statistic associated with differences in regression slopes is not greater than the critical value, so we do not conclude that the three regression slopes were different.

In SPSS, ANCOVA can be performed by the ANOVA subprogram. In the specification field of the ANOVA command, covariates are specified after the keyword WITH. The program given on page 337 performs the analysis described above.

TABLE 14.14. ANOVA table.

SOURCE OF VARIATION	DEGREES OF FREEDOM	SUM OF SQUARES	MEAN SUM OF SQUARES	VALUE OF F
Linear dependence	1	16346.3	16346.3	
Treatments	$3 - 1 = 2$	120.2	60.1	2.99
Differences in slopes	$3 - 1 = 2$	21.3	10.6	0.53
Within groups	$32 - (3 - 1) = 30$	602.5	20.1	
Total	$32 - 1 = 31$	17090.2		

The following form of the ANOVA table contains no test of equality of the regression slopes; they are assumed equal, and the quantity $SS_{lin\ dep} + SS_{within}$ is given as the residual sum of squares. The residual mean sum of squares is used as the denominator in forming F statistics.

ANOVA TABLE

Y	SALES
BY P	PACKAGING TECHNIQUE
WITH X	ADVERTISING EXPENDITURES

* *

SOURCE OF VARIATION	SUM OF SQUARES	DF	MEAN SQUARE	F	SIGNIF OF F
COVARIATES	16345.284	1	16345.284	732.777	.001
X	16345.284	1	16345.284	732.777	.001
MAIN EFFECTS	120.304	2	60.152	2.697	.085
P	120.304	2	60.152	2.697	.085
EXPLAINED	16465.588	3	5488.529	246.057	.001
RESIDUAL	624.567	28	22.306		
TOTAL	17090.155	31	551.295		

32 CASES WERE PROCESSED.
0 CASES (0 PCT) WERE MISSING.

Given the equality of regression slopes, however, our conclusions are the same: variation in the values of Y was due to the effect of the covariate X; we cannot conclude (at $\alpha = 5\%$) that the mean values of Y vary across the factor P.

```
10.     FILE NAME
10.005  NOVELTY
20.     VARIABLE LIST
20.005  LINE,P,X,Y
30.     INPUT FORMAT
30.005  FIXED (2F2.0,F3.1,F4.1)
40.     N OF CASES
40.005  32
50.     VAR LABELS
50.005  P PACKAGING TECHNIQUE/
50.010  X ADVERTISING EXPENDITURES/Y SALES
60.     ANOVA
60.005  Y BY P(1,3) WITH X
```

EXERCISES

1. Annoying Novelties, Inc., has been testing the relative effectiveness of three advertising media in three equivalent market areas. Sales, in thousands of dollars, are given in the following table for twelve months randomly chosen from the past three years. The values are organized by advertising medium.

	ADVERTISING MEDIUM		
MONTH	NEWSPAPER	RADIO	TELEVISION
1	79	63	72
2	85	84	82
3	94	73	85
4	90	80	73
5	82	97	72
6	78	75	93
7	91	81	93
8	89	87	93
9	86	85	89
10	83	92	91
11	94	74	92
12	90	80	92

   a. State formally the hypotheses being tested.

   b. These are the only advertising media being considered. Is the factor a fixed-effects one or a random-effects one?

   c. At the 5% significance level, are mean sales unequal across the three advertising media?

2. For the past seven months, a computer center has recorded the number of jobs submitted in each of four programming languages:

		LANGUAGE			
Month	BASIC	PASCAL	COBOL	FORTRAN	
1	1	4	14	2	
2	5	9	18	7	
3	2	5	9	3	
4	6	10	21	10	Hundreds
5	7	12	20	7	of jobs
6	3	5	12	5	
7	3	8	16	6	

a. We want to test the null hypothesis that the mean number of jobs is equal across the four languages, but first, calculate the variance of each of the four samples. What do these values indicate about the validity of the ANOVA procedure?

b. Note that for each sample the variance is proportional to the mean. Form a new data file by replacing each data value by its square root.

c. Find the four sample variances of the transformed values. Do the transformed samples satisfy the condition of equal variances?

d. Perform the hypothesis test described in (a), using the transformed values, at $\alpha = 5\%$. What conclusion do you reach?

3. Over a period of several months, a quality inspector for a chain of restaurants eats dinners at the chain's five outlets, rating the food and service at each meal on a 50-point scale. These are her ratings:

			RESTAURANT		
	14TH STREET	HIGHWAY 6	WESTVALE MALL	DOWNTOWN	AIRPORT ROAD
	30	41	38	37	40
	34	47	30	42	35
	26	44	34	45	43
Rating	33	39	35	39	41
		48	29	44	37
		43			40

a. According to the inspector's ratings, are the chain's five outlets of uniform quality? State formally the hypotheses being tested, and use $\alpha = 5\%$.

b. Do these five samples satisfy the condition of homogeneity of variance? Of what importance is this for the result of part (a)?

4. A consumer group investigates the price of a 1-pound loaf of Bake-rite wheat bread at six grocery stores in each of three cities. They find these prices:

CITY	PRICES (in cents)					
N. Umbrage	77	79	77	81	80	78
Weston	75	74	74	71	73	71
Midland	69	73	69	69	72	74

a. At the 5% significance level, can it be concluded that the mean price of Bake-rite wheat bread varies from city to city?

b. Is homogeneity of variance satisfied? What importance does this have for the result of part (a)?

5. Apply the Hartley and Cochran tests of homogeneity of variance to the three samples of values given in Exercise 1. At the 5% significance level, should we conclude from either test that the variances of the populations from which the samples were drawn are not the same? What assumption do we make to apply these tests?

6. In Exercise 2, the data values are transformed to ensure that homogeneity of variance is satisfied. Use the Hartley and Cochran tests on the original data to determine, at the 5% significance level, if this transformation is necessary.

7. Test the samples of values given in Exercise 1 for homogeneity of variance with the Bartlett test at the 5% significance level.

8. Use the Bartlett test at the 5% significance level to determine if the samples of values given in Exercise 2 do not satisfy the ANOVA assumption of equal population variances.

9. Create a raw data file and use the SPSS procedure ONEWAY to verify the results of Example 14.1 which involves books with covers of different colors.

10. Use SPSS to perform the analysis of variance of Exercise 1. Note that you must prepare and enter the given data as a raw data file.

11. Use SPSS to perform the analysis of variance and associated tests of Exercise 3. You must enter the given data as a raw data file.

12. Use ONEWAY to perform the hypothesis tests of Exercise 4. You must enter the data as a raw data file.

13. Use SPSS to perform the analysis described in Exercise 2. To transform the data as described, you can use the command COMPUTE. Note the effect on the sample variances of replacing each value by its square root.

14. Use the USNEWS file and the SPSS procedure ONEWAY to test, at the 5% significance level, the null hypothesis that the mean change in nonfarm employment does not change over the three regions. Be sure to state formally the hypotheses being tested.

15. Repeat Exercise 14 for mean change in construction activity across region. Use $\alpha = 5\%$.

16. In Exercises 14 and 15, can we continue to accept the null hypothesis of homogeneity of variance? Use ONEWAY and $\alpha = 10\%$. Why is this important?

17. In one-way analysis of variance, show that $\bar{X}_{.j}$ is an unbiased estimator of $\mu_{.j}$.

18. In one-way analysis of variance, show that

$$E(SS_{\text{between}}) = (k - 1)\sigma^2 + \sum_{j=1}^{k} n_j(\mu_j - \mu)^2.$$

19. Show that, in a one-way analysis of variance with k treatments, the number of possible comparisons between pairs of means is $k(k-1)/2$.

20. In Exercise 4, we concluded that the mean price of Bake-rite wheat bread varied from city to city. Use the Scheffé F test at the 5% significance level to determine which pairs of treatment means show significant differences.

21. In Exercise 3, the inspector found significant variation in quality across the chain's five restaurants. Use the Scheffé F test at the 5% significance level to determine which pairs of restaurants differ significantly in quality.

22. In Exercise 14, we found a significant difference in the mean values of X4 across the three regions. Use the Scheffé F test at $\alpha = 5\%$ to determine which pairs of treatment means exhibit significant differences. (*Hint:* You can use ONEWAY to find all the values you need to calculate the F statistics.)

23. Repeat Exercise 20 using Tukey's test to determine, at the 10% significance level, which pairs of treatment means show significant differences.

24. The following table contains the typing speeds, in words per minute, of 36 secretaries:

	H.S. DIPLOMA		A.S. DEGREE		B.S. DEGREE	
Civil Servant	29	31	23	62	17	32
	26	50	31	60	18	49
	42	25	18	20	50	58
Non-Civil Servant	17	62	35	83	17	28
	27	62	50	42	14	58
	50	29	62	19	29	62

Use two-way analysis of variance to test for significant row, column, and interaction effects on the dependent variable typing speed. Formally state the three pairs of hypotheses being tested, and state your conclusions in terms of the data file variables. Use $\alpha = 5\%$.

25. In a study of the relative effectiveness of several teaching methods, 45 students of low, moderate, or high aptitude participate in a short course taught in one of three distinct ways. Then all the students are tested on the material. These are their scores:

	METHOD		
APTITUDE	LECTURE	SMALL GROUP DISCUSSION	INDIVIDUAL STUDY
Low	45	63	58
	64	65	68
	58	80	75
	53	73	70
	49	78	74
Moderate	57	84	80
	75	88	76
	65	79	87
	68	80	88
	76	75	82
High	84	90	96
	91	87	99
	76	98	87
	87	95	92
	86	85	89

Use two-way analysis of variance to test for significant row, column, and interaction effects of the two factors on the students' scores, using $\alpha = 5\%$. State the three pairs of hypotheses being tested here, and state your conclusions in terms of the variables aptitude, method, and score.

26. Create a raw data file and use the SPSS procedure ANOVA to perform the tests described in Exercise 24.

27. Create a raw data file of the values given in Exercise 25, and use the SPSS procedure ANOVA to perform tests for row, column, and interaction effects.

28. In the USNEWS file, group the values of X4 into three classes with a RECODE command, then perform an analysis of variance to determine the effects of X1 and X4 on X3. Use $\alpha = 5\%$.

29. What assumptions must be satisfied to employ a

    a. completely randomized design?

    b. randomized block design?

30. In a two-way analysis of variance, what assumption must we make when we have only one entry per cell?

31. Consider these data, simplified from Exercise 25 by extracting the first entry in each class:

		METHOD		
APTITUDE	LECTURE	SMALL GROUPS DISCUSSION	INDIVIDUAL STUDY	
Low	45	63	58	
Moderate	57	84	80	
High	84	90	96	

Based only on these data, can we conclude, at the 5% significance level, that test scores vary across teaching method or aptitude?

32. Construct a 5 × 5 Latin square using the characters *A, B, C, D,* and *E.*

33. The Amalgamated Company measures productivity in a factory under these conditions:

    A : complete silence

    B : piped-in popular music

    C : piped-in country-western music

    D : piped-in classical music

    They control the influences of workday and job type by using a Latin square design, and accumulate these data, where the cell totals are in thousands of units. Perform the Latin square ANOVA on these data with $\alpha = 5\%$. What conclusions do you reach?

		JOB TYPE		
DAY	1	2	3	4
Monday	A(1.4)	B(1.6)	C(1.5)	D(1.7)
Tuesday	B(1.1)	A(1.0)	D(1.9)	C(1.5)
Wednesday	C(2.8)	D(2.7)	B(2.2)	A(2.1)
Thursday	D(3.4)	C(2.9)	A(2.5)	B(2.7)

34. To predict future equipment needs, a facilities manager evaluates three methods of data entry by means of this experiment. Eighteen newly hired trainees were given an aptitude test; scores on that test are the values of the variable *X* below. The trainees were divided into three groups, and for five weeks, each group entered data by one entry method. At the end of that time, their efficiency was evaluated. Their ratings are the values of *Y.*

				DATA ENTRY METHOD				
BATCH			REMOTE BATCH			TIME-SHARING		
TRAINEE	X	Y	TRAINEE	X	Y	TRAINEE	X	Y
A	32	62	G	21	72	M	38	95
B	46	66	H	24	85	N	45	88
C	27	64	I	18	61	O	52	104
D	35	48	J	32	87	P	48	86
E	31	57	K	35	69	Q	41	72
F	40	74	L	26	74	R	37	63

a. Plot Y_{ij} against X_{ij} for all the values.

b. For each group of values, find $\bar{X}_{.j}$, and calculate the values $Y_{ij}^* = Y_{ij} - B_j(X_{ij} - \bar{X}_j)$, where B_j is the slope of the regression line of the values X_{ij} and Y_{ij} for the jth group. Plot the adjusted values Y_{ij}^* against X_{ij}.

c. Find the slope B of the regression line of Y^* with X, and $\bar{X}$, the mean of all the X values. For each Y value, find $\hat{Y}_{ij} = Y_{ij}^* - B(X_{ij} - \bar{X})$, and plot a scatter diagram of $\hat{Y}$ with X.

d. What conclusions can you reach from parts (a), (b), and (c)?

e. Perform an analysis of variance of $\hat{Y}$ by data entry method, and compare the results with those of an analysis of covariance of Y by data entry method with aptitude.

35. A one-way classification with three treatments produced responses Y and associated observations X as shown in this table:

				TREATMENTS				
	1			2			3	
	X	Y		X	Y		X	Y
	6	41		3	42		8	39
	8	41		9	65		9	38
	2	23		10	70		12	61
	7	42		6	44		4	22
	9	56		4	29		7	26
	4	47		5	36		1	3
	8	65		8	42		8	33
	7	51		4	68		6	17
	6	41					9	14
	5	39					4	22
	4	57						

a. At the 5% significance level, must we conclude that the slope of the regression line of Y with X varies across the treatment groups? Use $\alpha = 5\%$.

b. Complete the analysis of covariance required to test for treatment effects on Y at the 5% significance level, while controlling for X.

36. Using the data given in Exercise 34, create a raw data file and use the SPSS procedure ANOVA to perform an analysis of covariance of efficiency score by data entry method while controlling for aptitude test score.

37. Create a raw file of the data given in Exercise 35 and use the SPSS procedure ANOVA to perform an analysis of covariance of Y by the treatment levels while controlling for X. Use the 5% significance level.

38. Using the USNEWS file and the SPSS procedure ANOVA, perform an analysis of covariance of X3 by X1 while controlling for X7. What conclusions, if any, do you reach?

Experimental Designs

INTRODUCTION

We have examined a number of ways of analyzing the data generated by an experiment, but have not dealt with creating the experiment. How does a researcher construct an experiment? What factors are important in its design? These are the fundamental questions in the area of experimental design.

Classification of Designs

The purpose of an experiment is to discover the effect of certain independent variables on a dependent variable. The outcome of the experiment will be measured in terms of variation in the value of the dependent variable.

Often, an effort will be made to control the possible effects of (independent) variables not intended to be part of the experiment, whose presence might influence the value of the dependent variable and thereby muddy the results. A *completely randomized block design* maintains such control by appropriate construction of the experiment, while the *analysis of covariance* model is a statistical technique to control the effects of extraneous variables or concomitant examples.

Another goal in experimental design is the maintenance of a balance between replicability of the research results and the ability to generalize them. Certainly it is desirable that the researcher be able to extend the results of his or her study to individuals/cases/situations which were not part of the experiment, but it should also be possible for others to replicate his or her results.

Before we define types of experimental designs which might meet these goals, we first distinguish two approaches to applied research:

Note that in *experimental research*, the experimenter *determines* values of the independent variables being examined, while in *ex post facto research*, he or she *observes* values of the independent variables, but does not influence them.

The statistical analyses applied to the data are not determined by the type of research employed. They will be the same in either case, and the following algorithm describes the analytical process:

   I. Observe the system to be studied.

  II. If the problem is defined, continue. ELSE go to I.

 III. If variables can be identified, continue. ELSE go to I.

 IV. If hypotheses can be formulated, continue. ELSE go to II.

  V. Establish the research design.

 VI. Determine the sampling procedures.

VII. Collect the data.

VIII. Perform the statistical analysis.

 IX. Interpret the results.

  X. If results correspond to the problem, continue. ELSE go to V.

 XI. Report the results.

When we identify ways to control the variation of extraneous independent variables, we have completely described the components of the system under study: the dependent variable, the independent variables whose effects on the dependent variable concern us, and other independent variables whose influence on the dependent variable we wish to control. There are four major methods for controlling the impact of these *control variables*:

a. holding such variables constant,

b. randomizing their effects,

c. matching experimental units with respect to values of the control variables, and

d. controlling their impact statistically.

We have used the technique of holding all the independent variables constant except the one being studied when we observed the drop in R^2 from the full to a reduced model in multiple linear regression. Randomization is possibly the "best" method of control; experimental units are assigned randomly to the various experimental treatments (values of the independent variable being studied) with respect to values of the control variables. There are several ways to match experimental units to control the impact of an independent variable. In one, the elements of the experimental group are stratified by values of the control variable, and these strata are then distributed among the treatments. In another, the sample is divided into equal parts corresponding to the number of values of the control variable and the experiment is repeated several times. Finally, control can be achieved statistically whenever the control variable is correlated with the dependent variable, since the control variable then accounts for part of the variation of the dependent variable. The researcher need only eliminate that portion of the variation from the analysis.

We can now define some types of experimental design.

DEFINITION

Classical experimental design:

a. Two groups are formed either by random assignment or by matching.

b. The groups are identical before the experiment begins.

c. During the experiment, one group (the *treatment group*) receives the experimental treatment and the other (the *control group*) does not.

d. After the experiment, differences in the values of the dependent variable from the control to the treatment group can thus be attributed to the experimental treatment.

Often, practical limitations—cost, time, etc.—prevent this design from being fully realized. In some cases, for example, a similar but nonequivalent control group is used. Or the researcher will construct a comparison group by dividing an intact group into treatment and comparison components, according to the value of some measure taken before administration of the experimental treatment. Designs with nonequivalent control groups are called *quasi-experimental designs*.

The simplest experimental design is called the

DEFINITION

Simple randomized design:

a. Only one independent variable and the dependent variable are included in the analysis.

b. Control is achieved by randomization, and possibly by holding other variables constant.

One-way analysis of variance is an example of such a design when the dependent variable is interval level. When the dependent variable is nominal, we have used the chi-square test.

The inclusion of more independent variables in the analysis leads to *factorial designs* in which

1. The researcher conducts several experiments simultaneously with the same experimental units, and

2. The possible effects of interactions between independent variables can be studied.

We make the following definition.

Factorial designs:

a. Two or more independent variables, called *factors*, are studied simultaneously.

b. No provisions are made for controlling the effects of other independent variables on the dependent variable other than randomization and possibly holding control variables constant.

c. Given the factors A and B (see Figure 15.1),

the following tests are performed.

H_o : row means are equal $(\mu_{A_1} = \mu_{A_2} = \ldots = \mu_{A_1})$

H_a : row means are unequal

H_o : column means are equal $(\mu_{B_1} = \mu_{B_2} = \ldots = \mu_{B_k})$

H_a : column means are unequal

H_o : cell means have equal trends across treatments (values of B)

H_a : cell means unequal trends across treatments

Factorial designs have been illustrated by multiple analysis of variance, but they may also be analyzed with multiple regression.

Simple randomized designs can be improved by using samples matched with respect to the *blocking variable*. Such designs are called

Randomized block designs:

a. The population is stratified by values of the blocking variable.

b. Subsamples are drawn from each strata in such a way that the distribution of values of the blocking variable in the entire sample grossly matches that of the population.

c. Given the factor A and blocking variable B, this hypothesis is tested (see Figure 15.2):

H_o : the column means are equal $(\mu_{A_1} = \mu_{A_2} = \ldots = \widetilde{\mu}_{A_k})$

H_a : the column means are unequal

FIGURE 15.1.   Factorial design. (A and B are factors).

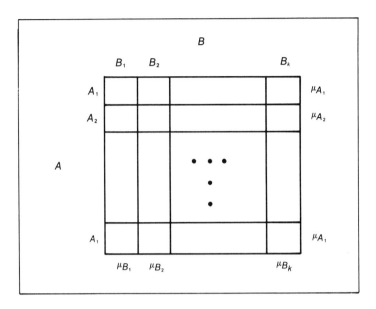

FIGURE 15.2.   Randomized block design. (A is a factor and B is a blocked factor).

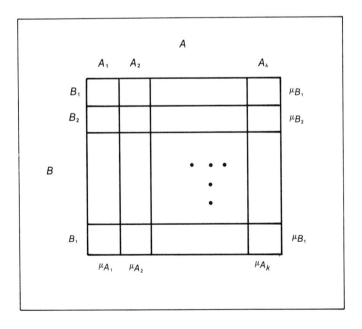

Whether the hypotheses of equal row means or interaction are to be tested will be a function of the particular experiment.

Such a design has three distinct advantages over a simple randomized design:

1. The stratification gives some control over the blocking variable.

2. Because of the stratification, the analysis is more sensitive to possible relations between the dependent variable and the factor.

3. Data for the independent variable can be studied separately.

Both analysis of variance and multiple regression can be used with the randomized block design.

If the effects of all treatments are temporary, the researcher may use a

---

DEFINITION

Repeated measure design: the experimental treatments are administered in order to each experimental unit.

The statistical techniques applied to data generated by such a design are the same as those used with a randomized block design.

Sometimes it is impossible to obtain experimental units for every cell of a factorial design, as in Figure 15.3.

Note that only one cell in each row contains data. Such experiments are generally analyzed as simple randomized designs, and are called *experimental designs with nested variables*. The factor B is *nested* in A in Figure 15.3.

Analysis of covariance extends the classical experimental designs described above by combining techniques of prediction with the analysis of variance. Prediction of the dependent variable from the control variables allows the adjustment of the values of the dependent variable. These adjusted values are then analyzed using the usual experimental designs.

FIGURE 15.3. Experimental design with nested variable B in A.

	A_1	A_2	A_3
B_1	not available	not available	
B_2	not available	not available	
B_3		not available	not available

B

Avoiding Pitfalls

In any research endeavor, it is wise to follow these guidelines:

1. Select a meaningful dependent variable.

2. When using a pre-test, do not use both select and control experimental units with the same measurements of the dependent variable.

If only one set of measurements of the dependent variable is used, when the selection is made it will alter, by what is called the *regression effect*, the results of the control mechanism.

3. Use the same instrument for pre- and post-testing.

4. Make sure that the treatment and control groups are comparable.

5. When the treatment and control groups are nonequivalent, use the appropriate statistical adjustments.

In particular, avoid the use of *raw gain scores*, the differences between pre- and post-test scores, unless the standard deviations of the pre- and post-test scores are equal.

6. Do not construct a matched control group after the treatment group has been selected.

If this is done, regression will act differently on the two groups, and will inflate apparent differences.

7. Be careful in the collection of data.

8. Avoid missing data. The analysis should be based on experimental units present in both control and experimental groups.

9. Do not generalize beyond the information extracted from the experiment and the results of the statistical analysis.

Experimental Goals

Upon completion of his or her research, the researcher should be able to answer the following questions:

1. Is there a significant difference in the value of the dependent variable which can be attributed to one or more of the factors?

2. Is this difference statistically significant?

3. Is the effect of factors on the dependent variable significant in terms of the problem?

4. Can the treatments of the experiment be applied in another setting with a reasonable expectation of similar results?

5. How likely is it that the observed effects are results of the treatments?

6. Are the results believable and interpretable?

EXERCISES

1. Discuss this statement: "Some of the most important variables in an experiment cannot be experimentally manipulated."

2. How can research be done if the experimenter cannot intervene, but must only observe the variables in their natural setting?

3. **a.** Does ex post facto research lend itself to the study of cause-and-effect relationships? Why or why not?

   b. Does experimental research lend itself to the study of cause-and-effect relationships? Why or why not?

4. Give an example illustrating that randomization is most effective with large samples.

5. Discuss this statement: "In a simple randomized design, the treatment samples are not simple random samples."

6. The matched-pairs design is a special case of randomized block design. The experimental units are paired by value of the blocking variable, then one experimental unit from each pair is assigned to each treatment group. Why is multiple regression analysis inappropriate with such a design?

7. When dealing with interval data, what statistical technique should be used for a repeated measures design with

   a. two groups?

   b. more than two groups?

8. How will a raw score analysis relate to the estimated treatment effect if the post-test score standard deviation is larger than that of the pre-test score?

9. Write a checklist to evaluate a journal article or research paper. (*Hint*: Write a list of what every paper should include.)

READINGS

Campbell, D.T., and Stanley, J.C.. *Experimental and Quasi-experimental Designs for Research*. Chicago: Rand McNally, 1966.

Winer, B. J. *Statistical Principles in Experimental Design*, 2nd ed. New York: McGraw-Hill, 1971.

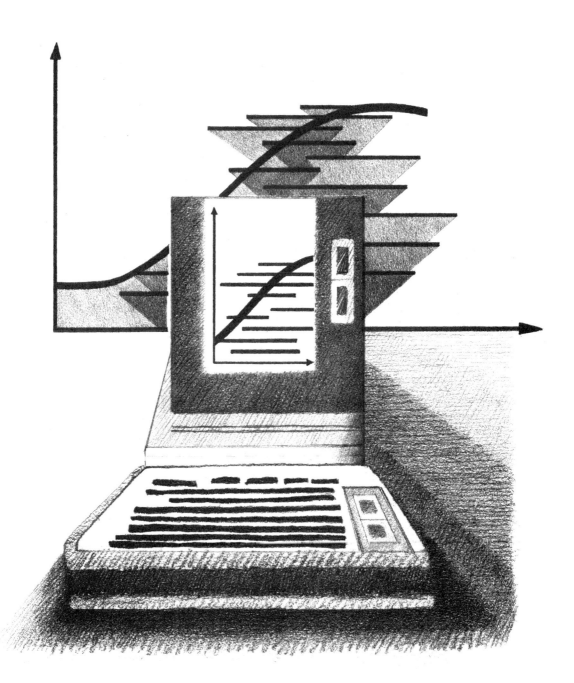

Nonparametric Tests

INTRODUCTION

In almost all our work so far, we have paid careful attention to the shapes of population distributions in assessing the applicability of our statistical analyses and tests. For example, to apply a t-test when using small samples, we must establish the normality of the underlying population(s); in linear regression we assume normality and homoscedasticity. That is, our tests thus far have been parametric, *depending on the characteristics of the underlying distributions.*

We now consider the class of nonparametric tests, *in which we need make few if any assumptions about the underlying distributions. In general, such tests are more widely applicable than parametric tests, but less powerful. We examine tests for randomness and goodness-of-fit, then nonparametric substitutes for several common parametric tests.*

Tests of Randomness

We sometimes wish to determine if a sequence of values is random, with no pattern or relationship among the values, as if each had been chosen at random from some range of values, or is not random, showing some pattern. Such investigations take the form of hypothesis tests with these hypotheses:

H_o: The values are random.

H_a: The values are not random.

The decision rule may be formulated in a variety of ways, of which we will examine four.

THE FREQUENCY TEST

Suppose that a sequence of n values is to be chosen, at random and one at a time, from a list of k candidate values. At each selection of a random value to be included in the sequence, each candidate value has probability k^{-1} of being selected. The number of times a given value appears has the binomial distribution $b(n, k^{-1})$, with mean $n \times k^{-1} = n/k$; we would expect each of the candidate values to appear in the sequence of n selected values approximately n/k times, *if the selection at each step is in fact random*. Significant deviations of the observed frequencies of the values from the predicted frequencies would suggest that the values are not random, but that some values are more likely than others.

We have used, in Chapter 11, tests based on a statistic with a chi-square distribution to compare predicted and observed frequencies, and we do so again here.

If the null hypothesis is true and the values are truly random, then the statistic

$$\chi^2 = \sum_{i=1}^{k} \frac{(O_i - n/k)^2}{n/k} = \frac{k}{n} \sum_{i=1}^{k} \left(O_i - \frac{n}{k}\right)^2 ,$$

where O_i is the observed frequency of each value, has a chi-square distribution with $k - 1$ degrees of freedom.

If the value of the chi-square statistic is near 0, the observed and expected frequencies must be similar, and we cannot conclude that the values are not random. If X^2 is large, the observed and expected frequencies must differ, and the values are not random. We may use the chi-square table in the usual way to find the critical value of the chi-square statistic above which we may reject H_o and accept H_a, concluding that the values are not random.

For example, consider this list of 50 digits from 0 to 4:

4 3 2 3 3 2 2 1 3 1 1 0 2 0 2 1 3 0

1 4 0 0 1 4 0 2 3 2 1 1 0 2 3 4 0 0

2 4 4 2 2 1 4 4 0 3 1 1 2 1

At the 5% significance level, we test the null hypothesis that these values are random against the alternative that they are not.

The digits are selected from a list of 5 possible values, 0, 1, 2, 3, and 4, so $k = 5$, while $n = 50$. Corresponding to $k - 1 = 4$ degrees of freedom, the critical value is $X^2 = 9.488$.

Each expected frequency is $50 \times \frac{1}{5} = 10$, while the five observed frequencies are

$$O_o = 10;\ O_1 = 12;\ O_2 = 12;\ O_3 = 8;\ O_4 = 8.$$

The value of the chi-square statistic is, then

$$
\begin{aligned}
X^2 &= \frac{5}{50} \sum_{i=0}^{4} (O_i - 10)^2 \\
&= \frac{1}{10} \left[(10-10)^2 + (12-10)^2 + (12-10)^2 + (8-10)^2 \right. \\
&\quad \left. + (8-10)^2 \right] \\
&= \frac{1}{10} \left[0+4+4+4+4 \right] \\
&= \frac{16}{10} = 1.6.
\end{aligned}
$$

This value is less than the critical value, so we do not reject the null hypothesis; we cannot conclude that these values were not randomly selected.

The SPSS subprogram NPAR TESTS will perform chi-square tests to compare expected and observed frequencies. With the numbers above in the raw data file NUMBERS, this program will perform the test in the previous example:

```
10.      FILE NAME
10.005   NUMBERS
20.      VARIABLE LIST
20.005   NUM
30.      INPUT FORMAT
30.005   FREEFIELD
40.      N OF CASES
40.005   50
50.      NPAR TESTS
50.005   CHI-SQUARE = NUM/EXPECTED = 10,10,10,10,10
```

Note the listing of the five expected frequencies, all equal to 10.

The output of the program coincides with our earlier work; the significance of 0.8088 is far above any significance level, and we do not conclude that the values are random.

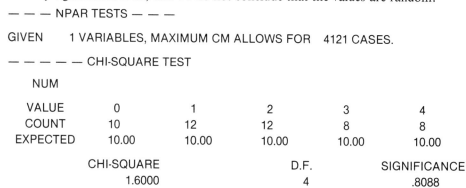

--- NPAR TESTS ---

GIVEN   1 VARIABLES, MAXIMUM CM ALLOWS FOR   4121 CASES.

----- CHI-SQUARE TEST

NUM

VALUE	0	1	2	3	4
COUNT	10	12	12	8	8
EXPECTED	10.00	10.00	10.00	10.00	10.00

CHI-SQUARE	D.F.	SIGNIFICANCE
1.6000	4	.8088

If the values are chosen from a much larger list of possible values (that is, if k is very large) the values can be grouped into classes, and the expected and observed frequencies of the classes of values compared. In this case, the associated number of degrees of freedom is one less than the number of classes. An SPSS program performing such a test would group the values with a RECODE command.

THE GAP TEST

A *gap* in a sequence of random numbers is the number of values *between* two identical values. For example, in the sequence 6 5 3 2 3 6 1 . . ., there is a gap of length 4 between the first and second occurrence of the digit 6, and a gap of length 1 between the first and second 3. In a sequence of digits chosen randomly from a set of k candidate values, the length of a gap after a given value is a random variable X with probability function

$$f(x) = P(X = x) = \left(1 - \frac{1}{k}\right)^x \cdot \frac{1}{k} , \qquad x = 0, 1, 2, \ldots .$$

That is, a gap of length x occurs after a particular value v when the next x values are not v, and the $(x + 1)$st value is v. This occurs with probability.

$$\left(1 - \frac{1}{k}\right)^x \cdot \frac{1}{k} .$$

If a sequence of random values contains N gaps, the expected number of gaps of length x is

$$N \cdot f(x) = N \cdot \left(1 - \frac{1}{k}\right)^x \cdot \frac{1}{k} .$$

To investigate the randomness of the sequence, we can compare the expected and observed numbers of gaps of each length using the chi-square statistic as in the frequency test. The associated number of degrees of freedom will be one less than the number of gap lengths considered.

Consider again the sequence of 50 digits from 0 to 4 tested in the previous example. There are $N = 45$ gaps in the sequence, whose values were chosen from the digits 0, 1, 2, 3, and 4. Since the probability of a gap of, say, length 2 is

$$\left(1 - \frac{1}{5}\right)^2 \cdot \frac{1}{5} = \frac{16}{125} = 0.28,$$

the expected number of gaps of length 2 is

$$\frac{16}{125} \times 45 = 5.76.$$

Finding the other expected gap frequencies in the same way, and counting the observed numbers of gaps, we obtain Table 16.1.

TABLE 16.1. The gap test of randomness.

| | DIGITS | | | | | GAPS | |
LENGTH	0	1	2	3	4	OBSERVED	EXPECTED
0	2	3	2	1	2	10	9.0
1	1	2	1	1		5	7.2
2	1	1	2			4	5.76
3	3	1	1	1	3	9	4.61
4		2	1			3	3.69
5	1	1	2	1		5	2.94
6						0	2.36
7			1	1		2	1.89
8	1					1	1.51
9			1			1	1.21
10		1		1		2	0.97
11	1					1	0.77
12			1			1	0.62
13						0	0.49
14						0	0.40
15						0	0.32
16						0	0.25
17						0	0.20
18				1		1	0.16
						45	

We now compute the chi-square statistic to test the null hypothesis that the sequence of values is random against the alternative that it is not:

$$X^2 = \sum_{i=0}^{18} \frac{(O_i - E_i)^2}{E_i} = \frac{(10 - 9.0)^2}{9.0} + \frac{(5 - 7.2)^2}{7.2} + \ldots$$

$$+ \frac{(1 - 0.16)^2}{0.16}$$

$$= 17.114.$$

Corresponding to $19 - 1 = 18$ degrees of freedom, the value of $x^2_{0.05}$ is 28.9; the value of the chi-square statistic is less than this critical value, so again we cannot reject the null hypothesis that the given sequence of values is random at the 5% significance level.

SPSS contains no procedure which performs the gap test.

THE POKER TEST

In Chapter 5 we performed several examples and exercises of this general form: "What is the probability that a five-card poker hand will contain two of a kind?" We can think of generating a random sequence of numbers as dealing from an infinite, well-shuffled deck, and apply reasoning similar to that of the card problems to develop another test of randomness.

Given a sequence which we wish to test for randomness, we break the sequence up into 5-tuples, and use a chi-square statistic to compare the expected and observed frequencies of particular types of "hands": two of a kind, three of a kind, two pair, etc. Because our "deck" is essentially infinite, finding the probabilities of the types of hands will differ from the card examples, as shown in this reuse of the first example of the chapter (p. 356).

Broken into 5-tuples, the sequence of numbers is this:

$$(4\ 3\ 2\ 3\ 3)\ (2\ 2\ 1\ 3\ 1)\ (1\ 0\ 2\ 0\ 2)\ (1\ 3\ 0\ 1\ 4)$$
$$(0\ 0\ 1\ 4\ 0)\ (2\ 3\ 2\ 1\ 1)\ (0\ 2\ 3\ 4\ 0)\ (0\ 2\ 4\ 4\ 2)$$
$$(2\ 1\ 4\ 4\ 0)\ (3\ 1\ 1\ 2\ 1)$$

When selecting digits randomly from the five alternatives 0, 1, 2, 3, and 4, the probability that an arbitrary 5-tuple will contain exactly one pair of identical values is:

$$1 \times \frac{1}{5} \times \frac{4}{5} \times \frac{3}{5} \times \frac{2}{5} \times \binom{5}{2} = \frac{4!5!}{5^4 2!3!} = \frac{48}{125} = 0.384.$$

There are ten 5-tuples in the sequence, so the expected number of hands containing exactly one pair of values is $10 \times 0.384 = 3.84$. Similar calculations, and counting the occurrences of the various arrangements, yield Table 16.2.

TABLE 16.2.　The poker test with five-number "hands" selected from five candidate values.

"HAND"	PROBABILITY	FREQUENCIES EXPECTED	OBSERVED
One pair	0.384	3.84	3
Two pair	0.288	2.88	4
Three of a kind	0.192	1.92	3
"Full house"	0.064	0.64	0
Four of a kind	0.032	0.32	0

We find the value of the chi-square statistic in the usual way:

$$X^2 = \sum_{i=1}^{5} \frac{(O_i - E_i)^2}{E_i} = \frac{(3-3.84)^2}{3.84} + \frac{(4-2.88)^2}{2.88} + \frac{(3-1.92)^2}{1.92}$$
$$\frac{(0-0.64)^2}{0.64} + \frac{(0-0.32)^2}{0.32} = 2.036.$$

The associated number of degrees of freedom is one less than the number of "hands" considered. In this case, that value is $5 - 1 = 4$, and the value of $x^2_{0.05}$ is then 9.488. The observed value of the chi-square statistic is less than this critical value, so we cannot conclude at the 5% significance level that the sequence is not random.

Again, this is not a test that SPSS will perform.

RUNS TEST

If a group of 12 men and women form a line in this order:

F   F   F   F   F   F   M   M   M   M   M   M,

we would conclude immediately that their arrangement is not random, as we would if this arrangement were observed:

M   F   M   F   M   F   M   F   M   F   M   F.

Sequences composed of only two types of symbols can be tested for randomness by counting the number of *runs*, or sequences of the same symbol, in the entire sequence. For example, the sequence

0   0   1   1   1   0   1

contains a run of two, a run of three, and two runs of one, a total of four.

If the number of runs in a sequence is small, lack of randomness through *clustering* is indicated, as in the first example above, while *mixing*—the second example—is reflected in too many runs. Since randomness may fail for either of these two mutually exclusive reasons, and the number of runs distinguishes them, we can perform a two-sided test for randomness or either of two one-sided tests:

Two-sided test:

H_o: The sequence is random.

H_a: The sequence is not random.

One-sided tests:
lower-tail

H_o: The sequence is random.

H_a: The sequence tends to cluster.

upper-tail

H_o: The sequence is random.

H_a: The sequence tends to mix.

A sequence to be tested consists of n_1 symbols of one type and n_2 symbols of a second; the number of possible runs has a minimum of 2 and a maximum of $2n_1$ if $n_1 = n_2$ or $2n_1 + 1$ if $n_1 < n_2$. Table A-9 provides critical values of the two-sided test at the 5% significance level (or the one-sided tests at $\alpha = 2.5\%$) for n_1 and $n_2 \leq 20$.

For example, this sequence of binary digits contains 6 runs and is composed of $n_1 = $ ten 0's and $n_2 = $ twelve 1's:

0   0   0   1   1   1   1   0   0   0   0   1   1   0   0   0   1   1   1   1   1   1

From Table A-9, the critical values for the two-sided test at the 5% significance level are 7 and 17; the observed number of runs is less than 7, so we conclude that the sequence is not random.

More strongly, since the number of runs is less than the critical value for the lower-tail test at the 2.5% significance level, we conclude that the sequence tends to cluster.

For values of n_1 and n_2 greater than those accommodated by the table, we let $N = n_1 + n_2$ be the total number of symbols in the sequence and U the number of runs observed and apply this result:

As N increases, the sampling distribution of U approaches the normal distribution with mean $\mu_u = 1 + (2n_1n_2)/N$ and standard deviation

$$S_u = \left[\frac{2n_1n_2(2n_1n_2 - N)}{N^2(N - 1)} \right]^{1/2}.$$

The critical values of U for the two-sided test at the α significance level are:

$$U^* = \mu_u \pm z_{\alpha/2} S_u.$$

Runs Above and Below a Central Value

A sequence whose entries take on more than two values may be tested for randomness with a runs test in two different ways. In the first, a dichotomy is imposed on the values in the sequence by creating a new sequence whose entries reflect whether each entry in the original sequence is above (A) or below (B) some central value in the population from which the original entries were selected, typically the mean or median. The sequence of A's and B's can be tested with the runs test already given. Such a test is said to be based on *runs above and below* the mean or median.

For example, to test this sequence of 20 digits, from 0 to 9, we create the corresponding sequence of symbols A and B depending on whether each entry in the sequence is above or below the population median of 4.5:

3	4	4	1	8	0	1	0	9	5	7	6	8	2	5	9	3	0	8	7
B	*B*	*B*	*B*	*A*	*B*	*B*	*B*	*A*	*A*	*A*	*A*	*A*	*B*	*A*	*A*	*B*	*B*	*A*	*A*

There are 8 runs in the sequence of 10 A's and 10 B's; from Table A-9, the critical values for the two-sided test at the 5% significance level are 6 and 16. The observed number of runs lies between the critical values, so we cannot conclude that the original sequence is not random.

The SPSS subprogram NPAR TESTS will perform tests based on runs above and below a central value or on runs in dichotomous data when the specification RUNS is used. The general form of the command is this:

$$\text{NPAR TESTS; RUNS} \quad \left\{ \begin{array}{c} \text{MEAN} \\ \text{MEDIAN} \\ \text{MODE} \\ \text{or} \\ \text{value} \end{array} \right\} \quad = \text{variable list/}$$

The procedure will then examine runs above and below the specified statistic or value for the variables given in the variable list. For dichotomous data, specify a value between the two values of the variable.

For example, this program tests the randomness of the values of the variables X5 and X7 from the USNEWS file about their respective medians.

```
10.     GET FILE
10.005  USNEWS
20.     NPAR TESTS
20.005  RUNS (MEDIAN) = X5,X7
```

— — — — — RUNS TEST
X5   AVERAGE INCOME OF FACTORY WORKERS

CASES	TEST VALUE	RUNS	LT	GE
73	15226.0000	39	36	37

Z	2-TAILED P
.3553	.7224

— — — — — RUNS TEST
X7   CHANGE IN CONSTRUCTION ACTIVITY

CASES	TEST VALUE	RUNS	LT	GE
74	4.1500	31	37	37

Z	2-TAILED P
− 1.6387	.1013

From the output, we see that it is reasonable to conclude that the values of X5 are randomly distributed about their median, and the same can be said about the values of X7.

RUNS UP AND DOWN

In dichotomizing a sequence, as was done in the previous section, information about the sequence is inevitably lost. Some of this can be preserved by considering, not runs above and below a value, but directional runs, runs up and down. Again, we create a new sequence of symbols from the original sequence, but here, every entry except the first receives $a+$ or $a-$ depending on whether the entry is greater or less than its predecessor. The runs test is then applied to the sequence of $+$'s and $-$'s. (If an entry replicates its predecessor, a 0 is entered. 0's are ignored in creating and counting runs.)

Applying this technique to the previous example, we generate this sequence of $+$'s and $-$'s:

```
3   4   4   1   8   0   1   0   9   5   7   6   8   2   5   9   3   0   8   7
    +   0   −   +   −   +   −   +   −   +   −   +   −   +   +   −   −   +   −
```

In the sequence of 9 $+$'s and 9 $-$'s, we find $U = 16$ runs. From Table A-9, the critical values for the two-sided test at $a = 5\%$ are 5 and 15. U is not between these critical values, so we can conclude from this test that the original sequence of values is not random. In particular, since U is greater than the upper critical value, we can conclude that the sequence is not random because the direction changes too often.

This procedure is not performed by SPSS.

THE KOLMOGOROV GOODNESS-OF-FIT TEST

One of the many applications of the chi-square distribution is the goodness-of-fit test, which is used to investigate whether a set of values might have come from a specified distribution. Another technique, developed by the Russian mathematician A. N. Kolmogorov, tests for goodness-of-fit by comparing the empirical *cumulative distribution function* (CDF) with the hypothesized CDF, using the largest absolute difference of the two functions as the test statistic, usually called D. As with the chi-square test, the null hypothesis is that the values come from the proposed distribution, while the alternative is that they come from some other distribution.

For the western cities, consider the 16 non-missing values of X5, income of the average factory worker, from the USNEWS data file. We will use the Kolmogorov test to compare the null hypothesis that these values come from a normal distribution with mean 15000 and standard deviation 2300 with the alternative hypothesis that they come from some other distribution at the 5% significance level.*

At each of the sixteen data values, the cumulative distribution function $F(x)$ of the hypothesized normal distribution has the value

$$F(x) = P(X \leq x), \text{ where } X \text{ is } N(15000, 2300).$$

For example, at 14410, one of the data values, the value of $F(x)$ is

$$F(14410) = P(X \leq 14410) = P\left(\frac{(X - 15000)}{2300} \leq \frac{14410 - 15000)}{2300} \right)$$

$$= P(Z \leq -0.26) = 0.5000 - 0.1026 = 0.3974.$$

The value of the cumulative *empirical* distribution at any data point x is

$$S(x) = \frac{N(x)}{N},$$

where $N(x)$ is the number of values less than or equal to x, and N is the total number of data values (in our example, 16). At 14410, then, the value of $S(x)$ is

$$S(14410) = \frac{9}{16} = 0.5625.$$

Similar calculations provide the remainder of the entries in Table 16.3, in which the values of X5 have been placed in ascending order.

The relationship between S and F can be seen by graphing the functions on common axes (Figure 16.1). Note that F is continuous and increasing, while S is an increasing step function.

The Kolmogorov test statistic D is the smallest number greater than or equal to the absolute difference $|S(x) - F(x)|$ for all values of x. It can be seen from Figure 16.1 that F and S differ most in the vicinity of $x = 14818$. In particular, at $x = 14818$, this difference is $|0.7500 - 0.4681| = 0.2819$.

In some situations the statistic D is not so easily found. Consider a portion of this graph in a Kolmogorov test as shown in Figure 16.2.

---

*This is the same problem addressed in Chapter 11, but reduced in size by considering only the western cities.

TABLE 16.3.   The Kolmogorov test of goodness-of-fit.

X5	S(X)	F(X)
11275	0.0625	0.0529
12708	0.1250	0.1660
12871	0.1875	0.1762
13459	0.2500	0.2514
14083	0.3125	0.3446
14142	0.3750	0.3557
14228	0.4375	0.3669
14291	0.5000	0.3783
14410	0.5625	0.3974
14524	0.6250	0.4168
14768	0.6875	0.4602
14818	0.7500	0.4681
15953	0.8125	0.6591
17639	0.8750	0.8749
17985	0.9375	0.9032
19307	1.0000	0.9693

The largest difference between F and S is indicated, but it is not simply $|S(x_o) - F(x_o)|$; rather, it is the smallest number larger than $|S(x_1) - F(x)|$ as x approaches x_o. Since F is continuous, D is $|S(x_1) - F(x_o)|$.

Critical values for the two-sided Kolmogorov test, corresponding to the sample size, are given in Table A-10; for our sample of 16 values and a significance level of 10%, the critical

FIGURE 16.1.   The empirical and theoretical distributions in the Kolmogorov test.

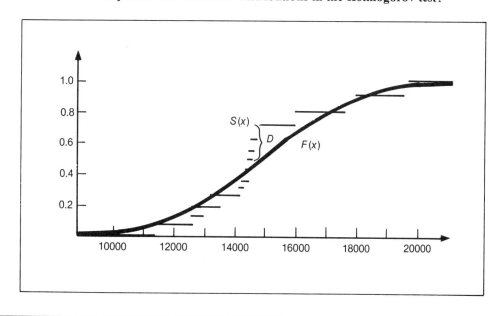

FIGURE 16.2. Finding D in the Kolmogorov test.

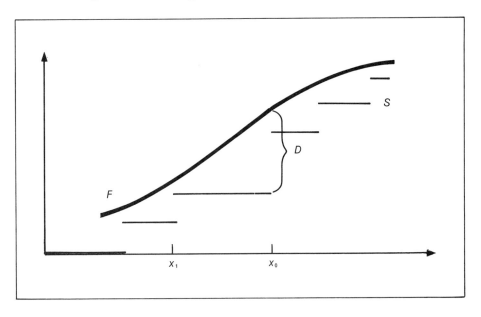

value is 0.327. The test statistic, 0.2819, is less than the critical value, so we cannot conclude that the sample values did not come from the hypothesized normal distribution.

The SPSS subprocedure NPAR TESTS will perform the Kolmogorov test of goodness-of-fit with certain distributions when the specification K-S is used (for Kolmogorov-Smirnov). The general form of the command is this:

$$\text{NPAR TESTS; K-S} \left(\begin{array}{c} \text{UNIFORM} \\ \text{NORMAL} \\ \text{or} \\ \text{POISSON} \end{array} \right) \text{ parameters)} = \text{variable list,}$$

where the parameters, if given, are the minimum and maximum values for the uniform distribution, the mean and standard deviation for the normal, and the mean for the Poisson. If omitted, these parameters will be calculated from the sample.

This program uses NPAR TESTS to test if the values of X5 from the USNEWS file might have come from a normal distribution with mean 15000 and standard deviation 2300:

```
10.     GET FILE
10.005 USNEWS
20.     NPAR TESTS
20.005 KS (NORMAL,15000,2300) = X5/
```

— — — — — KOLMOGOROV — SMIRNOV GOODNESS OF FIT TEST

X5   AVERAGE INCOME OF FACTORY WORKERS

TEST DIST. — NORMAL (MEAN =   15000.0000 STD DEV =   2300.0000)

CASES	MAX(ABS DIFF)	MAX (+ DIFF)	(MAX − DIFF)
73	.0718	.0421	− .0718

K-S Z	2-TAILED P
.613	.847

From the 2-tailed probability, much larger than any significance level, we see that we cannot reject the null hypothesis.

Two important observations should be made about the Kolmogorov test. First, the distribution of the test statistic D is not dependent on the nature of the underlying population distribution. Second, while the chi-square test of goodness-of-fit requires grouping possible values of a continuous distribution into classes, no such grouping is required here. Thus, it can be argued that the Kolmogorov test is better suited to situations involving continuous distributions.

THE KOLMOGOROV-SMIRNOV TWO-SAMPLE TEST

In a procedure very much like the Kolmogorov test, we can examine whether two independent samples might have come from identical distributions. This test was developed by another Russian, N. V. Smirnov, and generally carries the names of both men.

Given two independent samples, we construct the two empirical cumulative distribution functions $S_1(x)$ and $S_2(x)$, and use the statistic $D = $ maximum $|S_1(x) - S_2(x)|$ to test these hypotheses:

H_o: The samples come from identical populations.

H_a: The populations are not identical.

The calculation of D is more straightforward here than in the Kolmogorov one-sample test, as shown by this example in which we compare the values of X5 from the USNEWS data file for 10 eastern and the 16 western cities of the previous section.

Table 16.4 shows the empirical cumulative distribution functions $S_1(x)$ for the first 10 eastern cities and $S_2(x)$ for the western, and all the relevant differences between them.

The value of D is simply the largest of the absolute differences, in this case, 0.2875. From Table A-10b, the critical value when the sizes of the samples are 10 and 16 is ½ for the 5% significance level. We cannot conclude that the distributions are different.

For sample sizes n_1 and n_2 larger than those given in the table, this formula approximates the critical value of the statistic D at the 5% significance level:

$$(1.36) \sqrt{\frac{n_1 + n_2}{n_1 \, n_2}}$$

Kolmogorov-Smirnov tests can be performed by the SPSS subprogram NPAR TESTS with the specification K-S. The general form of the command is this:

NPAR TEST; K-S = dependent variable list BY independent variable

(value 1, value 2)

TABLE 16.4.   **Empirical cumulative distribution functions for eastern cities $S_1(x)$, and for western cities, $S_2(x)$.**

EASTERN	X5 WESTERN	$S_1(x)$	$S_2(x)$	$|S_1(x) - S_2(x)|$
10905		0.10	0	0.1000
	11275	0.10	0.0625	0.0375
11299		0.20	0.0625	0.1375
	12708	0.20	0.1250	0.0750
	12871	0.20	0.1875	0.0125
13002		0.30	0.1875	0.1125
13432		0.40	0.1875	0.2125
	13459	0.40	0.2500	0.1500
13979		0.50	0.2500	0.2500
	14083	0.50	0.3125	0.1875
14132		0.60	0.3125	0.2875
	14142	0.60	0.3750	0.2250
	14228	0.60	0.4375	0.1625
	14291	0.60	0.5000	0.1000
	14410	0.60	0.5625	0.0375
	14524	0.60	0.6250	0.0250
	14768	0.60	0.6875	0.0875
	14818	0.60	0.7500	0.1500
15784		0.70	0.7500	0.0500
	15953	0.70	0.8125	0.1125
16178		0.80	0.8125	0.0125
	17639	0.80	0.8750	0.0750
	17985	0.80	0.9375	0.1375
18362		0.90	0.9375	0.0375
18740		1.00	0.9375	0.0625
	19307	1.00	1.0000	0

The values of variables named in the dependent variable list are divided into two groups according to the given values of the independent variable, and these two groups of values are compared with the Kolmogorov-Smirnov test.

The example given here, including all the cases for which X1 is 1 or 3, will be performed by this program:

```
10.     GET FILE
10.005  USNEWS
20.     NPAR TESTS
20.005  K-S = X5 BY X1(1,3)/
```

The output of the program is this:

```
— — — — — KOLMOGOROV-SMIRNOV 2-SAMPLE TEST

    X5        AVERAGE INCOME OF FACTORY WORKERS
  BY X1       CITY REGION

WARNING — DUE TO SMALL SAMPLE SIZE, PROBABILITY TABLES
SHOULD BE CONSULTED.
```

					MAX(ABS DIFF)	MAX(+ DIFF)	MAX(− DIFF)
=	1.	=	3.		.2338	.2338	− .2315
	27		16				

K-S Z	2-TAILED P
.741	.642

Again, we cannot reject the null hypothesis to conclude that the distributions of average income of factory workers are different in the eastern and western regions.

Nonparametric Substitutes for Some Familiar Parametric Tests

The conditions necessary for procedures such as the *t*-test or analysis of variance do not always hold in practice. For such situations we have a selection of nonparametric tests which do not require conditions on the underlying distributions.

THE MANN-WHITNEY TEST

The Mann-Whitney test considers the null hypothesis that two independent samples come from identical population distributions against the alternative that the underlying populations are different. The test is most sensitive to a difference in the *locations* of the two populations, so it is often thought of as a nonparametric substitute for the independent sample *t*-test (its effectiveness is not affected by the nature of the underlying distributions), and it is assumed that the underlying populations differ only in location.

To perform this test, we take two independent samples, then order *all* the sampled values together as one large group, creating order statistics, and assigning the rank 1 to the smallest, 2 to the next smallest, and so on. Identical values are given the mean rank of the positions they would occupy were they slightly different. The test statistic is based on these ranks.

Let n_1 and n_2 be the sizes of the two samples, and let R_1 and R_2 be the sums of their ranks, respectively. Then for sample sizes of at least 8, if the null hypothesis is true and the two population means are equal, both R_1 and R_2, as random variables, have approximately normal distributions with means

$$\mu_{R_1} = \frac{n_1 \ (n_1 + n_2 + 1)}{2} \text{ and } \mu_{R_2} = \frac{n_2(n_1 + n_2 + 1)}{2}$$

and standard deviations

$$\sigma_{R_1} = \sigma_{R_2} = \left[\frac{n_1 n_2(n_1 + n_2 + 1)}{12}\right]^{1/2}.$$

Either R_1 or R_2 may be used as the test statistic, and the critical values for a two-sided test of the form

$$H_o: \mu_1 = \mu_2$$

$$H_a: \mu_1 \neq \mu_2,$$

with significance level α, will be $\mu_R \pm z_{\alpha/2} \sigma_R$. If μ_1 is not equal to μ_2, the values of sample 1 will tend to cluster at either the beginning or the end of the ranking, thus making the corresponding rank sum R_1 either small or large. If R_1 is not between the critical values, we may reject the null hypothesis and conclude that the population means are different.

When ties occur, the distributions of R_1 and R_2 are not exactly as described above, and correction factors are sometimes used. The difference is very small, however, unless the number of ties is large. (This is true of all tests based on sample ranks.)

For example, consider these two sets of test scores, from students instructed by two distinct methods, but given the same exam:

Group 1:   65   73   98   67   59   86   83   73   79   90

Group 2:   78   64   67   82   94   55   74   77   81   71

We order all the values and assign ranks. Note how ties are resolved:

| GROUP | | | GROUP | | |
1	2	RANK	1	2	RANK
	55	1		78	12
59		2	79		13
	64	3		81	14
65		4		82	15
67	67	5.5	83		16
	71	7	86		17
73		8.5	90		18
73		8.5		94	19
	74	10	98		20
	77	11			

The sum of ranks for group 1 is $R_1 = 112.5$ (R_2 is 97.5). The expected value of R_1 if H_o is true is

$$\frac{10(10 + 10 + 1)}{2} = 105,$$

and σ_R is:

$$\left[\frac{n_1 n_2(n_1 + n_2 + 1)}{12}\right]^{1/2} = \left[\frac{10 \times 10(10 + 10 + 1)}{12}\right]^{1/2} = \left(\frac{2100}{12}\right)^{1/2} = 13.23.$$

At the 10% significance level, the critical values of R_1 are

$$105 \pm 1.645 \times 13.23 = 105 \pm 21.76 = (83.24, 126.76.)$$

Since the test statistic R_1 does not lie outside the critical values, we cannot reject the null hypothesis; we cannot conclude that the test scores for the two groups differ significantly.

One-sided tests for the difference between two means can also be performed using the Mann-Whitney technique. Suppose, for example, that the values of sample 1 are in general less than those of sample 2. Then the rank sum R_1 will be reduced. If it is small enough, we may conclude that $\mu_1 < \mu_2$, and the critical value for this lower-tail test is $\mu_R - z_\alpha \sigma_R$. Similarly, the critical value for the upper-tail test with significance level α is $\mu_R + z_\alpha \sigma_R$.

The Mann-Whitney test can be performed by the SPSS subprogram NPAR TESTS, initiated by a command of this form:

NPAR TESTS; M-W = dependent variable list BY independent

variable (value 1, value 2)/

Again, the variables named in the dependent variable list are divided into two groups depending on the corresponding values of the independent variable, and these groups are administered the Mann-Whitney test. (NPAR TESTS do not use the rank sum R_1 as test statistic, but rather the value $U = R_1 - n_1 (n_1 + 1)/2$; the test based on U is entirely equivalent to the test we have described.)

This program, for example, tests the null hypothesis that the values of X3, unemployment, from the USNEWS file, are identically distributed in the eastern and western regions:

```
10.      GET FILE
10.005  USNEWS
20.      NPAR TESTS
20.005  M-W = X3 BY X1(1,3)/
```

The output is this:

— — — — MANN—WHITNEY U – WILCOXON RANK SUM W TEST

X3	UNEMPLOYMENT
BY X1	CITY REGION

X1 =	1.	X1 =	3.
MEAN RANK	NUMBER	MEAN RANK	NUMBER
21.54	25	21.44	17

		CORRECTED FOR TIES	
U	W	Z	2-TAILED P
211.5	364.5	.0256	.9795

We cannot conclude that the distributions of the values of X3 are different in the eastern and western regions. In particular, we cannot conclude that their locations are different.

THE WILCOXON MATCHED-PAIRS SIGNED-RANK TEST

When considering samples of matched-pair data (X_i, Y_i), we have used the *t*-test to examine the corresponding sample of *matched-pair differences* $D_i = X_i - Y_i$. In the Wilcoxon test for the difference of the central locations of two populations based on a sample of matched pairs, we construct the test statistic from the ranks and directions of the matched pair differences.

Consider, in an experiment of this kind, the population of all possible or hypothetical matched-pair differences, and let the median of this population be M_D. We assume that the differences are symmetrically distributed around M_D, and state our hypotheses in terms of this parameter. If the two populations have the same central tendencies, then M_D will be near 0; if the populations differ in location, we will have $M_D \neq 0$. These are the hypotheses of the two-sided Wilcoxon test:

$$H_o: M_D = 0$$

$$H_a: M_D \neq 0.$$

One-sided tests can also be performed, and will be mentioned later.

In Chapter 11, this example was solved with a matched-pairs t-test. We perform it here with the Wilcoxon test. Two tire-tread materials are being compared. Pairs of tires of the two materials are mounted on test cars and their lifetimes measured. The results are shown in Table 16.5.

TABLE 16.5.   Comparison of two tire-tread materials.

PAIR	1	2	3	4	5	6	7	8	9	10	11	12
Life of material A (1000's of miles)	42.6	43.1	41.9	45.2	43.6	41.0	42.6	43.7	42.2	46.3	42.4	41.8
Life of material B (1000's of miles)	40.3	42.7	42.0	42.1	43.4	39.5	41.5	43.4	41.4	43.1	41.6	40.1

From the values in Table 16.5, we assemble Table 16.6, which includes the matched-pair differences, their ranks, and their signs.

TABLE 16.6.   Tire-tread differences, ranks, and signs.

PAIR	A	B	$D = A - B$	D	RANK OF D	SIGN OF D
1	42.6	40.3	2.3	2.3	10	+
2	43.1	42.7	0.4	0.4	4	+
3	41.9	42.0	− 0.1	0.1	1	−
4	45.2	42.1	3.1	3.1	11	+
5	43.6	43.4	0.2	0.2	2	+
6	41.0	39.5	1.5	1.5	8	+
7	42.6	41.5	1.1	1.1	7	+
8	43.7	43.4	0.3	0.3	3	+
9	42.2	41.4	0.8	0.8	5.5	+
10	46.3	43.1	3.2	3.2	12	+
11	42.4	41.6	0.8	0.8	5.5	+
12	41.8	40.1	1.7	1.7	9	+

If a difference is 0, it is disregarded, and the sample size is reduced by 1; note again how ties are resolved in ranking the absolute values of the differences.

From Table 16.6, we compute two statistics, the sum of the ranks of the positive difference T^+, and the sum of the ranks of the negative difference T^-. Here

$$T^+ = 77; \ T^- = 1.$$

Either can be used as the test statistic, and from Table A-11 we see that values this extreme would occur under the null hypothesis with almost negligible probability. Thus, we can reject the null hypothesis and conclude that there is a significant difference in the durability of the two materials.

For tests in which the alternative hypothesis is of the form H_a: $M_D > 0$, we reject the null hypothesis and accept this alternative if T^+ is large enough (and T^- small enough); we accept the alternative H_a: $M_D < 0$ if T^- is sufficiently large.

For sample sizes larger than those treated in Table A-11, we can use the fact that as the sample size increases, the sampling distributions of both T^+ and T^- under H_o approach the normal distribution with mean $n(n + 1)/4$ and standard deviation $[n(n + 1)(2n + 1)/24]^{1/2}$.

The Wilcoxon test can be performed by NPAR TESTS when this command is used:

NPAR TESTS; WILCOXON = variable list [WITH variable list]/or ALL/

The test will be performed for all variables named in the first list with all those named in the second; if the keyword WITH is omitted, the test is performed for all pairs of variables in the variable list.

This program uses the USNEWS file to compare X2, change in department store sales, with X7, change in construction activity

```
10.      GET FILE
10.005   USNEWS
20.      NPAR TESTS
20.005   WILCOXON = X2 WITH X7/
```

and produces this output:

— — — — WILCOXON MATCHED-PAIRS SIGNED-RANKS TEST

| X2 | CHANGE IN DEPARTMENT STORE SALES | | | | |
| WITH X7 | CHANGE IN CONSTRUCTION ACTIVITY | | | | |

| | | 50 – RANKS | 20 + RANKS | | |
CASES	TIES	MEAN	MEAN	Z	2-TAIL
70	0	38.87	27.07	– 4.102	.0018

We conclude that there is a significant difference in the values of X2 and X7 in the cities in the study.

THE KRUSKAL-WALLIS TEST

In the parametric case, we used one-way analysis of variance to test the equality of several population means using independent samples. In this way, ANOVA was an extension of the independent samples t-test for the equality of two population means. The Mann-Whitney test is a nonparametric substitute for the t-test, and compared the central tendencies of two populations; we now extend that idea to three or more populations with the Kruskal-Wallis test, which uses the ranks of the elements of independent samples from $k \geq 3$ populations to test these hypotheses:

H_o: The population distributions are identical.

H_a: The population distributions are not identical.

The test is most sensitive to differences in the locations, in particular the medians, of the populations, so it is generally thought of as a test of the equality of the population medians:

H_o: The population medians are equal.

H_a: At least one median differs from the others.

We demonstrate the Kruskal-Wallis test with an exercise from Chapter 14: A computer center has, for seven months, recorded the numbers of jobs run in four popular languages. The numbers of jobs are shown in Table 16.7.

TABLE 16.7.   Demonstration of the Kruskal-Wallis test.

MONTH	BASIC	PASCAL	COBOL	FORTRAN	
1	1	5	14	2	
2	5	9	18	7	
3	2	5	9	3	Hundreds
4	6	11	21	10	of
5	7	12	20	7	jobs
6	3	5	12	5	
7	3	8	16	6	

We wish to test the null hypothesis that the median numbers of jobs in each language are the same against the alternative that at least two of the medians differ.

The Kruskal-Wallis test is based on the ranks of all the observations, so we begin by constructing a table of ranks (Table 16.8). Ties again receive an average rank value.

TABLE 16.8.   Table of ranks for the Kruskal-Wallis test.

MONTH	BASIC	PASCAL	COBOL	FORTRAN
1	1	9	24	2.5
2	9	18.5	26	15
3	2.5	9	18.5	5
4	12.5	21	28	20
5	15	22.5	27	15
6	5	9	22.5	9
7	5	17	25	12.5

If for each sample n_i is the sample size and R_i is the sum of the ranks of the sample, the Kruskal-Wallis test statistic H is given by this formula:

$$H = \frac{12}{n(n + 1)} \left[\sum_{i=1}^{k} \frac{R_i^2}{n_i} \right] - 3(n + 1),$$

where k is the number of samples, and $n = \sum_{i=1}^{k} n_i$ is the total number of observations. In our example, $R_1 = 50$, $R_2 = 106$, $R_3 = 171$, and $R_4 = 79$, so the value of H is:

$$H = \frac{12}{28(28+1)} \left[\sum_{i=1}^{4} \frac{R_i^2}{n_i} \right] - 3(28+1)$$

$$= \frac{12}{28 \times 29} \left[\frac{50^2}{7} + \frac{106^2}{7} + \frac{171^2}{7} + \frac{79^2}{7} \right] - 3 \times 29$$

$$= \frac{12}{812} \times \frac{49218}{7} - 87 = 16.91.$$

If the null hypothesis is true and the population medians are equal, the sampling distribution of H is a chi-square distribution with $k - 1$ degrees of freedom. In our example, corresponding to $4 - 1 = 3$ degrees of freedom and a significance level of 5%, the critical value from Table A-6 is $x_{0.05}^2 = 7.815$. The observed value of H is greater than the critical value, so we may reject H_o and conclude that the population medians are unequal.

Note that the ability to reject H_o, as in analysis of variance, does not allow us to say with the same confidence that one particular median is greater than the others. Further tests must be performed to establish such a result.

The following command will cause the subprogram NPAR TESTS to perform the Kruskal-Wallis test:

NPAR TESTS; K-W = dependent variable list BY independent

variable (value 1, value 2)/

Here, the Kruskal-Wallis test is performed on every variable named in the dependent variable list, divided into groups by all values of the independent variable between, and including, those given.

This program, using the USNEWS file once again, performs the Kruskal-Wallis test on the values of X5, as grouped by the three regions (the values of X1).

```
10.      GET FILE
10.005  USNEWS
20.      NPAR TESTS
20.005  K-W = X5 BY X1(1,3)/
```

This output is produced:

— — — — — KRUSKAL-WALLIS 1-WAY ANOVA

X5	AVERAGE INCOME OF FACTORY WORKERS		
BY X1	CITY REGION		

X1	1	2	3
NUMBER	27	30	16
MEAN RANKS	32.48	43.10	33.19

CORRECTED FOR TIES

CASES	CHI-SQUARE	SIGNIFICANCE	CHI-SQUARE	SIGNIFICANCE
73	4.221	.121	4.221	.121

Though the significance of the chi-square statistic approaches 10%, we cannot conclude on the basis of this test that there is a difference in workers' average income among the three regions.

The Spearman Rank Correlation Coefficient

The ranks of elements of samples can also be used to construct a measure of correlation.

Consider a sample of n pairs of values (X_i, Y_i). We order the X values and the Y values separately, and rank the values in the usual way. Corresponding to the original pair of values, we have n pairs of *ranks*, and the Spearman rank correlation coefficient, named for the English psychologist C. Spearman, is computed from the sum of the squares of the differences between these ranks.

If there is a perfect correlation between the X and Y values, then the ranks in each pair will be identical in this fashion:

	RANKS	
	X	*Y*
	1	1
	2	2
	.	.
	.	.
	.	.
	$n-1$	$n-1$
	n	n

Each of the rank differences D_i will be 0, as will the sum of their squares.

If there is a perfect inverse correlation between the X and Y values, then the ranks of one will ascend while those of the other descend, in this way:

	RANKS	
	X	*Y*
	1	n
	2	$n-1$
	.	.
	.	.
	.	.
	$n-1$	2
	n	1

This situation produces the largest sum of the squares of the rank differences:

$$\sum_{i=1}^{n} D_i^2 = (n-1)^2 + [(n-1) - 2]^2 + \ldots + [2 - (n-1)]^2 + (1-n)^2$$

$$= 2[(n-1)^2 + (n-3)^2 + \ldots]$$

$$= \frac{n(n^2-1)}{3}.$$

The value of $\sum_{i=1}^{n} D_i^2$ must fall between 0, indicating perfect direct correlation, and $n(n^2 - 1)/3$, indicating perfect inverse correlation. We prefer that measures of correlation fall between -1 and 1, and a function of $\sum_{i=1}^{n} D_i^2$ which does this in a familiar way is defined to be the Spearman rank correlation coefficient R_s:

$$R_s = 1 - \frac{6 \sum_{i=1}^{n} D_i^2}{n(n^2 - 1)} \; .$$

A formula more commonly used to calculate R_s is this:

$$R_s = \frac{12 \sum_{i=1}^{n} R_{X_i} R_{Y_i}}{n(n^2 - 1)} - \frac{3(n + 1)}{n - 1} \; ,$$

where R_{X_i} and R_{Y_i} are the ranks of the values X_i and Y_i. This formula is also equivalent to those above:

$$R_s = \frac{12 \sum_{i=1}^{n} (R_{X_i} - \bar{R}_X)(R_{Y_i} - \bar{R}_Y)}{n(n^2 - 1)} \; .$$

That is, if the original data values are themselves ranks, R_s is equivalent to the Pearson correlation coefficient.

As an example, consider the values shown here in Table 16.9 (from Chapter 12) of floor area and natural gas use in January for fifteen midwestern houses.

TABLE 16.9.

	VALUES		RANKS		
AREA	GAS IN 1000s OF CU FT		AREA	GAS	D_i
1.4	31.6		3.5	3	0.5
1.7	40.3		6.5	6	0.5
2.1	52.0		13	13	0
1.9	41.5		10	7	3
2.4	50.7		15	12	3
1.8	43.8		8	8	0
1.7	44.0		6.5	9	-2.5
1.9	48.2		10	11	-1
1.3	30.5		2	1	1
1.2	30.7		1	2	-1
2.0	55.2		12	14	-2
1.9	47.6		10	10	0
2.2	57.2		14	15	-1
1.4	36.2		3.5	5	-1.5
1.6	34.5		5	4	1

The sum of the squares of the rank differences is $\sum_{i=1}^{15} D_i^2 = 36.0$, and the value of the Spearman coefficient is

$$R_s = 1 - \frac{6 \sum_{i=1}^{15} D_i^2}{15(15^2 - 1)} = 1 - \frac{6 \times 36}{15 \times 224} = 1 - 0.064 = 0.936.$$

The value of R_s is near 1, indicating as before a strong correlation between gas use and house floor area.

The Spearman rank correlation coefficient is computed by the SPSS subprogram NONPAR CORR. The specification field of the NONPAR CORR command is identical in format and function to that of PEARSON CORR. If the gas use and house area values make up the raw data file FUEL, this program performs the calculations given above:

```
10.       FILE NAME
10.005  FUEL
20.       VARIABLE LIST
20.005  AREA, GAS
30.       INPUT FORMAT
30.005  FREEFIELD
40.       N OF CASES
40.005  15
50.       NONPAR CORR
50.005  GAS WITH AREA
```

This output is produced:

$- - -$ SPEARMAN CORRELATION $- - -$

VARIABLE PAIR	COEF.	SIG.	N
GAS AREA	.9354	.001	15

Note the small significance associated with the value of the coefficient R_s. We can safely conclude that there is a correlation between the gas use and area figures in the data file.

EXERCISES

1. These 50 values were selected from the digits 5, 6, 7, 8, and 9:

```
6 6 8 5 7 7 6 8 6 5
5 6 9 6 8 8 9 7 8 6
7 6 5 8 5 7 8 5 8 8
7 8 9 9 5 7 8 6 9 6
5 6 7 7 5 7 7 6 5 5
```

Use the frequency test to choose between these hypotheses at the 5% significance level:

H_o: The values are random.

H_a: The values are not random.

2. Consider this sequence of digits:

```
1 5 6 0 3 8 2 1 4 6 2 1 5 1 2
0 4 9 4 2 3 7 2 1 1 6 3 8 9 0
2 1 5 0 1 4 4 3 6 2 1 0 0 3 2
1 0 9 4 1 5 2 2 1 3 3 6 1 0 4
```

Use the frequency test to determine, at the 5% significance level, if these values have been randomly selected.

3. Consider the 73 values of X3, unemployment, in the USNEWS file. Group the values into four categories from 2.0 to 10.0 and use the frequency test to determine if the values are randomly selected from this range of values. Use $\alpha = 10\%$.

4. Create a raw data file and use the SPSS subprogram NPAR TESTS with the specification CHI-SQUARE to perform the test described in Exercise 2. What conclusion do you draw?

5. Use the SPSS subprogram NPAR TESTS to perform the frequency test described in Exercise 1. What conclusion do you draw?

6. Are there sequences of values that are clearly not random, but which will not be detected as such by the frequency test? If so, show some examples.

7. Use the RECODE command with subprogram NPAR TESTS to perform the test of Exercise 3. Does the result of this test surprise you?

8. Repeat Exercise 7, but using the RECODE command to group the values of X3 into eight classes from 2.0 to 10.0. What differences in the results do you find? (*Note*: Do not forget to adjust the expected values.)

9. Using the gap test, can you conclude at the 5% significance level that the values of Exercise 1 were not randomly selected?

10. Using the gap test, can you conclude at the 5% significance level that the values given in Exercise 2 were not randomly selected?

11. Again grouping the values of X3 from the USNEWS file into four classes from 2.0 to 10.0, use the gap test to determine at the 5% significance level if these values have not been randomly selected from the interval [2.0, 10.0].

12. Consider this sequence of 50 digits:

```
0 1 2 3 4 0 1 2 3 4 0 1 2 3 4 0 1 2 3 4 0 1 2 3 4

0 1 2 3 4 0 1 2 3 4 0 1 2 3 4 0 1 2 3 4 0 1 2 3 4
```

a. Verify that the value of the chi-square statistic in a frequency test of randomness for these values is zero. What conclusion would be drawn from the frequency test?

b. Why does the frequency test not reveal that the sequence is not random? Are there other situations in which the frequency test will fail to reveal nonrandomness?

c. Does the gap test conclude that this sequence is not random? Why?

d. Are there situations in which the gap test will fail to reveal the nonrandomness of a sequence? If so, describe some such sequences.

13. Use the poker test to test the randomness of the sequence of values given in Exercise 1 at the 5% significance level. Note that the probabilities of the several "hands" are the same as those given in Table 16.2.

14. Use the poker test to determine if the sequence of values given in Exercise 2 was randomly selected. Note that the probabilities of the various "hands" are not those given in the chapter, but must be recalculated.

15. Verify the probabilities and expected values given in Table 16.2. How would these change if each "hand" were a 4-tuple?

16. Can the poker test be used in conjunction with grouped values? Do this with the values of X3 from the USNEWS file, grouping them into four classes from 2.0 to 10.0. Use $\alpha = 5\%$. (*Hint*: Note that each value is now being selected from four alternatives, and disregard the last three values of X3.)

17. **a.** Assemble a table of probabilities for a poker test on a sequence of digits from 1 to 7 which uses 4-tuples (rather than 5-tuples).

    b. Use the table of part (a) to test this sequence of digits for randomness.

    $$
    \begin{array}{ccccccccccc}
    1 & 3 & 1 & 4 & 7 & 6 & 2 & 5 & 3 & 2 & 6 & 1 \\
    4 & 1 & 3 & 7 & 6 & 6 & 4 & 3 & 3 & 1 & 4 & 2 \\
    1 & 2 & 5 & 5 & 3 & 2 & 6 & 5 & 4 & 4 & 2 & 3 \\
    1 & 5 & 1 & 1 & 2 & 2 & 6 & 3 & 1 & 4 & 2 & 5 \\
    \end{array}
    $$

18. Are there sequences which are clearly not random but whose nonrandomness will not be detected by the poker test? If so, describe an example.

19. Use the runs test to determine, at the 5% significance level, if it can be concluded that these binary sequences are not random.

    a. 0 0 1 1 1 0 1 1 1 0 0 0 1 1 0 1 1 0 1 1 1

    b. 0 1 0 0 1 1 0 0 0 1 1 1 0 0 0 0 1 1 1 1 0 0 0 0 0 1 1 1 1 1

    c. 1 0 1 1 0 0 1 1 1 0 0 0 1 1 1 1 0 0 0 0 1 1 1 0 0 0 1 1 0 0 0 1 0

20. At the 2.5% significance level, can we conlude that this sequence of symbols tends to cluster?

    $$A\ A\ A\ B\ A\ A\ A\ A\ B\ B\ A\ A\ A\ A\ B\ A\ A\ A\ A\ B\ B\ B\ B$$

21. At the 2.5% significance level, can we conclude that this sequence of symbols tends to mix?

    $$+\ -\ -\ +\ -\ +\ +\ +\ -\ +\ -\ +\ +\ -\ +\ -\ -\ -\ +\ -\ +\ +\ -\ -\ +\ +\ -$$

22. Are there sequences composed of only two characters in which a pattern is clearly visible but whose nonrandomness is not revealed by the runs test? If so, describe or exhibit such a sequence.

23. In a sequence composed of n_1 0's and n_2 1's, show that the maximum number of runs possible is $2n_1$ if $n_1 = n_2$ and $2n_1 + 1$ if $n_1 < n_2$.

24. Use the normal approximation to the statistic U in a runs test of the randomness of this binary sequence:

    $$
    \begin{array}{cccccccccccccccccccc}
    0 & 0 & 0 & 0 & 1 & 1 & 1 & 1 & 1 & 0 & 0 & 1 & 1 & 1 & 0 & 0 & 0 & 0 & 0 & 0 \\
    0 & 1 & 1 & 1 & 1 & 1 & 0 & 0 & 0 & 0 & 0 & 1 & 1 & 1 & 1 & 1 & 1 & 0 & 0 & 0 \\
    0 & 0 & 0 & 1 & 1 & 1 & 1 & 1 & 0 & 0 & 0 & 0 & 0 & 0 & 0 & 1 & 1 & 1 & 1 \\
    \end{array}
    $$

25. Using 4.5 as the central value, use the runs test above and below that value to examine the sequence given in Exercise 2. At the 5% significance level, can we conclude that the sequence is not random?

26. Pick a central value and apply the runs test above and below that value to the sequence in Exercise 1. At the 5% significance level, what conclusion can you draw?

27. Using the value 3.5 to divide the range of values, use the runs test above and below that value to examine the sequence given in Exercise 17 (b). At the 5% significance level, what conclusion do you draw?

28. Use the SPSS subprogram NPAR TESTS with the specification RUNS to perform the test described in Exercise 20.

29. Use NPAR TESTS to perform the test described in Exercise 21.

30. Use NPAR TESTS to perform the test described in Exercise 24.

31. Consider the 74 values of X4, change in nonfarm employment, in the USNEWS data file. Use the SPSS subprogram NPAR TESTS to perform parts (a) and (b).

   a. Can we conclude, at the 10% significance level, that the values of X4 are not randomly distributed above and below their mean?

   b. Can we conclude, at the 10% significance level, that the values of X4 are not randomly distributed above and below their median?

   c. Under what conditions might the responses to parts (a) and (b) be different?

32. Are there sequences composed of more than two values which are clearly not random but whose nonrandomness will not be detected by a runs test above and below their mean? If so, describe an example.

33. a. Are there sequences composed of more than two values which are clearly not random but whose nonrandomness will not be detected by a runs test above and below their median? If so, describe an example.

   b. What difficulties can you see with a test of randomness based on runs above and below the mode?

34. Use NPAR TESTS to determine which, if any, of the seven variables in the USNEWS file are not randomly distributed about their means at the 10% significance level. (Note that this can be done in a very short program.)

35. Perform a test of runs up and down of the randomness of the sequence of values given in Exercise 1. At the 5% significance level, does this test show that the sequence is not random?

36. Perform a test of runs up and down at the 5% significance level on the binary sequence given in Exercise 24. Compare the value of the test statistic and your conclusion to those of Exercise 24. What does this comparison reveal?

37. Perform a test of runs up and down at the 5% significance level on the sequence of values given in Exercise 2. What conclusion can you draw?

38. It is suggested that these values were randomly selected from a population which is normally distributed with mean 100 and standard deviation 20.

105	64	121	109	96
84	110	91	101	120
116	72	78	95	118
116	94	135	127	60

At the 5% significance level, does the Kolmogorov test cause this hypothesis to be rejected?

39. Use the Kolmogorov test to determine if the null hypothesis that these values are a sample from an exponential distribution with mean 8 can be rejected at the 5% significance level.

$$9 \quad 1 \quad 5 \quad 5 \quad 6 \quad 7 \quad 3 \quad 2 \quad 10 \quad 4$$

40. Use the Kolmogorov test to determine if it can be concluded that these values were not selected from a uniform distribution on the interval [0, 1]. Use $\alpha = 5\%$, and note that this is equivalent to asking if the values were randomly selected from [0, 1].

$$
\begin{array}{ccccc}
0.3214 & 0.6509 & 0.2321 & 0.6817 & 0.1521 \\
0.6134 & 0.2152 & 0.7786 & 0.9054 & 0.1147 \\
0.3211 & 0.0121 & 0.5311 & 0.8203 & 0.4640 \\
\end{array}
$$

41. Use the SPSS subprogram NPAR TESTS with the specification K-S to perform the test described in Exercise 38.

42. Use the subprogram NPAR TESTS to perform the Kolmogorov test described in Exercise 40.

43. It is suggested that the values of X2, change in department store sales in the USNEWS file, have a normal distribution with mean 9.0 and standard deviation 10.0. Use the subprogram NPAR TESTS to determine if this claim can be disproved at the 10% significance level.

44. Use the subprogram NPAR TESTS and the Kolmogorov test to examine the null hypothesis that the values of X6 in the USNEWS file have a normal distribution. Use $\alpha = 10\%$.

45. Use the Kolmogorov-Smirnov test with a significance level of 5% to test the null hypothesis that these values came from a distribution identical to the underlying distribution of the sample given in Exercise 38 against the alternative hypothesis that the underlying distributions are not the same.

$$
\begin{array}{ccccc}
101 & 63 & 57 & 59 & 86 \\
110 & 115 & 92 & 104 & 66 \\
73 & 79 & 102 & 69 & 90 \\
\end{array}
$$

46. Compare this sample of values with those given in Exercise 39.

$$3 \quad 7 \quad 2 \quad 5 \quad 5 \quad 3 \quad 8 \quad 9 \quad 4 \quad 2$$

Using the Kolmogorov-Smirnov test, can it be concluded at the 10% significance level that the two samples have different underlying distributions?

47. Use the SPSS subprogram NPAR TESTS with the specification K-S to perform the test described in Exercise 45.

48. Use NPAR TESTS to perform the test described in Exercise 46.

49. It is suggested that the distributions of the variable X2 for the eastern and western regions are the same. Use NPAR TESTS to determine if this claim can be disproven at the 10% significance level.

50. Use the subprogram NPAR TESTS to determine if it can be shown at the 10% significance level that the distributions of the values of X7 from the USNEWS file are different for the eastern and western regions.

51. Consider the two samples of values given in Exercises 38 and 45. Assuming that the underlying populations of the two samples differ only in location, if at all, use the Mann-Whitney test to choose between these hypotheses at the 5% significance level:

$$H_o: \mu_A = \mu_B$$

$$H_a: \mu_A \neq \mu_B$$

52. Consider the two samples of values given in Exercises 39 and 46. Use the Mann-Whitney test at the 10% significance level to determine if we can conclude that the underlying population means are different. What assumption are we making here?

53. Use the subprogram NPAR TESTS with the specification M-W to perform the test described in Exercise 51. (*Hint*: In creating the raw data file, it will be necessary to include another variable to indicate which sample each value is from.)

54. Use NPAR TESTS to perform the test described in Exercise 52.

55. It is suggested that mean income of factory workers is different in the western region of the United States from in the eastern region. Based on the data in the USNEWS file and the subprogram NPAR TESTS, use the Mann-Whitney test to test this contention at the 5% significance level. What assumption is being made?

56. Use the subprogram NPAR TESTS to perform a Mann-Whitney test to examine whether the mean change in factory workers' income is different in the eastern and central regions of the United States. Use, of course, X6 from the USNEWS file.

57. Using the Wilcoxon test, determine whether it can be concluded at the 10% significance level that the elements of these matched pairs represent populations whose medians differ.

Sample 1:	24	26	28	29	24	29	31	27	21
Sample 2:	30	31	27	36	25	26	33	29	30

58. Perform the test described in Exercise 17, Chapter 10, using the Wilcoxon test. Note that we no longer require the assumption of normality.

59. Use the subprogram NPAR TESTS with the specification WILCOXON to perform the test described in Exercise 57.

60. Use NPAR TESTS with the specification WILCOXON to perform the test described in Exercise 58.

61. Use NPAR TESTS and the Wilcoxon test to test the null hypothesis that the change in nonfarm employment and change in construction activity are equal in the United States against the alternative that they are not at the 10% significance level. Refer to X4 and X7 in the USNEWS file.

62. Repeat Exercise 61 for the variables X2 and X6 in the USNEWS data file. What conclusion can you draw?

63. A soft drink firm test-markets a new beverage in cans of four different colors in six sales regions. Their sales, in cases of beverage per one thousand population, for the six areas for a 1-week period are these:

	COLOR OF CAN			
AREA	*RED*	*YELLOW*	*BLUE*	*GREEN*
1	24	26	18	16
2	40	38	21	33
3	28	32	24	22
4	16	21	32	19
5	22	26	16	24
6	35	32	26	20

Use the Kruskal-Wallis test to determine if the median sales levels are not all the same at the 5% significance level. In examining the medians, what assumption do we make?

64. A new shaving cream is being introduced, and the manufacturers are now testing three new advertising campaigns in a total of 21 markets. In the third week after introduction, their sales, in cases per thousand population, form this table:

| | CAMPAIGN | |
A	B	C
38	26	40
42	30	38
27	18	30
60	42	38
36	24	41
54	30	45
40	26	39

Use the Kruskal-Wallis test to determine, at the 5% significance level, if the median sales levels for the three campaigns are different.

65. Create a raw data set, then use the SPSS subprogram NPAR TESTS to perform the Kruskal-Wallis test described in Exercise 63.

66. Use NPAR TESTS to perform the Kruskal-Wallis test described in Exercise 64.

67. Consider the values of the variable X6, change in factory workers' income, in the USNEWS file. Use NPAR TESTS with the specification K-W to determine, at the 10% level of significance, if there is a difference in the median values of X6 in the eastern, central, and western regions of the United States.

68. Repeat Exercise 67 with the variable X2 from the USNEWS file.

69. Consider the pairs of values given in Exercise 57. Compute the Spearman rank correlation coefficient for these pairs. What does this value tell you?

70. For the data given in Exercise 63, compute the Spearman rank correlation coefficient between the sales figures for the red and blue cans in the six sales areas.

71. Create a raw data file, and use the SPSS subprogram NONPAR CORR to compute the Spearman coefficient described in Exercise 69. What does the significance of this value tell you?

72. Use the subprogram NONPAR CORR to perform the calculations required in Exercise 70. What do you learn from the significance of R_s?

73. Use the subprogram NONPAR CORR to find the Spearman rank correlation coefficient between the values X2 and X4 in the USNEWS data file. What interpretation do you place on your results?

74. Use NONPAR CORR to find the Spearman rank correlation coefficient of the values of the variables X6 and X7 in the USNEWS file. Interpret your results.

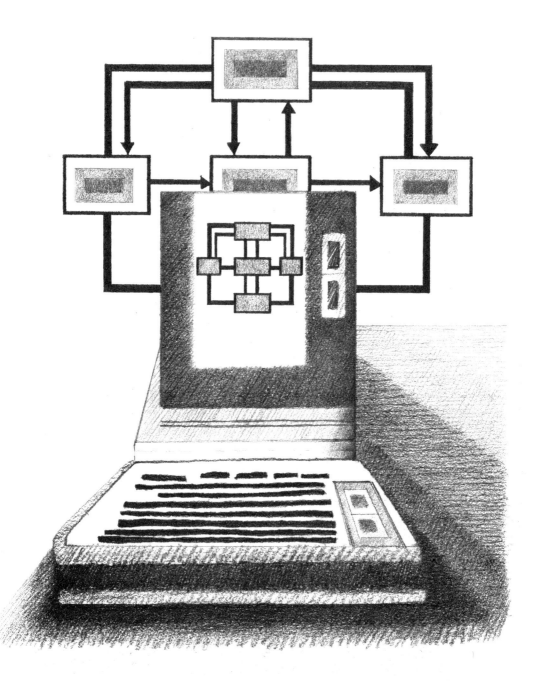

Statistical Applications Hardware

INTRODUCTION

In previous chapters we have worked with a great number of statistical processes and tests, and we have assumed the availability of electronic aids to perform the associated computations. Now we consider those aids; we look in turn at the several types of digital electronic computing devices—nonprogrammable and programmable hand-held calculators, minicomputer systems, microcomputers, and large computer systems—which can be used to perform statistical tasks.

Nonprogrammable Calculators

The two types of computing devices with which you are probably the most familiar are simple hand-held calculators and large-scale computer systems. Figure 17.1 (a) and (b) illustrate the fundamental difference between these two devices.

While the computer offers greater flexibility of input and output, the primary difference between these two types of devices is the presence in the computer of a *memory*, a section of the computer in which can be stored both *data*, values which will be used in or produced by

FIGURE 17.1.   **(a) Block diagram of a nonprogrammable calculator. (b) block diagram of a general-purpose digital computer. Data flow: --->; control flow →.**

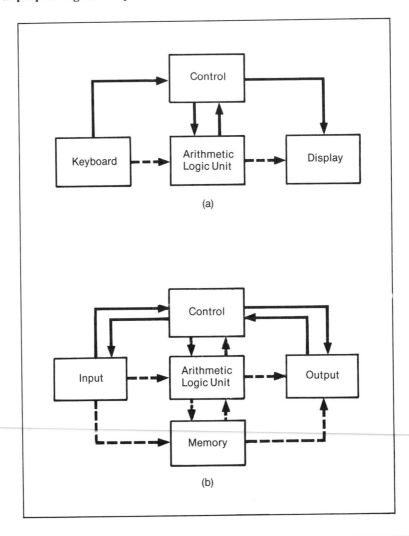

computations, and the instructions which make up those computations, the *program*. The instructions of the program are executed at a rate of thousands or millions per second, without waiting for the human intervention that a nonprogrammable device requires, and, after execution, the program continues to "live" in the computer's memory, so that it can be executed again, using the same or different data. With a nonprogrammable calculator, the operations which make up a given calculation must be entered from the keyboard each time the calculation is performed.

Even with these limitations, however, calculators are convenient and inexpensive computational aids. They are traditionally divided by performance into three categories: standard four-function, expanded function, and special purpose.

The standard four-function calculator is attractive primarily because of its price, which can be as low as $5. It can be used for simple statistical calculations, but for all except the most elementary, paper and pencil will be required to record intermediate results, and the very common operation of taking a square root (as in finding standard deviation, or a correlation coefficient) must be done by an approximation technique like Newton's Method:

To find $\sqrt{N}$, choose an initial estimate a_1. Then let $a_{n+1} = (a_n + N/a_n)/2$ for $n = 1, 2, \ldots$. Continue this process of successive approximations until a_n and a_{n+1} agree to your satisfaction. Then $a_n = a_{n+1} = \sqrt{N}$.

Approximation techniques can also be used for other functions, but these methods are tedious, and the accuracy of most four-function devices is no more than eight digits.

Expanded-function calculators generally provide eight digits of display, one or more registers for the storage of intermediate results, a constant facility for using the same value in repeated calculations, and keys for the operations of square and square root. Often their displays and internal workings require less power, so that battery life is increased. Such units sell for from $10 to $50; typical units also have keys for finding π, percent, and change in percent.

The highest level of sophistication in nonprogrammable calculators is reached by the special-purpose calculators. These are designed for specific applications, including science, statistics, business, navigation, and engineering, and generally cost between $20 and $100. A statistics-oriented calculator in the $50 range would provide these capabilities:

1. Six or more fully addressable memories.

2. All the functions of an expanded function calculator.

3. Logarithmic, exponential, and trigonometric function keys.

4. A display of at least ten digits with scientific notation.

5. Single-key computation of linear regression statistics—$\bar{X}$ and $\bar{Y}$, intercept and slope of the regression line, R, R^2, and standard deviations—after the data have been keyed in.

At a slightly higher price, calculators with more capabilities of interest to the statistician are available:

1. Random values from various probability distributions.

2. Probability function values from the classical probability distributions.

3. Keys for finding sums of squares and sums of cross products.

Programmable Calculators

Programmable calculators, like other programmable devices, have the ability to retain and repeat a sequence of instructions, a program. The simplest of these is the *learn mode* or *straight-line* programmable, which stores a sequence of instructions as they are entered from the keyboard when the mode switch is in the "learn" position. The sequence of operations can then be executed repeatedly, without reentering the computation steps though branching is not possible in these simple programs, and they cannot be edited.

Learn-mode calculators are available with three keyboard entry systems: algebraic, Reverse Polish, and plus-equals. Table 17.1 shows the differences among the three systems in entering and evaluating the expression

$$\frac{(4+3) \times 6}{2} - 9.$$

TABLE 17.1.   Three keyboard entry systems of learn-mode calculators.

ALGEBRAIC		REVERSE POLISH		PLUS-EQUALS	
Key	*Display*	*Key*	*Display*	*Key*	*Display*
Clear	—	Clear	—	Clear	—
4	4	4	4	4	4
+	4	↑	4	+	4
3	3	3	3	$\overline{\overline{3}}$)	3
×	7	+	7	×	7
÷	42	6	6	6	6
2	2	×	42	÷	42
−	21	2	2	$\overline{\overline{2}}$)	21
9	9	÷	21	9	9
=	12	9	9	(≡)	12
		−	12		

Reverse Polish entry often requires fewer keystrokes than does algebraic entry, which is attractive for long problems, and devices using plus-equals entry are sometimes less expensive as this method can be implemented at lower cost than the others.

More expensive programmable calculators offer many of the programming features of computers, such as comparisons and branching within a program (if a value is, say, greater than 1, perform one set of instructions; if it is not, perform a different set), and storage of programs on an external medium. Small magnetic cards are generally used, and a program recorded on a card can be read back into the calculator and executed at a later time.

Finally, the decreasing cost of microelectronics has blurred the distinction between programmable calculators and small computers. Printers which produce output on adding machine paper can be connected to the same devices mentioned above, and manufacturers such as Hewlett-Packard, Wang, and Radio Shack offer calculators programmable in higher-level languages with extended input/output capabilities. Some of these systems offer an array of peripheral devices comparable to those of a microcomputer system, and on them the user can implement a great variety of statistical computations.

Mini- and Microcomputer Systems

Moving up in size and complexity from programmable calculators, one encounters first microcomputers, then minicomputer systems. Historically, however, minis have been with us since the mid 1960's, while micros are a more recent development. We first consider minicomputers.

A computer word is the number of bits treated as a unit by the machine's hardware. In minicomputers a word generally contains from 12 to 32 bits, and is thereby capable of holding from two to four character codes per word. The electronics of the central processing unit (CPU) are distributed among several integrated circuits, which most often occupy all of one circuit board.

Minicomputer systems are most often operated interactively, with processing occurring in "real time" as a user sits at a terminal, rather than in the batch mode as is possible to use with larger systems. Operating systems—the software which supervises the operation of the system and allocates its computational and storage resources among the requested tasks—are less costly for such systems than for batch systems, and most of these offer time-sharing—the ability to serve several users simultaneously—as well. Such systems are available for under $25,000, though increased *I/O* sophistication can raise their prices considerably. Software support includes high-level languages such as BASIC, FORTRAN, and PASCAL, and many applications packages including some for statistical analysis. Digital Electronic Corporation's (DEC) PDP-11, for example, offers a statistical program package comparable to SPSS with tutorial and browse modes.

One of the first commercially successful minicomputers was the PDP-8, introduced by DEC in 1968. Shortly after came the Hewlett-Packard 2116, the Data General Nova, and many others. Today, machines electronically equivalent to the PDP-8 and PDP-11 are offered as kits by Heathkit. All these systems offer programming in high-level languages, a variety of input and output devices, mass storage on floppy disks, and a selection of application software packages, including statistical packages.

The CPU of a minicomputer, that board with all the integrated circuits on it, will alone cost more than $1,000. Recently, however, microelectronic technology has advanced to the point that all the components of the CPU can be manufactured in a single integrated circuit, called the microprocessing unit (MPU), and selling for under $10. A typical MPU is represented by Figure 17.2.

The data buffer holds data or instructions which are being gathered from memory or placed in memory, as directed by the instruction control. Arithmetic and logic operations dictated by the instruction currently held in the instruction register are performed by the arithmetic and logic unit (ALU), while the other registers and accumulators are used by the instruction control to control the flow of execution of the program and hold the results of computations. All this takes place on a chip of silicon no more than half an inch on a side.

The microcomputer is made up of the MPU, interfaces for connecting to peripheral devices, and a power supply, while adding peripheral devices such as a CRT terminal, printer, cassette tape recorder, or disk drive and software (programs) completes the microcomputer system. Such systems now range in price from less than $500 up, and many are available as kits.

For statistical use, a microcomputer system should include:

1. *I/O* interfaces to accommodate future system expansion.

FIGURE 17.2.   MPU unit. Data flow: →.

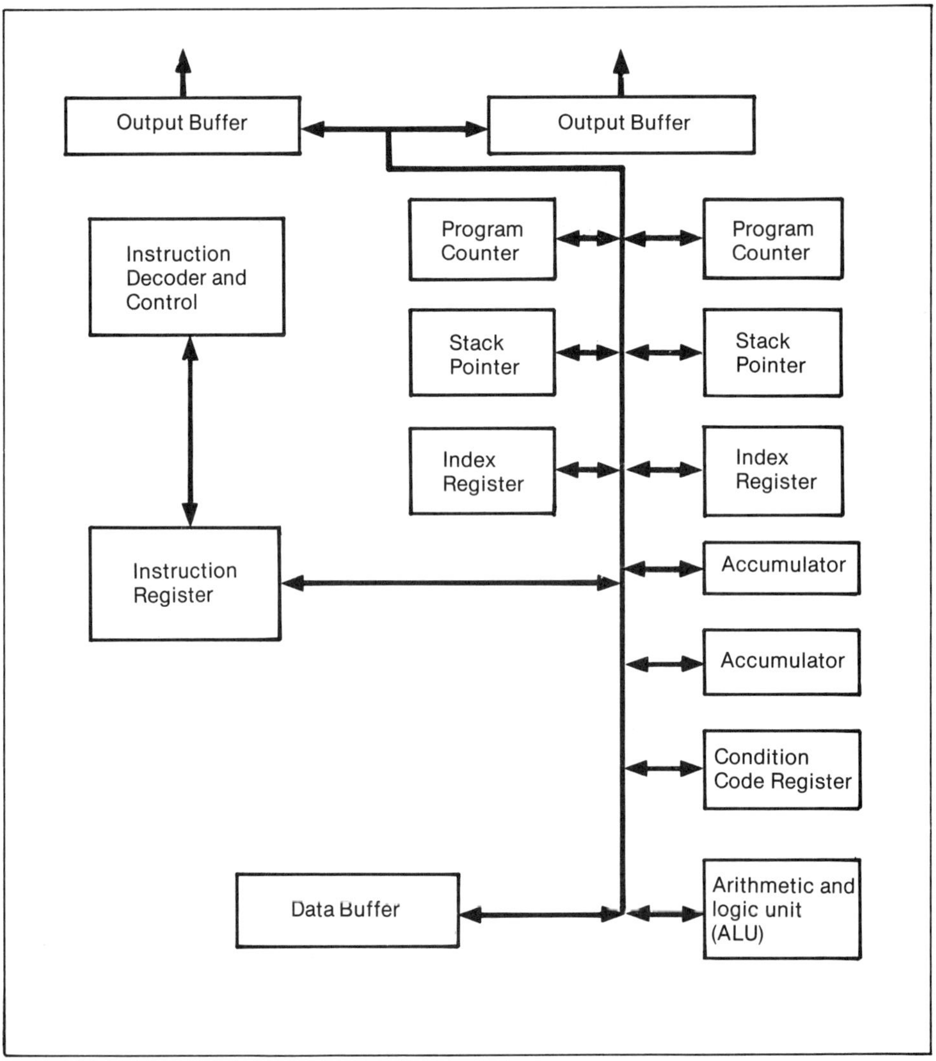

2. At least 16K (16×1024) words of memory.

3. Supporting software.

The user should pay particular attention to the last point. Without the software with which to operate a system—the operating system, editors, compilers, etc. —it is so much elegant hardware. If you do not want to write all these programs yourself, find what software is available and what it costs if it is not included in your system.

Popular microcomputer systems include the Commodore Company's PET, Radio Shack's TRS-80, the Apple II, the IBM Personal Computer, Ohio Scientific's line of microcomputers, and others. Their microprocessors generally support 8 bit words, with at least 16K of memory and software for BASIC and for the assembler language of the MPU. The IBM Personal Computer supports a 16 bit MPU. Additional *I/O* connections allow the use of one or more cassette tape recorders, two or more diskette drives, a plotter or line printer, and other peripherals. Applications packages are available in a wide variety of areas including basic math, business, graphics, and word processing. The statistical packages (usually in BASIC for $20 to $40) calculate the usual descriptive statistics, find Pearson product-moment correlations, and perform simple linear regressions and one-sample *t*-tests. More advanced packages are also available.

1. Their memory capacity is generally limited to 64K 8 bit words.

2. Access times between the MPU and mass storage devices such as a floppy disk are long.

Large Computer Systems

While all the systems we have discussed so far are dedicated to one task at a time, a large-scale computer system is often serving several users at essentially the same time. This is accomplished through multiprogramming (sharing of memory among users), multiprocessing (two or more CPU's dedicated to each job), expanded memory capabilities (thousands or millions of K bytes of storage), increased word size (32 to 64 bits/word, with 6 or 8 bits/byte), or more sophisticated software. Generally, time-sharing, batch, and remote batch-processing capabilities are present, and the system can support almost any combination of terminals, line printers, high-speed tape drives, disk drives, and card readers.

Large-scale systems generally support several high-level programming languages such as FORTRAN, ALGOL (originally intended as a language for the publication of algorithms), BASIC, COBOL (COmmon Business Oriented Language), and PL/1 (a general purpose language which combines many of the features of FORTRAN and COBOL). Additional special purpose languages (e.g., SPSS, SAS, BMD), file management systems, text editors, and others are also available for such systems. You have probably been doing your work in this course in just such an environment, on a machine built by IBM, Control Data Corporation, Honeywell, Univac, Burroughs, or another major manufacturer. Vendor support and service for such systems is generally contracted for when the system is purchased or leased, and represents a substantial fraction of the system cost.

EXERCISES

1. Outline the historical development of computers. (Reference: Saul Rosen, "Electronic Computers: A Historical Survey." *Computing Surveys,* March 1969, pp. 7–36.)

2. What is a stored-program computer? What are the advantages of the stored-program feature?

3. What differences distinguish a minicomputer from a large-scale computer system?

4. What peripheral devices are particularly suited to a mini- or a microcomputer?

5. What features should be considered in the selection of a
   a. microcomputer
   b. programmable calculator
   c. nonprogrammable calculator
   for statistical applications?

6. Define the following:
   a. microcomputer
   b. programmable calculator
   c. microprocessor

7. Compare the use of a mini-/microcomputer to that of a large-scale computer system for statistical processing.

8. What factors are responsible for the recent rapid increase in the number of microcomputers in use in the United States?

9. With respect to Exercise 8, why has there not been a similar growth in the use of microcomputers for statistical processing?

10. Discuss the advantages and disadvantages of using:

    a. core memory

    b. disk drives

    c. tape drives

    for storing data for statistical analysis.

Statistical Applications Software

INTRODUCTION

Having examined the several types of computing hardware which can be employed in statistical analyses, we now consider statistical software, the programs that direct the hardware to perform the operations. We will sketch answers to these questions: What considerations are important in the design of a statistical software package? What design constructs are possible? How can packages be compared? How can a researcher select the package which best fits his or her needs?

In many if not most situations, the researcher will not be able to choose from the entire array of all available software packages, but will have to select from a more limited list (possibly only one) of packages available on the computer system he or she will use. The questions above remain important, however, for choosing wisely from this limited selection and for using the selected package efficiently.

Design Constructs

An applications software package, in statistics or in any other area, may take one of these four forms:

1. A set of independent programs.

2. A set of subroutines to be called by a user-generated program.

3. A single, multipurpose program.

4. An applications compiler.

Further, a package constructed on any of these patterns may be designed for batch use, with programs submitted on decks of cards, or for interactive use by users at terminals. Batch systems are generally less expensive, while interactive systems are more expensive but easier to use. Many packages now include both capabilities.

The design of a package must take into account a number of considerations, including:

1. The intended application of the package (e.g., primarily the behavioral sciences, some other area, or a number of fields?)

2. The background of users (undergraduates or professional statisticians?)

3. The level of experience of the users (How familiar will they be with computers and programming?)

4. The hardware the package will run on.

We now consider the four major design constructs in more detail.

Sets of Programs

In a package of this kind, each program is complete in itself, and each performs just one statistical operation on the data file: computing descriptive statistics, for example, or performing a *t*-test. Each program will use the same input formatting, style of documentation, and coding language, and some effort will also be made to standardize the output formats.

The advantages of such a package are these: the programs can be easily used by a non-programmer, and they generally offer some flexibility of input format. In many cases, however, the user will want to perform many analyses on a particular data file, or the same analysis on several files. A package of individual programs is clumsy and time consuming in such situations.

Sets of Subroutines

When the package consists of a set of subroutines, the user must write a main program to read in the data file and call the subroutines to perform statistical analyses. This main program will be written in a high-level language like FORTRAN or APL, while the subroutines will be entirely input-free. There are several advantages to such a package:

1. Statistical processing can be tailored to each job.

2. Repetitive runs of several programs can be done easily.

3. Routines not in the package can be supplied by the user.

On the other hand, a set of subroutines can be used only by someone familiar with programming in the language of the subroutines; many statistical users will not have such training. Packages of this type generally have less documentation provided than might be required by the general user. Therefore, a high level of sophistication in computing may be required to obtain statistical results.

LARGE, MULTIPLE-USE PROGRAMS

Packages that are single, large programs can be divided into two subcategories. In the first, the entire program, which can perform a variety of statistical and processing tasks, is maintained in the computer's memory. Such packages tend to occupy vast amounts of storage, and they can require large amounts of time for *I/O*. They are, however, easily used by the nonprogrammer. In the second, the first section of code loaded into main memory produces a menu of available operations, then generates calls to the parts of the program which perform tasks the user requests, either in a specified code or in some English-like series of statements. Only those parts of the program related to the current tasks are maintained in memory, which can then be used more efficiently. Such systems are well-adapted to interactive use, and some also provide tutorial sessions in the use of the package.

APPLICATIONS COMPILERS

Packages of this type are, in fact, programming languages in which statistical tasks, or large steps in statistical tasks, are initiated by single statements in the language. They are useful to the researcher who wants to perform a nonstandard analysis quickly, but they again require that the user be a competent programmer. Very few such systems have yet been built. APL is an example of a statistical applications compiler.

Problem Areas

Many if not most statistical packages now in wide use were written by programmers rather than statisticians. Many were not constructed in modules, were designed for the hardware of the 1960's or earlier, and lack complete documentation. These and other conditions have led to the following problem areas in the design and use of statistical packages:

1. Statistical packages are often misunderstood. inefficiently used, even misused.

The ease with which some packages can be used may encourage the unwary to apply the analyses in the package to situations and data inconsistent with the techniques available; for example, assuming that a group of values has come from a normally distributed population when this has not been established. Documentation should show not only how to cause the analyses to be performed but also the situations to which the analyses are applicable, the assumptions on which the techniques are based, and the conclusions that can be drawn. This information should be available not only in the written documentation of the package but also during package use.

   2.   Overly stringent requirements of system hardware or software.

Some packages are written for specific machines, and can be used only with difficulty, if at all, on any other. Some require specialized devices, a plotter, for example, that may not be available.

   3.   Lack of transportability.

It should be possible to execute a package on a wide variety of machines, particularly since hardware changes at computer facilities occur with surprising frequency. Acquisition of a new machine should not render all purchased software unusable.

   4.   Inflexible and complicated control languages.

In some earlier packages, the statements used to control the operation of the package seem almost intentionally obscure, with stringent requirements for the placement of instructions on each line and instructions which, at first glance, seem arbitrary sequences of characters. The control language of a package should be easily accessible to those who use it. In particular, English-like words should be used, and their placement on the lines should not be overly critical.

   5.   Appropriateness of algorithms.

An algorithm is a series of steps which together describe the operations the computer must perform to accomplish a given task. The algorithms that perform statistical analyses must be appropriate to the demands of the tasks. For example, in computing the variance of a very large list of relatively uniform values by the formula

$$S^2 = \frac{1}{n-1} \sum_{i=1}^{n} X_i^2 - \frac{n}{n-1} \bar{X}^2,$$

the difference between the summation and $n\bar{X}^2$ may be very small compared to the value of the summation. In such a case, the variance may be distorted by the limited precision of the computer, and it would be better to use the formula provided by the definition of sample variance to find S^2.

   6.   Sophistication of programming language design.

The language in which the package is written is clearly relevant to its performance. There are many issues in programming language design which relate to package performance, for which we will provide one example.

Subroutines may be classified as *open* or *closed*. A closed subroutine (or subprocedure, or subprogram) is compiled and stored in the computer's memory separately from the main program that calls it. A call statement in the main program causes a jump in the flow of execution into the subroutine, and at the completion of the subroutine, execution returns to the point in the main program immediately after the call. With such a linking mechanism, used, for example, in FORTRAN, a subroutine need not reside in memory simultaneously with its main program except when the subroutine is executed.

The code of an open subroutine, in contrast, is a part of the code of the main program (PASCAL is an example of such a language); the subroutine is compiled and resides in memory with its main program. Many modern computers are designed to execute such languages efficiently, but may encounter difficulties with closed subroutines. If many users are engaging the computer, there may be insufficient memory remaining to bring in a closed subroutine. Thus, that program will wait until other executions are completed and sufficient space in memory is freed to bring in the subroutine.

7. Lack of modularity.

Many early packages are composed of large blocks of code not broken down into modular sections. Such code is difficult to maintain or modify, and often lacks transportability because machine-dependent statements are hard to locate.

8. Documentation availability.

Often, particularly with packages for mini- and microcomputers, it is difficult, if not impossible, to obtain duplicate copies of the package manual. When several people are likely to use a given package, each person should be able to have a copy of the manual at reasonable cost.

Desirable Package Features

It is obvious that the single most desirable characteristic of any package of programs is *ease of use*; the two primary areas here are input of the data file and instructing the package to perform the desired tasks. A package that does not allow direct, easily understood operation in these areas can be more an impediment than an aid. Other useful characteristics of a statistical program package are these:

1. Availability of both batch and interactive modes.

2. An interactive process to generate successive models, with each model a better approximation of the situation represented by the data. Graphical displays are desirable here.

3. An English-like command structure.

4. Intelligible error messages during execution to avoid dependence on the manual.

5. Results in English sentences whenever possible, rather than just tables of values.

6. The ability to generate new values based on existing values.

7. The creation of histograms, and of graphs of one variable against another.

8. Manuals showing examples of data files, runs using the package, and interpretation of results.

9. The ability to manipulate data files.

This last characteristic deserves further comment. In addition to sorting of files, the attachment of labels to variables, and the selection of particular cases, it is also often useful to *invert* a file, that is, to rearrange the records in a new order based on the values of a particular variable. For example, consider the simple data file in Table 18.1.

TABLE 18.1.   Simple data file.

RECORD NUMBER	CAR TYPE	YEAR	COST	COLOR	REGION
1	Ford	1967	$3250	Blue	Southwest
2	Ford	1977	$9000	Yellow	Southeast
3	Buick	1972	$6500	Blue	West
4	Dodge	1977	$7200	Green	Southeast

The following inverted file could be used to find the average cost of a car by region.

REGION	RECORDS
Southeast	2,4
Southwest	1
West	3

The programmer who is planning to write a statistical applications package would take all the above considerations into account as would the buyer of a prepared package. The overall requirements remain the same.

Evaluation Checklist

Literally hundreds of statistical software packages exist, so to obtain the best value when you are selecting one, you should perform an analysis of competing systems using criteria which can be objectively applied to each system. The criteria listed below have been found useful in such analyses, for both statistical and nonstatistical packages.

1. Cost elements
    In addition to the cost of the package itself, there are many other factors that may affect the entire cost associated with the package, including

    a. Conversion to the new package.

    b. Training support personnel.

    c. System modifications to accept the new package.

    d. New forms, manuals, and other operating aids.

    e. Maintenance.

    f. Additional charges for multiple machines or locations.

2. Availability
    Is the package operational? What is its operational history? Seek references and contact other users of the candidate package.

3. Equipment configuration
   Ascertain the minimal and optimal configurations of hardware for the package, and where your machine fits into that range.

4. Software environment
   Most installations have standards with respect to programming languages, operating systems, and utilities, and the usages of the package should conform to those standards.

5. Quality of system design
   Packages built up of *modules* can be more easily maintained and modified; expandable packages can be increased in size to handle new processing tasks; and fast, accurate algorithms will handle more tasks more precisely.

6. Documentation
   Documentation should be available for all levels of the package, including:

   a. A general overview of and introduction to the package.

   b. A systems manual describing the package processes, and input and output.

   c. Program-level documentation, with descriptions and flowcharts, record layouts, and annotated program listings.

   d. Operator instructions.

   e. User instructions, with example sessions.

7. Installation support
   In installing the package, support should be available for preparation, training personnel, and the installation itself.

8. Maintenance
   Who assumes responsibility for maintenance, and how much support is provided?

9. Enhancement
   The supplier may be planning enhancements of the package. The user should be aware of the cost, availability, and scheduling of these changes.

10. Supplier integrity
    Is the supplier committed to support the package, and the maintenance of it? Does the supplier have the personnel and financial resources to assure such a commitment?

Every December, *Datamation* magazine publishes a survey of users of packages which is conducted by the magazine in cooperation with Datapro Research Corporation. The survey covers all types of packages for large-scale computers and minicomputers. Users are asked to grade packages they have used in the previous year in seven categories: overall satisfaction, throughput/efficiency, ease of installation, ease of use, documentation, vendor's technical support, and training. Top-rated packages are listed in an "Honor Roll"; in recent years, both SPSS and SAS have been placed on the list.

Table 18.2 displays some salient characteristics of several popular statistical packages.

TABLE 18.2. Characteristics of statistical packages.

NAME	MACHINE	INTENDED FOR CORE SIZE	LANGUAGE	TYPE OF PACKAGE SET OF PROGRAMS	SET OF SUBROUTINES	LARGE PROGRAM	COMPILER	QUALITY OF DOCUMENTATION[a]	TYPE OF USER[a]
BMD	IBM 360	128K to 512K byte	FORTRAN IV (G and H levels)	X				Average manuals	Professional statistician
SSP/360	IBM 360	variable	FORTRAN IV		X			Good	Fortran programmer
SPSS	CDC 600 CDC 3200-3600 PDP 10 IBM 360/40	190K bytes	FORTRAN IV			X		Above average	General
Data-Text	IBM 360/50 CDC 6000	250K bytes	ASA FORTRAN IV			X	X	Good	General Professional statistician
OMNITAB II	Machine independent	90K words	ANSI and FORTRAN			X		Very good	Programmer, Professional statistician
SAS	IBM 360	?	FORTRAN	X				Very good	General
FOSOL	IBM 360/50 CDC 6000	?	FORTRAN COMPASS				X	Very good	Professional statistician, Programmer, General

[a] The last two columns only reflect the authors' opinions.

Other Package Sources

In addition to commercial packages such as those mentioned in Table 18.2, many colleges and universities have generated packages for statistical and pedogogical applications, which are often easily accessible to the nonprogrammer. Here are four examples:

The CADA (Computer-Assisted Data Analysis) monitor of the University of Iowa is a conversational, interactive software package intended for students with limited backgrounds and for more sophisticated users.

The University of Chicago's School of Management offers IDA (Instructional Data Analysis), also intended for students and others needing modern statistical techniques with a minimum of mathematical and technical prerequisites.

From the Department of Quantitative Methods at Georgia State University comes the Georgia State Time-Sharing System, developed for the use of students in the decision sciences there.

Charles T. Clark and A.W. Hunt of the University of Texas at Austin offer a set of programs called STATPAK specifically designed to be used in instruction with *Introduction to Business and Economic Statistics* by Stockton and Clark. This package is designed for batch use only.

Statistical programs are also available in texts which combine instruction in statistics and programming. These generally contain discussion of statistical techniques and listings of programs to perform the analyses. Examples are: *FORTRAN Programming for the Behavioral Sciences* by Donald J. Veldman, *Introduction to Statistics and Computer Programming* by Carl F. Kossack and Claudia I. Henschke, *APL-Stat* by James B. Ramsey and Gerald L. Musgrave.

STATISTICAL PACKAGES FOR MICROCOMPUTERS

While statistical packages have long been available for larger machines, their development for microcomputers has only recently begun. The major microcomputer manufacturers now offer sets of programs in BASIC to perform the fundamental statistical analyses, and other packages are available through independent software suppliers. Advertisements from these vendors can be found in magazines devoted to microcomputers such as *Byte, Creative Computing, On Computing,* and *Interface Age.*

STATISTICAL PACKAGE SOURCES

BMD (Biomedical Computer Programs)
University of California Press
2223 Fulton St.
Berkeley, CA 94720

SSP/360 (Scientific Subroutine Package)
IBM Corporation
Data Processing Division
112 East Post Road
White Plains, NY 10601

EXERCISES

1. Evaluate two or more statistical packages available at your computer center using the criteria of the *Datamation* survey.

2. Evaluate two or more statistical packages using the checklist given in this chapter.

3. Compare two or more statistical packages available at your computer center with respect to ease of use when performing:

   a. Step-wise regression.

   b. Set-wise regression.

   c. Tests of the drop in predictive efficiency between full and reduced regression models.

4. Compare two or more statistical packages with respect to analyzing

   a. Completely randomized designs.

   b. Latin square designs.

   c. Factorial designs.

5. Evaluate the statistical packages available at your computer center with respect to nonparametric procedures.

6. From the literature, develop a review of a selection of statistical software packages for large-scale computer systems. (*Hint:* These journals may prove to be useful: *ACM Computing Surveys, Communications of the ACM,* and the *Journal of the American Statistical Association*.)

7. Review a selection of statistical software packages for microcomputers.

SPSS (Statistical Package for the Social Sciences)
National Opinion Research Center
University of Chicago
6030 S. Ellis Avenue
Chicago, IL 60637

Data-Text
D.J. Armor
Department of Social Relations
Harvard University
Cambridge, MA 02138

OMNITAB II
National Technical Information Service
5285 Park Royal Road
Springfield, VA 22151

SAS (Statistical Analysis System)
J.H. Goodnight
Institute of Statistics
North Carolina State University at Raleigh
P.O. Box 5457
Raleigh, NC 27607

FOSOL
D.A. Florian
Quantitative Methods Department
Sangamon State University
Springfield, IL 62700

Appendices

Appendix A: Statistical Tables

TABLE A-1. Binomial probabilities

n	x	.01	.05	.10	.20	.30	.40	.50	.60	.70	.80	.90	.95	.99	x
2	0	980	902	810	640	490	360	250	160	090	040	010	002	0+	0
	1	020	095	180	320	420	480	500	480	420	320	180	095	020	1
	2	0+	002	010	040	090	160	250	360	490	640	810	902	980	2
3	0	970	857	729	512	343	216	125	064	027	008	001	0+	0+	0
	1	029	135	243	384	441	432	375	288	189	096	027	007	0+	1
	2	0+	007	027	096	189	288	375	432	441	384	243	135	029	2
	3	0+	0+	001	008	027	064	125	216	343	512	729	857	970	3
4	0	961	815	656	410	240	130	062	026	008	002	0+	0+	0+	0
	1	039	171	292	410	412	346	250	154	076	026	004	0+	0+	1
	2	001	014	049	154	265	346	375	346	265	154	049	014	001	2
	3	0+	0+	004	026	076	154	250	346	412	410	292	171	039	3
	4	0+	0+	0+	002	008	026	062	130	240	410	656	815	961	4
5	0	951	774	590	328	168	078	031	010	002	0+	0+	0+	0+	0
	1	048	204	328	410	360	259	156	077	028	006	0+	0+	0+	1
	2	001	021	073	205	309	346	312	230	132	051	008	001	0+	2
	3	0+	001	008	051	132	230	312	346	309	205	073	021	001	3
	4	0+	0+	0+	006	028	077	156	259	360	410	328	204	048	4
	5	0+	0+	0+	0+	002	010	031	078	168	328	590	774	951	5
6	0	941	735	531	262	118	047	016	004	001	0+	0+	0+	0+	0
	1	057	232	354	393	303	187	094	037	010	002	0+	0+	0+	1
	2	001	031	098	246	324	311	234	138	060	015	001	0+	0+	2
	3	0+	002	015	082	185	276	312	276	185	082	015	002	0+	3
	4	0+	0+	001	015	060	138	234	311	324	246	098	031	001	4
	5	0+	0+	0+	002	010	037	094	187	303	393	354	232	057	5
	6	0+	0+	0+	0+	001	004	016	047	118	262	531	735	941	6
7	0	932	698	478	210	082	028	008	002	0+	0+	0+	0+	0+	0
	1	066	257	372	367	247	131	055	017	004	0+	0+	0+	0+	1
	2	002	041	124	275	318	261	164	077	025	004	0+	0+	0+	2
	3	0+	004	023	115	227	290	273	194	097	029	003	0+	0+	3
	4	0+	0+	003	029	097	194	273	290	227	115	023	004	0+	4
	5	0+	0+	0+	004	025	077	164	261	318	275	124	041	002	5
	6	0+	0+	0+	0+	004	017	055	131	247	367	372	257	066	6
	7	0+	0+	0+	0+	0+	002	008	028	082	210	478	698	932	7
8	0	923	663	430	168	058	017	004	001	0+	0+	0+	0+	0+	0
	1	075	279	383	336	198	090	031	008	001	0+	0+	0+	0	1
	2	003	051	149	294	296	209	109	041	010	001	0+	0+	0+	2
	3	0+	005	033	147	254	279	219	124	047	009	0+	0+	0+	3
	4	0+	0+	005	046	136	232	273	232	136	046	005	0+	0+	4
	5	0+	0+	0+	009	047	124	219	279	254	147	033	005	0+	5
	6	0+	0+	0+	001	010	041	109	209	296	294	149	051	003	6
	7	0+	0+	0+	0+	001	008	031	090	198	336	383	279	075	7
	8	0+	0+	0+	0+	0+	001	004	017	058	168	430	663	923	8

TABLE A-1. Binomial probabilities (Continued)

n	x	.01	.05	.10	.20	.30	.40	p .50	.60	.70	.80	.90	.95	.99	x
9	0	914	630	387	134	040	010	002	0+	0+	0+	0+	0+	0+	0
	1	083	299	387	302	156	060	018	004	0+	0+	0+	0+	0+	1
	2	003	063	172	302	267	161	070	021	004	0+	0+	0+	0+	2
	3	0+	008	045	176	267	251	164	074	021	003	0+	0+	0+	3
	4	0+	001	007	066	172	251	246	167	074	017	001	0+	0+	4
	5	0+	0+	001	017	074	167	246	251	172	066	007	001	0+	5
	6	0+	0+	0+	003	021	074	164	251	267	176	045	008	0+	6
	7	0+	0+	0+	0+	004	021	070	161	267	302	172	063	003	7
	8	0+	0+	0+	0+	0+	004	018	060	156	302	387	299	083	8
	9	0+	0+	0+	0+	0+	0+	002	010	040	134	387	630	914	9
10	0	904	599	349	107	028	006	001	0+	0+	0+	0+	0+	0+	0
	1	091	315	387	268	121	040	010	002	0+	0+	0+	0+	0+	1
	2	004	075	194	302	233	121	044	011	001	0+	0+	0+	0+	2
	3	0+	010	057	201	267	215	117	042	009	001	0+	0+	0+	3
	4	0+	001	011	088	200	251	205	111	037	006	0+	0+	0+	4
	5	0+	0+	001	026	103	201	246	201	103	026	001	0+	0+	5
	6	0+	0+	0+	006	037	111	205	251	200	088	011	001	0+	6
	7	0+	0+	0+	001	009	042	117	215	267	201	057	010	0+	7
	8	0+	0+	0+	0+	001	011	044	121	233	302	194	075	004	8
	9	0+	0+	0+	0+	0+	002	010	040	121	268	387	315	091	9
	10	0+	0+	0+	0+	0+	0+	001	006	028	107	349	599	904	10
11	0	895	569	314	086	020	004	0+	0+	0+	0+	0+	0+	0+	0
	1	099	329	384	236	093	027	005	001	0+	0+	0+	0+	0+	1
	2	005	087	213	295	200	089	027	005	001	0+	0+	0+	0+	2
	3	0+	014	071	221	257	177	081	023	004	0+	0+	0+	0+	3
	4	0+	001	016	111	220	236	161	070	017	002	0+	0+	0+	4
	5	0+	0+	002	039	132	221	226	147	057	010	0+	0+	0+	5
	6	0+	0+	0+	010	057	147	226	221	132	039	002	0+	0+	6
	7	0+	0+	0+	002	017	070	161	236	220	111	016	001	0+	7
	8	0+	0+	0+	0+	004	023	081	177	257	221	071	014	0+	8
	9	0+	0+	0+	0+	001	005	027	089	200	295	213	087	005	9
	10	0+	0+	0+	0+	0+	001	005	027	093	236	384	329	099	10
	11	0+	0+	0+	0+	0+	0+	0+	004	020	086	314	569	895	11
12	0	886	540	282	069	014	002	0+	0+	0+	0+	0+	0+	0+	0
	1	107	341	377	206	071	017	003	0+	0+	0+	0+	0+	0+	1
	2	006	099	230	283	168	064	016	002	0+	0+	0+	0+	0+	2
	3	0+	017	085	236	240	142	054	012	001	0+	0+	0+	0+	3
	4	0+	002	021	133	231	213	121	042	008	001	0+	0+	0+	4
	5	0+	0+	004	053	158	227	193	101	029	003	0+	0+	0+	5
	6	0+	0+	0+	016	079	177	226	177	079	016	0+	0+	0+	6
	7	0+	0+	0+	003	029	101	193	227	158	053	004	0+	0+	7
	8	0+	0+	0+	001	008	042	121	213	231	133	021	002	0+	8
	9	0+	0+	0+	0+	001	012	054	142	240	236	085	017	0+	9

TABLE A-1. Binomial probabilities (Continued)

n	x	.01	.05	.10	.20	.30	.40	p .50	.60	.70	.80	.90	.95	.99	x
12	10	0+	0+	0+	0+	0+	002	016	064	168	283	230	099	006	10
	11	0+	0+	0+	0+	0+	0+	003	017	071	206	377	341	107	11
	12	0+	0+	0+	0+	0+	0+	0+	002	014	069	282	540	886	12
13	0	878	513	254	055	010	001	0+	0+	0+	0+	0+	0+	0+	0
	1	115	351	367	179	054	011	002	0+	0+	0+	0+	0+	0+	1
	2	007	111	245	268	139	045	010	001	0+	0+	0+	0+	0+	2
	3	0+	021	100	246	218	111	035	006	001	0+	0+	0+	0+	3
	4	0+	003	028	154	234	184	087	024	003	0+	0+	0+	0+	4
	5	0+	0+	006	069	180	221	157	066	014	001	0+	0+	0+	5
	6	0+	0+	001	023	103	197	209	131	044	006	0+	0+	0+	6
	7	0+	0+	0+	006	044	131	209	197	103	023	001	0+	0+	7
	8	0+	0+	0+	001	014	066	157	221	180	069	006	0+	0+	8
	9	0+	0+	0+	0+	003	024	087	184	234	154	028	003	0+	9
	10	0+	0+	0+	0+	001	006	035	111	218	246	100	021	0+	10
	11	0+	0+	0+	0+	0+	001	010	045	139	268	245	111	007	11
	12	0+	0+	0+	0+	0+	0+	002	011	054	179	367	351	115	12
	13	0+	0+	0+	0+	0+	0+	0+	001	010	055	254	513	878	13
14	0	869	488	229	044	007	001	0+	0+	0+	0+	0+	0+	0+	0
	1	123	359	356	154	041	007	001	0+	0+	0+	0+	0+	0+	1
	2	008	123	257	250	113	032	006	001	0+	0+	0+	0+	0+	2
	3	0+	026	114	250	194	085	022	003	0+	0+	0+	0+	0+	3
	4	0+	004	035	172	229	155	061	014	001	0+	0+	0+	0+	4
	5	0+	0+	008	086	196	207	122	041	007	0+	0+	0+	0+	5
	6	0+	0+	001	032	126	207	183	092	023	002	0+	0+	0+	6
	7	0+	0+	0+	009	062	157	209	157	062	009	0+	0+	0+	7
	8	0+	0+	0+	002	023	092	183	207	126	032	001	0+	0+	8
	9	0+	0+	0+	0+	007	041	122	207	196	086	008	0+	0+	9
	10	0+	0+	0+	0+	001	014	061	155	229	172	035	004	0+	10
	11	0+	0+	0+	0+	0+	003	022	085	194	250	114	026	0+	11
	12	0+	0+	0+	0+	0+	001	006	032	113	250	257	123	008	12
	13	0+	0+	0+	0+	0+	0+	001	007	041	154	356	359	123	13
	14	0+	0+	0+	0+	0+	0+	0+	001	007	044	229	488	869	14
15	0	860	463	206	035	005	0+	0+	0+	0+	0+	0+	0+	0+	0
	1	130	366	343	132	031	005	0+	0+	0+	0+	0+	0+	0+	1
	2	009	135	267	231	092	022	003	0+	0+	0+	0+	0+	0+	2
	3	0+	031	129	250	170	063	014	002	0+	0+	0+	0+	0+	3
	4	0+	005	043	188	219	127	042	007	001	0+	0+	0+	0+	4
	5	0+	001	010	103	206	186	092	024	003	0+	0+	0+	0+	5
	6	0+	0+	002	043	147	207	153	061	012	001	0+	0+	0+	6
	7	0+	0+	0+	014	081	177	196	118	035	003	0+	0+	0+	7
	8	0+	0+	0+	003	035	118	196	177	081	014	0+	0+	0+	8
	9	0+	0+	0+	001	012	061	153	207	147	043	002	0+	0+	9

Reprinted from Frederick Mosteller, Robert E. K. Rourke, and George B. Thomas, Jr., *Probability with statistical applications,* Second Edition, ©1970, Addison-Wesley Publishing Company, Reading, Mass., pp. 475–477. Reprinted with permission.

TABLE A-2. The Poisson distribution

A random variable X is said to have the Poisson distribution if (for some $\mu > 0$; see Definitions on pg. 107).

$$F_X(x) = \begin{cases} e^{-\mu} \dfrac{\mu^x}{x!}, & x = 0, 1, 2, \dots \\ 0, & \text{otherwise.} \end{cases}$$

The following table gives values of $F_X(x)$ (for various values of μ).

x	.1	.2	.3	4	μ .5	.6	.7	.8	.9	1.0
0	.9048	.8187	.7408	.6703	.6065	.5488	.4966	.4493	.4066	.3679
1	.0905	.1637	.2222	.2681	.3033	.3293	.3476	.3595	.3659	.3679
2	.0045	.0164	.0333	.0536	.0758	.0988	.1217	.1438	.1647	.1839
3	.0002	.0011	.0033	.0072	.0126	.0198	.0284	.0383	.0494	.0613
4	.0000	.0001	.0002	.0007	.0016	.0030	.0050	.0077	.0111	.0153
5	.0000	.0000	.0000	.0001	.0002	.0004	.0007	.0012	.0020	.0031
6	.0000	.0000	.0000	.0000	.0000	.0000	.0001	.0002	.0003	.0005
7	.0000	.0000	.0000	.0000	.0000	.0000	.0000	.0000	.0000	.0001

x	1.1	1.2	1.3	1.4	μ 1.5	1.6	1.7	1.8	1.9	2.0
0	.3329	.3012	.2725	.2466	.2231	.2019	.1827	.1653	.1496	.1353
1	.3662	.3614	.3543	.3452	.3347	.3230	.3106	.2975	.2842	.2707
2	.2014	.2169	.2303	.2417	.2510	.2584	.2640	.2678	.2700	.2707
3	.0738	.0867	.0998	.1128	.1255	.1378	.1496	.1607	.1710	.1804
4	.0203	.0260	.0324	.0395	.0471	.0551	.0636	.0723	.0812	.0902
5	.0045	.0062	.0084	.0111	.0141	.0176	.0216	.0260	.0309	.0361
6	.0008	.0012	.0018	.0026	.0035	.0047	.0061	.0078	.0098	.0120
7	.0001	.0002	.0003	.0005	.0008	.0011	.0015	.0020	.0027	.0034
8	.0000	.0000	.0001	.0001	.0001	.0002	.0003	.0005	.0006	.0009
9	.0000	.0000	.0000	.0000	.0000	.0000	.0001	.0001	.0001	.0002

x	2.1	2.2	2.3	2.4	μ 2.5	2.6	2.7	2.8	2.9	3.0
0	.1225	.1108	.1003	.0907	.0821	.0743	.0672	.0608	.0550	.0498
1	.2572	.2438	.2306	.2177	.2052	.1931	.1815	.1703	.1596	.1494
2	.2700	.2681	.2652	.2613	.2565	.2510	.2450	.2384	.2314	.2240
3	.1890	.1966	.2033	.2090	.2138	.2176	.2205	.2225	.2237	.2240
4	.0992	.1082	.1169	.1254	.1336	.1414	.1488	.1557	.1622	.1680
5	.0417	.0476	.0538	.0602	.0668	.0735	.0804	.0872	.0940	.1008
6	.0146	.0174	.0206	.0241	.0278	.0319	.0362	.0407	.0455	.0504
7	.0044	.0055	.0068	.0083	.0099	.0118	.0139	.0163	.0188	.0216
8	.0011	.0015	.0019	.0025	.0031	.0038	.0047	.0057	.0068	.0081
9	.0003	.0004	.0005	.0007	.0009	.0011	.0014	.0018	.0022	.0027
10	.0001	.0001	.0001	.0002	.0002	.0003	.0004	.0005	.0006	.0008
11	.0000	.0000	.0000	.0000	.0000	.0001	.0001	.0001	.0002	.0002
12	.0000	.0000	.0000	.0000	.0000	.0000	.0000	.0000	.0000	.0001

x	3.1	3.2	3.3	3.4	μ 3.5	3.6	3.7	3.8	3.9	4.0
0	.0450	.0408	.0369	.0334	.0302	.0273	.0247	.0224	.0202	.0183
1	.1397	.1304	.1217	.1135	.1057	.0984	.0915	.0850	.0789	.0733
2	.2165	.2087	.2008	.1929	.1850	.1771	.1692	.1615	.1539	.1465
3	.2237	.2226	.2209	.2186	.2158	.2125	.2087	.2046	.2001	.1954
4	.1734	.1781	.1823	.1858	.1888	.1912	.1931	.1944	.1951	.1954
5	.1075	.1140	.1203	.1264	.1322	.1377	.1429	.1477	.1522	.1563
6	.0555	.0608	.0662	.0716	.0771	.0826	.0881	.0936	.0989	.1042
7	.0246	.0278	.0312	.0348	.0385	.0425	.0466	.0508	.0551	.0595
8	.0095	.0111	.0129	.0148	.0169	.0191	.0215	.0241	.0269	.0298
9	.0033	.0040	.0047	.0056	.0066	.0076	.0089	.0102	.0116	.0132
10	.0010	.0013	.0016	.0019	.0023	.0028	.0033	.0039	.0045	.0053
11	.0003	.0004	.0005	.0006	.0007	.0009	.0011	.0013	.0016	.0019
12	.0001	.0001	.0001	.0002	.0002	.0003	.0003	.0004	.0005	.0006
13	.0000	.0000	.0000	.0000	.0001	.0001	.0001	.0001	.0002	.0002
14	.0000	.0000	.0000	.0000	.0000	.0000	.0000	.0000	.0000	.0001

TABLE A-2. The Poisson distribution (Continued)

$$F_X(x) = e^{-\mu} \frac{\mu^x}{x!}$$

x	4.1	4.2	4.3	4.4	4.5	4.6	4.7	4.8	4.9	5.0
0	.0166	.0150	.0136	.0123	.0111	.0101	.0091	.0082	.0074	.0067
1	.0679	.0630	.0583	.0540	.0500	.0462	.0427	.0395	.0365	.0337
2	.1393	.1323	.1254	.1188	.1125	.1063	.1005	.0948	.0894	.0842
3	.1904	.1852	.1798	.1743	.1687	.1631	.1574	.1517	.1460	.1404
4	.1951	.1944	.1933	.1917	.1898	.1875	.1849	.1820	.1789	.1755
5	.1600	.1633	.1662	.1687	.1708	.1725	.1738	.1747	.1753	.1755
6	.1093	.1143	.1191	.1237	.1281	.1323	.1362	.1398	.1432	.1462
7	.0640	.0686	.0732	.0778	.0824	.0869	.0914	.0959	.1002	.1044
8	.0328	.0360	.0393	.0428	.0463	.0500	.0537	.0575	.0614	.0653
9	.0150	.0168	.0188	.0209	.0232	.0255	.0280	.0307	.0334	.0363
10	.0061	.0071	.0081	.0092	.0104	.0118	.0132	.0147	.0164	.0181
11	.0023	.0027	.0032	.0037	.0043	.0049	.0056	.0064	.0073	.0082
12	.0008	.0009	.0011	.0014	.0016	.0019	.0022	.0026	.0030	.0034
13	.0002	.0003	.0004	.0005	.0006	.0007	.0008	.0009	.0011	.0013
14	.0001	.0001	.0001	.0001	.0002	.0002	.0003	.0003	.0004	.0005
15	.0000	.0000	.0000	.0000	.0001	.0001	.0001	.0001	.0001	.0002

x	5.1	5.2	5.3	5.4	5.5	5.6	5.7	5.8	5.9	6.0
0	.0061	.0055	.0050	.0045	.0041	.0037	.0033	.0030	.0027	.0025
1	.0311	.0287	.0265	.0244	.0225	.0207	.0191	.0176	.0162	.0149
2	.0793	.0746	.0701	.0659	.0618	.0580	.0544	.0509	.0477	.0446
3	.1348	.1293	.1239	.1185	.1133	.1082	.1033	.0985	.0938	.0892
4	.1719	.1681	.1641	.1600	.1558	.1515	.1472	.1428	.1383	.1339
5	.1753	.1748	.1740	.1728	.1714	.1697	.1678	.1656	.1632	.1606
6	.1490	.1515	.1537	.1555	.1571	.1584	.1594	.1601	.1605	.1606
7	.1086	.1125	.1163	.1200	.1234	.1267	.1298	.1326	.1353	.1377
8	.0692	.0731	.0771	.0810	.0849	.0887	.0925	.0962	.0998	.1033
9	.0392	.0423	.0454	.0486	.0519	.0552	.0586	.0620	.0654	.0688
10	.0200	.0220	.0241	.0262	.0285	.0309	.0334	.0359	.0386	.0413
11	.0093	.0104	.0116	.0129	.0143	.0157	.0173	.0190	.0207	.0225
12	.0039	.0045	.0051	.0058	.0065	.0073	.0082	.0092	.0102	.0113
13	.0015	.0018	.0021	.0024	.0028	.0032	.0036	.0041	.0046	.0052
14	.0006	.0007	.0008	.0009	.0011	.0013	.0015	.0017	.0019	.0022
15	.0002	.0002	.0003	.0003	.0004	.0005	.0006	.0007	.0008	.0009
16	.0001	.0001	.0001	.0001	.0001	.0002	.0002	.0002	.0003	.0003
17	.0000	.0000	.0000	.0000	.0000	.0001	.0001	.0001	.0001	.0001

x	6.1	6.2	6.3	6.4	6.5	6.6	6.7	6.8	6.9	7.0
0	.0022	.0020	.0018	.0017	.0015	.0014	.0012	.0011	.0010	.0009
1	.0137	.0126	.0116	.0106	.0098	.0090	.0082	.0076	.0070	.0064
2	.0417	.0390	.0364	.0340	.0318	.0296	.0276	.0258	.0240	.0223
3	.0848	.0806	.0765	.0726	.0688	.0652	.0617	.0584	.0552	.0521
4	.1294	.1249	.1205	.1162	.1118	.1076	.1034	.0992	.0952	.0912
5	.1579	.1549	.1519	.1487	.1454	.1420	.1385	.1349	.1314	.1277
6	.1605	.1601	.1595	.1586	.1575	.1562	.1546	.1529	.1511	.1490
7	.1399	.1418	.1435	.1450	.1462	.1472	.1480	.1486	.1489	.1490
8	.1066	.1099	.1130	.1160	.1188	.1215	.1240	.1263	.1284	.1304
9	.0723	.0757	.0791	.0825	.0858	.0891	.0923	.0954	.0985	.1014
10	.0441	.0469	.0498	.0528	.0558	.0588	.0618	.0649	.0679	.0710
11	.0245	.0265	.0285	.0307	.0330	.0353	.0377	.0401	.0426	.0452
12	.0124	.0137	.0150	.0164	.0179	.0194	.0210	.0227	.0245	.0264
13	.0058	.0065	.0073	.0081	.0089	.0098	.0108	.0119	.0130	.0142
14	.0025	.0029	.0033	.0037	.0041	.0046	.0052	.0058	.0064	.0071
15	.0010	.0012	.0014	.0016	.0018	.0020	.0023	.0026	.0029	.0033
16	.0004	.0005	.0006	.0006	.0007	.0008	.0010	.0011	.0013	.0014
17	.0001	.0002	.0002	.0002	.0003	.0003	.0004	.0004	.0005	.0006
18	.0000	.0001	.0001	.0001	.0001	.0001	.0001	.0002	.0002	.0002
19	.0000	.0000	.0000	.0000	.0000	.0000	.0000	.0001	.0001	.0001

TABLE A-2. The Poisson distribution (Continued)

$$F_X(x) = e^{-\mu}\frac{\mu^x}{x!}$$

x	7.1	7.2	7.3	7.4	7.5	7.6	7.7	7.8	7.9	8.0
0	.0008	.0007	.0007	.0006	.0006	.0005	.0005	.0004	.0004	.0003
1	.0059	.0054	.0049	.0045	.0041	.0038	.0035	.0032	.0029	.0027
2	.0208	.0194	.0180	.0167	.0156	.0145	.0134	.0125	.0116	.0107
3	.0492	.0464	.0438	.0413	.0389	.0366	.0345	.0324	.0305	.0286
4	.0874	.0836	.0799	.0764	.0729	.0696	.0663	.0632	.0602	.0573
5	.1241	.1204	.1167	.1130	.1094	.1057	.1021	.0986	.0951	.0916
6	.1468	.1445	.1420	.1394	.1367	.1339	.1311	.1282	.1252	.1221
7	.1489	.1486	.1481	.1474	.1465	.1454	.1442	.1428	.1413	.1396
8	.1321	.1337	.1351	.1363	.1373	.1382	.1388	.1392	.1395	.1396
9	.1042	.1070	.1096	.1121	.1144	.1167	.1187	.1207	.1224	.1241
10	.0740	.0770	.0800	.0829	.0858	.0887	.0914	.0941	.0967	.0993
11	.0478	.0504	.0531	.0558	.0585	.0613	.0640	.0667	.0695	.0722
12	.0283	.0303	.0323	.0344	.0366	.0388	.0411	.0434	.0457	.0481
13	.0154	.0168	.0181	.0196	.0211	.0227	.0243	.0260	.0278	.0296
14	.0078	.0086	.0095	.0104	.0113	.0123	.0134	.0145	.0157	.0169
15	.0037	.0041	.0046	.0051	.0057	.0062	.0069	.0075	.0083	.0090
16	.0016	.0019	.0021	.0024	.0026	.0030	.0033	.0037	.0041	.0045
17	.0007	.0008	.0009	.0010	.0012	.0013	.0015	.0017	.0019	.0021
18	.0003	.0003	.0004	.0004	.0005	.0006	.0006	.0007	.0008	.0009
19	.0001	.0001	.0001	.0002	.0002	.0002	.0003	.0003	.0003	.0004
20	.0000	.0000	.0001	.0001	.0001	.0001	.0001	.0001	.0001	.0002
21	.0000	.0000	.0000	.0000	.0000	.0000	.0000	.0000	.0001	.0001

x	8.1	8.2	8.3	8.4	8.5	8.6	8.7	8.8	8.9	9.0
0	.0003	.0003	.0002	.0002	.0002	.0002	.0002	.0002	.0001	.0001
1	.0025	.0023	.0021	.0019	.0017	.0016	.0014	.0013	.0012	.0011
2	.0100	.0092	.0086	.0079	.0074	.0068	.0063	.0058	.0054	.0050
3	.0269	.0252	.0237	.0222	.0208	.0195	.0183	.0171	.0160	.0150
4	.0544	.0517	.0491	.0466	.0443	.0420	.0398	.0377	.0357	.0337
5	.0882	.0849	.0816	.0784	.0752	.0722	.0692	.0663	.0635	.0607
6	.1191	.1160	.1128	.1097	.1066	.1034	.1003	.0972	.0941	.0911
7	.1378	.1358	.1338	.1317	.1294	.1271	.1247	.1222	.1197	.1171
8	.1395	.1392	.1388	.1382	.1375	.1366	.1356	.1344	.1332	.1318
9	.1256	.1269	.1280	.1290	.1299	.1306	.1311	.1315	.1317	.1318
10	.1017	.1040	.1063	.1084	.1104	.1123	.1140	.1157	.1172	.1186
11	.0749	.0776	.0802	.0828	.0853	.0878	.0902	.0925	.0948	.0970
12	.0505	.0530	.0555	.0579	.0604	.0629	.0654	.0679	.0703	.0728
13	.0315	.0334	.0354	.0374	.0395	.0416	.0438	.0459	.0481	.0504
14	.0182	.0196	.0210	.0225	.0240	.0256	.0272	.0289	.0306	.0324
15	.0098	.0107	.0116	.0126	.0136	.0147	.0158	.0169	.0182	.0194
16	.0050	.0055	.0060	.0066	.0072	.0079	.0086	.0093	.0101	.0109
17	.0024	.0026	.0029	.0033	.0036	.0040	.0044	.0048	.0053	.0058
18	.0011	.0012	.0014	.0015	.0017	.0019	.0021	.0024	.0026	.0029
19	.0005	.0005	.0006	.0007	.0008	.0009	.0010	.0011	.0012	.0014
20	.0002	.0002	.0002	.0003	.0003	.0004	.0004	.0005	.0005	.0006
21	.0001	.0001	.0001	.0001	.0001	.0002	.0002	.0002	.0002	.0003
22	.0000	.0000	.0000	.0000	.0001	.0001	.0001	.0001	.0001	.0001

x	9.1	9.2	9.3	9.4	9.5	9.6	9.7	9.8	9.9	10
0	.0001	.0001	.0001	.0001	.0001	.0001	.0001	.0001	.0001	.0000
1	.0010	.0009	.0009	.0008	.0007	.0007	.0006	.0005	.0005	.0005
2	.0046	.0043	.0040	.0037	.0034	.0031	.0029	.0027	.0025	.0023
3	.0140	.0131	.0123	.0115	.0107	.0100	.0093	.0087	.0081	.0076
4	.0319	.0302	.0285	.0269	.0254	.0240	.0226	.0213	.0201	.0189
5	.0581	.0555	.0530	.0506	.0483	.0460	.0439	.0418	.0398	.0378
6	.0881	.0851	.0822	.0793	.0764	.0736	.0709	.0682	.0656	.0631
7	.1145	.1118	.1091	.1064	.1037	.1010	.0982	.0955	.0928	.0901
8	.1302	.1286	.1269	.1251	.1232	.1212	.1191	.1170	.1148	.1126
9	.1317	.1315	.1311	.1306	.1300	.1293	.1284	.1274	.1263	.1251

TABLE A-2. The Poisson distribution (Continued)

$$F_X(x) = e^{-\mu}\frac{\mu^x}{x!}$$

μ

x	9.1	9.2	9.3	9.4	9.5	9.6	9.7	9.8	9.9	10
10	.1198	.1210	.1219	.1228	.1235	.1241	.1245	.1249	.1250	.1251
11	.0991	.1012	.1031	.1049	.1067	.1083	.1098	.1112	.1125	.1137
12	.0752	.0776	.0799	.0822	.0844	.0866	.0888	.0908	.0928	.0948
13	.0526	.0549	.0572	.0594	.0617	.0640	.0662	.0685	.0707	.0729
14	.0342	.0361	.0380	.0399	.0419	.0439	.0459	.0479	.0500	.0521
15	.0208	.0221	.0235	.0250	.0265	.0281	.0297	.0313	.0330	.0347
16	.0118	.0127	.0137	.0147	.0157	.0168	.0180	.0192	.0204	.0217
17	.0063	.0069	.0075	.0081	.0088	.0095	.0103	.0111	.0119	.0128
18	.0032	.0035	.0039	.0042	.0046	.0051	.0055	.0060	.0065	.0071
19	.0015	.0017	.0019	.0021	.0023	.0026	.0028	.0031	.0034	.0037
20	.0007	.0008	.0009	.0010	.0011	.0012	.0014	.0015	.0017	.0019
21	.0003	.0003	.0004	.0004	.0005	.0006	.0006	.0007	.0008	.0009
22	.0001	.0001	.0002	.0002	.0002	.0002	.0003	.0003	.0004	.0004
23	.0000	.0001	.0001	.0001	.0001	.0001	.0001	.0001	.0002	.0002
24	.0000	.0000	.0000	.0000	.0000	.0000	.0000	.0001	.0001	.0001

μ

x	11	12	13	14	15	16	17	18	19	20
0	.0000	.0000	.0000	.0000	.0000	.0000	.0000	.0000	.0000	.0000
1	.0002	.0001	.0000	.0000	.0000	.0000	.0000	.0000	.0000	.0000
2	.0010	.0004	.0002	.0001	.0000	.0000	.0000	.0000	.0000	.0000
3	.0037	.0018	.0008	.0004	.0002	.0001	.0000	.0000	.0000	.0000
4	.0102	.0053	.0027	.0013	.0006	.0003	.0001	.0001	.0000	.0000
5	.0224	.0127	.0070	.0037	.0019	.0010	.0005	.0002	.0001	.0001
6	.0411	.0255	.0152	.0087	.0048	.0026	.0014	.0007	.0004	.0002
7	.0646	.0437	.0281	.0174	.0104	.0060	.0034	.0018	.0010	.0005
8	.0888	.0655	.0457	.0304	.0194	.0120	.0072	.0042	.0024	.0013
9	.1085	.0874	.0661	.0473	.0324	.0213	.0135	.0083	.0050	.0029
10	.1194	.1048	.0859	.0663	.0486	.0341	.0230	.0150	.0095	.0058
11	.1194	.1144	.1015	.0844	.0663	.0496	.0355	.0245	.0164	.0106
12	.1094	.1144	.1099	.0984	.0829	.0661	.0504	.0368	.0259	.0176
13	.0926	.1056	.1099	.1060	.0956	.0814	.0658	.0509	.0378	.0271
14	.0728	.0905	.1021	.1060	.1024	.0930	.0800	.0655	.0514	.0387
15	.0534	.0724	.0885	.0989	.1024	.0992	.0906	.0786	.0650	.0516
16	.0367	.0543	.0719	.0866	.0960	.0992	.0963	.0884	.0772	.0646
17	.0237	.0383	.0550	.0713	.0847	.0934	.0963	.0936	.0863	.0760
18	.0145	.0256	.0397	.0554	.0706	.0830	.0909	.0936	.0911	.0844
19	.0084	.0161	.0272	.0409	.0557	.0699	.0814	.0887	.0911	.0888
20	.0046	.0097	.0177	.0286	.0418	.0559	.0692	.0798	.0866	.0888
21	.0024	.0055	.0109	.0191	.0299	.0426	.0560	.0684	.0783	.0846
22	.0012	.0030	.0065	.0121	.0204	.0310	.0433	.0560	.0676	.0769
23	.0006	.0016	.0037	.0074	.0133	.0216	.0320	.0438	.0559	.0669
24	.0003	.0008	.0020	.0043	.0083	.0144	.0226	.0328	.0442	.0557
25	.0001	.0004	.0010	.0024	.0050	.0092	.0154	.0237	.0336	.0446
26	.0000	.0002	.0005	.0013	.0029	.0057	.0101	.0164	.0246	.0343
27	.0000	.0001	.0002	.0007	.0016	.0034	.0063	.0109	.0173	.0254
28	.0000	.0000	.0001	.0003	.0009	.0019	.0038	.0070	.0117	.0181
29	.0000	.0000	.0001	.0002	.0004	.0011	.0023	.0044	.0077	.0125
30	.0000	.0000	.0000	.0001	.0002	.0006	.0013	.0026	.0049	.0083
31	.0000	.0000	.0000	.0000	.0001	.0003	.0007	.0015	.0030	.0054
32	.0000	.0000	.0000	.0000	.0001	.0001	.0004	.0009	.0018	.0034
33	.0000	.0000	.0000	.0000	.0000	.0001	.0002	.0005	.0010	.0020
34	.0000	.0000	.0000	.0000	.0000	.0000	.0001	.0002	.0006	.0012
35	.0000	.0000	.0000	.0000	.0000	.0000	.0000	.0001	.0003	.0007
36	.0000	.0000	.0000	.0000	.0000	.0000	.0000	.0001	.0002	.0004
37	.0000	.0000	.0000	.0000	.0000	.0000	.0000	.0000	.0001	.0002
38	.0000	.0000	.0000	.0000	.0000	.0000	.0000	.0000	.0000	.0001
39	.0000	.0000	.0000	.0000	.0000	.0000	.0000	.0000	.0000	.0001

From *Handbook of Probability and Statistics With Tables*, 2nd ed. By Burlington and May. Copyright © 1970 by McGraw-Hill, Inc. Used with permission.

TABLE A-3. Values of $e^{-\mu}$

μ	$e^{-\mu}$	μ	$e^{-\mu}$	μ	$e^{-\mu}$	μ	$e^{-\mu}$
0.00	1.000000	2.60	.074274	5.10	.006097	7.60	.000501
0.10	.904837	2.70	.067206	5.20	.005517	7.70	.000453
0.20	.818731	2.80	.060810	5.30	.004992	7.80	.000410
0.30	.740818	2.90	.055023	5.40	.004517	7.90	.000371
0.40	.670320	3.00	.049787	5.50	.004087	8.00	.000336
0.50	.606531	3.10	.045049	5.60	.003698	8.10	.000304
0.60	.548812	3.20	.040762	5.70	.003346	8.20	.000275
0.70	.496585	3.30	.036883	5.80	.003028	8.30	.000249
0.80	.449329	3.40	.033373	5.90	.002739	8.40	.000225
0.90	.406570	3.50	.030197	6.00	.002479	8.50	.000204
1.00	.357879	3.60	.027324	6.10	.002243	8.60	.000184
1.10	.332871	3.70	.024724	6.20	.002029	8.70	.000167
1.20	.301194	3.80	.022371	6.30	.001836	8.80	.000151
1.30	.272532	3.90	.020242	6.40	.001661	8.90	.000136
1.40	.246597	4.00	.018316	6.50	.001503	9.00	.000123
1.50	.223130	4.10	.016573	6.60	.001360	9.10	.000112
1.60	.201897	4.20	.014996	6.70	.001231	9.20	.000101
1.70	.182684	4.30	.013569	6.80	.001114	9.30	.000091
1.80	.165299	4.40	.012277	6.90	.001008	9.40	.000083
1.90	.149569	4.50	.011109	7.00	.000912	9.50	.000075
2.00	.135335	4.60	.010052	7.10	.000825	9.60	.000068
2.10	.122456	4.70	.009095	7.20	.000747	9.70	.000061
2.20	.110803	4.80	.008230	7.30	.000676	9.80	.000056
2.30	.100259	4.90	.007447	7.40	.000611	9.90	.000050
2.40	.090718	5.00	.006738	7.50	.000553	10.00	.000045
2.50	.082085						

Reprinted with permission of Duxbury Press.

TABLE A-4. Standard Normal curve areas

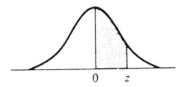

z	.00	.01	.02	.03	.04	.05	.06	.07	.08	.09
0.0	.0000	.0040	.0080	.0120	.0160	.0199	.0239	.0279	.0319	.0359
0.1	.0398	.0438	.0478	.0517	.0557	.0596	.0636	.0675	.0714	.0753
0.2	.0793	.0832	.0871	.0910	.0948	.0987	.1026	.1064	.1103	.1141
0.3	.1179	.1217	.1255	.1293	.1331	.1368	.1406	.1443	.1480	.1517
0.4	.1554	.1591	.1628	.1664	.1700	.1736	.1772	.1808	.1844	.1879
0.5	.1915	.1950	.1985	.2019	.2054	.2088	.2123	.2157	.2190	.2224
0.6	.2257	.2291	.2324	.2357	.2389	.2422	.2454	.2486	.2517	.2549
0.7	.2580	.2611	.2642	.2673	.2704	.2734	.2764	.2794	.2823	.2852
0.8	.2881	.2910	.2939	.2967	.2995	.3023	.3051	.3078	.3106	.3133
0.9	.3159	.3186	.3212	.3238	.3264	.3289	.3315	.3340	.3365	.3389
1.0	.3413	.3438	.3461	.3485	.3508	.3531	.3554	.3577	.3599	.3621
1.1	.3643	.3665	.3686	.3708	.3729	.3749	.3770	.3790	.3810	.3830
1.2	.3849	.3869	.3888	.3907	.3925	.3944	.3962	.3980	.3997	.4015
1.3	.4032	.4049	.4066	.4082	.4099	.4115	.4131	.4147	.4162	.4177
1.4	.4192	.4207	.4222	.4236	.4251	.4265	.4279	.4292	.4306	.4319
1.5	.4332	.4345	.4357	.4370	.4382	.4394	.4406	.4418	.4429	.4441
1.6	.4452	.4463	.4474	.4484	.4495	.4505	.4515	.4525	.4535	.4545
1.7	.4554	.4564	.4573	.4582	.4591	.4599	.4608	.4616	.4625	.4633
1.8	.4641	.4649	.4656	.4664	.4671	.4678	.4686	.4693	.4699	.4706
1.9	.4713	.4719	.4726	.4732	.4738	.4744	.4750	.4756	.4761	.4767
2.0	.4772	.4778	.4783	.4788	.4793	.4798	.4803	.4808	.4812	.4817
2.1	.4821	.4826	.4830	.4834	.4838	.4842	.4846	.4850	.4854	.4857
2.2	.4861	.4864	.4868	.4871	.4875	.4878	.4881	.4884	.4887	.4890
2.3	.4893	.4896	.4898	.4901	.4904	.4906	.4909	.4911	.4913	.4916
2.4	.4918	.4920	.4922	.4925	.4927	.4929	.4931	.4932	.4934	.4936
2.5	.4938	.4940	.4941	.4943	.4945	.4946	.4948	.4949	.4951	.4952
2.6	.4953	.4955	.4956	.4957	.4959	.4960	.4961	.4962	.4963	.4964
2.7	.4965	.4966	.4967	.4968	.4969	.4970	.4971	.4972	.4973	.4974
2.8	.4974	.4975	.4976	.4977	.4977	.4978	.4979	.4979	.4980	.4981
2.9	.4981	.4982	.4982	.4983	.4984	.4984	.4985	.4985	.4986	.4986
3.0	.4987	.4987	.4987	.4988	.4988	.4989	.4989	.4989	.4990	.4990

This table is abridged from Table 1 of *Statistical Tables and Formulas,* by A. Hald (New York: John Wiley & Sons, Inc., 1952). Reproduced by permission of A. Hald and the publishers, John Wiley & Sons, Inc.

TABLE A-5. The t-distribution

The following table gives values of y such that (for various values of n, γ)

$$P[X \leq y] = \int_{-\infty}^{y} f_X(x)\, dx = \gamma$$

γ / n	·60	·75	·90	·95	·975	·99	·995	·9975	·999	·9995
1	·325	1·000	3·078	6·314	12·706	31·821	63·657	127·32	318·31	636·62
2	·289	·816	1·886	2·920	4·303	6·965	9·925	14·089	22·327	31·598
3	·277	·765	1·638	2·353	3·182	4·541	5·841	7·453	10·214	12·924
4	·271	·741	1·533	2·132	2·776	3·747	4·604	5·598	7·173	8·610
5	·267	·727	1·476	2·015	2·571	3·365	4·032	4·773	5·893	6·869
6	·265	·718	1·440	1·943	2·447	3·143	3·707	4·317	5·208	5·959
7	·263	·711	1·415	1·895	2·365	2·998	3·499	4·029	4·785	5·408
8	·262	·706	1·397	1·860	2·306	2·896	3·355	3·833	4·501	5·041
9	·261	·703	1·383	1·833	2·262	2·821	3·250	3·690	4·297	4·781
10	·260	·700	1·372	1·812	2·228	2·764	3·169	3·581	4·144	4·587
11	·260	·697	1·363	1·796	2·201	2·718	3·106	3·497	4·025	4·437
12	·259	·695	1·356	1·782	2·179	2·681	3·055	3·428	3·930	4·318
13	·259	·694	1·350	1·771	2·160	2·650	3·012	3·372	3·852	4·221
14	·258	·692	1·345	1·761	2·145	2·624	2·977	3·326	3·787	4·140
15	·258	·691	1·341	1·753	2·131	2·602	2·947	3·286	3·733	4·073
16	·258	·690	1·337	1·746	2·120	2·583	2·921	3·252	3·686	4·015
17	·257	·689	1·333	1·740	2·110	2·567	2·898	3·222	3·646	3·965
18	·257	·688	1·330	1·734	2·101	2·552	2·878	3·197	3·610	3·922
19	·257	·688	1·328	1·729	2·093	2·539	2·861	3·174	3·579	3·883
20	·257	687	1·325	1·725	2·086	2·528	2·845	3·153	3·552	3·850
21	·257	·686	1·323	1·721	2·080	2·518	2·831	3·135	3·527	3·819
22	·256	·686	1·321	1·717	2·074	2·508	2·819	3·119	3·505	3·792
23	·256	·685	1·319	1·714	2·069	2·500	2·807	3·104	3·485	3·767
24	·256	·685	1·318	1·711	2·064	2·492	2·797	3·091	3·467	3·745
25	·256	·684	1·316	1·708	2·060	2·485	2·787	3·078	3·450	3·725
26	·256	·684	1·315	1·706	2·056	2·479	2·779	3·067	3·435	3·707
27	·256	·684	1·314	1·703	2·052	2·473	2·771	3·057	3·421	3·690
28	·256	·683	1·313	1·701	2·048	2·467	2·763	3·047	3·408	3·674
29	·256	·683	1·311	1·699	2·045	2·462	2·756	3·038	3·396	3·659
30	·256	·683	1·310	1·697	2·042	2·457	2·750	3·030	3·385	3·646
40	·255	·681	1·303	1·684	2·021	2·423	2·704	2·971	3·307	3·551
60	·254	·679	1·296	1·671	2·000	2·390	2·660	2·915	3·232	3·460
120	·254	·677	1·289	1·658	1·980	2·358	2·617	2·860	3·160	3·373
∞	·253	·674	1·282	1·645	1·960	2·326	2·576	2·807	3·090	3·291

TABLE A-6. Critical values of the chi square distribution

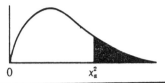

d.f.	$\chi^2{}_{0.995}$	$\chi^2{}_{0.990}$	$\chi^2{}_{0.975}$	$\chi^2{}_{0.950}$	$\chi^2{}_{0.900}$
1	0.0000393	0.0001571	0.0009821	0.0039321	0.0157908
2	0.0100251	0.0201007	0.0506356	0.102587	0.210720
3	0.0717212	0.114832	0.215795	0.351846	0.584375
4	0.206990	0.297110	0.484419	0.710721	1.063623
5	0.411740	0.554300	0.831211	1.145476	1.61031
6	0.675727	0.872085	1.237347	1.63539	2.20413
7	0.989265	1.239043	1.68987	2.16735	2.83311
8	1.344419	1.646482	2.17973	2.73264	3.48954
9	1.734926	2.087912	2.70039	3.32511	4.16816
10	2.15585	2.55821	3.24697	3.94030	4.86518
11	2.60321	3.05347	3.81575	4.57481	5.57779
12	3.07382	3.57056	4.40379	5.22603	6.30380
13	3.56503	4.10691	5.00874	5.89186	7.04150
14	4.07468	4.66043	5.62872	6.57063	7.78953
15	4.60094	5.22935	6.26214	7.26094	8.54675
16	5.14224	5.81221	6.90766	7.96164	9.31223
17	5.69724	6.40776	7.56418	8.67176	10.0852
18	6.26481	7.01491	8.23075	9.39046	10.8649
19	6.84398	7.63273	8.90655	10.1170	11.6509
20	7.43386	8.26040	9.59083	10.8508	12.4426
21	8.03366	8.89720	10.28293	11.5913	13.2396
22	8.64272	9.54249	10.9823	12.3380	14.0415
23	9.26042	10.19567	11.6885	13.0905	14.8479
24	9.88623	10.8564	12.4011	13.8484	15.6587
25	10.5197	11.5240	13.1197	14.6114	16.4734
26	11.1603	12.1981	13.8439	15.3791	17.2919
27	11.8076	12.8786	14.5733	16.1513	18.1138
28	12.4613	13.5648	15.3079	16.9279	18.9392
29	13.1211	14.2565	16.0471	17.7083	19.7677
30	13.7867	14.9535	16.7908	18.4926	20.5992
40	20.7065	22.1643	24.4331	26.5093	29.0505
50	27.9907	29.7067	32.3574	34.7642	37.6886
60	35.5346	37.4848	40.4817	43.1879	46.4589
70	43.2752	45.4418	48.7576	51.7393	55.3290
80	51.1720	53.5400	57.1532	60.3915	64.2778
90	59.1963	61.7541	65.6466	69.1260	73.2912
100	67.3276	70.0648	74.2219	77.9295	82.3581

TABLE A-6.   Critical values of the chi square distribution (Continued)

$\chi^2_{0.100}$	$\chi^2_{0.050}$	$\chi^2_{0.025}$	$\chi^2_{0.010}$	$\chi^2_{0.005}$	d.f.
2.70554	3.84146	5.02389	6.63490	7.87944	1
4.60517	5.99147	7.37776	9.21034	10.5966	2
6.25139	7.81473	9.34840	11.3449	12.8381	3
7.77944	9.48773	11.1433	13.2767	14.8602	4
9.23635	11.0705	12.8325	15.0863	16.7496	5
10.6446	12.5916	14.4494	16.8119	18.5476	6
12.0170	14.0671	16.0128	18.4753	20.2777	7
13.3616	15.5073	17.5346	20.0902	21.9550	8
14.6837	16.9190	19.0228	21.6660	23.5893	9
15.9871	18.3070	20.4831	23.2093	25.1882	10
17.2750	19.6751	21.9200	24.7250	26.7569	11
18.5494	21.0261	23.3367	26.2170	28.2995	12
19.8119	22.3621	24.7356	27.6883	29.8194	13
21.0642	23.6848	26.1190	29.1413	31.3193	14
22.3072	24.9958	27.4884	30.5779	32.8013	15
23.5418	26.2962	28.8454	31.9999	34.2672	16
24.7690	27.5871	30.1910	33.4087	35.7185	17
25.9894	28.8693	31.5264	34.8053	37.1564	18
27.2036	30.1435	32.8523	36.1908	38.5822	19
28.4120	31.4104	34.1696	37.5662	39.9968	20
29.6151	32.6705	35.4789	38.9321	41.4010	21
30.8133	33.9244	36.7807	40.2894	42.7956	22
32.0069	35.1725	38.0757	41.6384	44.1813	23
33.1963	36.4151	39.3641	42.9798	45.5585	24
34.3816	37.6525	40.6465	44.3141	46.9278	25
35.5631	38.8852	41.9232	45.6417	48.2899	26
36.7412	40.1133	43.1944	46.9630	49.6449	27
37.9159	41.3372	44.4607	48.2782	50.9933	28
39.0875	42.5569	45.7222	49.5879	52.3356	29
40.2560	43.7729	46.9792	50.8922	53.6720	30
51.8050	55.7585	59.3417	63.6907	66.7659	40
63.1671	67.5048	71.4202	76.1539	79.4900	50
74.3970	79.0819	83.2976	88.3794	91.9517	60
85.5271	90.5312	95.0231	100.425	104.215	70
96.5782	101.879	106.629	112.329	116.321	80
107.565	113.145	118.136	124.116	128.299	90
118.498	124.342	129.561	135.807	140.169	100

From ''Tables of the Percentage Points of the χ^2-Distribution.'' *Biometrika*, Vol. 32 (1941), pp. 188–189, by Catherine M. Thompson. Reproduced by permission of Professor D. V. Lindley.

TABLE A-7. Critical values of the F distribution*

5 per cent (roman type) and 1 per cent (bold-face type) points for the distribution of F*

Degrees of freedom for lesser mean square (rows) — *Degrees of freedom for greater mean square* (columns)

df	1	2	3	4	5	6	7	8	9	10	11	12	14	16	20	24	30	40	50	75	100	200	500	∞
1	161	200	216	225	230	234	237	239	241	242	243	244	245	246	248	249	250	251	252	253	253	254	254	254
	4052	**4999**	**5403**	**5625**	**5764**	**5859**	**5928**	**5981**	**6022**	**6056**	**6082**	**6106**	**6142**	**6169**	**6208**	**6234**	**6258**	**6286**	**6302**	**6323**	**6334**	**6352**	**6361**	**6366**
2	18.51	19.00	19.16	19.25	19.30	19.33	19.36	19.37	19.38	19.39	19.40	19.41	19.42	19.43	19.44	19.45	19.46	19.47	19.47	19.48	19.49	19.49	19.50	19.50
	98.49	**99.01**	**99.17**	**99.25**	**99.30**	**99.33**	**99.34**	**99.36**	**99.38**	**99.40**	**99.41**	**99.42**	**99.43**	**99.44**	**99.45**	**99.46**	**99.47**	**99.48**	**99.48**	**99.49**	**99.49**	**99.49**	**99.50**	**99.50**
3	10.13	9.55	9.28	9.12	9.01	8.94	8.88	8.84	8.81	8.78	8.76	8.74	8.71	8.69	8.66	8.64	8.62	8.60	8.58	8.57	8.56	8.54	8.54	8.53
	34.12	**30.81**	**29.46**	**28.71**	**28.24**	**27.91**	**27.67**	**27.49**	**27.34**	**27.23**	**27.13**	**27.05**	**26.92**	**26.83**	**26.69**	**26.60**	**26.50**	**26.41**	**26.35**	**26.27**	**26.23**	**26.18**	**26.14**	**26.12**
4	7.71	6.94	6.59	6.39	6.26	6.16	6.09	6.04	6.00	5.96	5.93	5.91	5.87	5.84	5.80	5.77	5.74	5.71	5.70	5.68	5.66	5.65	5.64	5.63
	21.20	**18.00**	**16.69**	**15.98**	**15.52**	**15.21**	**14.98**	**14.80**	**14.66**	**14.54**	**14.45**	**14.37**	**14.24**	**14.15**	**14.02**	**13.93**	**13.83**	**13.74**	**13.69**	**13.61**	**13.57**	**13.52**	**13.48**	**13.46**
5	6.61	5.79	5.41	5.19	5.05	4.95	4.88	4.82	4.78	4.74	4.70	4.68	4.64	4.60	4.56	4.53	4.50	4.46	4.44	4.42	4.40	4.38	4.37	4.36
	16.26	**13.27**	**12.06**	**11.39**	**10.97**	**10.67**	**10.45**	**10.27**	**10.15**	**10.05**	**9.96**	**9.89**	**9.77**	**9.68**	**9.55**	**9.47**	**9.38**	**9.29**	**9.24**	**9.17**	**9.13**	**9.07**	**9.04**	**9.02**
6	5.99	5.14	4.76	4.53	4.39	4.28	4.21	4.15	4.10	4.06	4.03	4.00	3.96	3.92	3.87	3.84	3.81	3.77	3.75	3.72	3.71	3.69	3.68	3.67
	13.74	**10.92**	**9.78**	**9.15**	**8.75**	**8.47**	**8.26**	**8.10**	**7.98**	**7.87**	**7.79**	**7.72**	**7.60**	**7.52**	**7.39**	**7.31**	**7.23**	**7.14**	**7.09**	**7.02**	**6.99**	**6.94**	**6.90**	**6.88**
7	5.59	4.74	4.35	4.12	3.97	3.87	3.79	3.73	3.68	3.63	3.60	3.57	3.52	3.49	3.44	3.41	3.38	3.34	3.32	3.29	3.28	3.25	3.24	3.23
	12.25	**9.55**	**8.45**	**7.85**	**7.46**	**7.19**	**7.00**	**6.84**	**6.71**	**6.62**	**6.54**	**6.47**	**6.35**	**6.27**	**6.15**	**6.07**	**5.98**	**5.90**	**5.85**	**5.78**	**5.75**	**5.70**	**5.67**	**5.65**
8	5.32	4.46	4.07	3.84	3.69	3.58	3.50	3.44	3.39	3.34	3.31	3.28	3.23	3.20	3.15	3.12	3.08	3.05	3.03	3.00	2.98	2.96	2.94	2.93
	11.26	**8.65**	**7.59**	**7.01**	**6.63**	**6.37**	**6.19**	**6.03**	**5.91**	**5.82**	**5.74**	**5.67**	**5.56**	**5.48**	**5.36**	**5.28**	**5.20**	**5.11**	**5.06**	**5.00**	**4.96**	**4.91**	**4.88**	**4.86**
9	5.12	4.26	3.86	3.63	3.48	3.37	3.29	3.23	3.18	3.13	3.10	3.07	3.02	2.98	2.93	2.90	2.86	2.82	2.80	2.77	2.76	2.73	2.72	2.71
	10.56	**8.02**	**6.99**	**6.42**	**6.06**	**5.80**	**5.62**	**5.47**	**5.35**	**5.26**	**5.18**	**5.11**	**5.00**	**4.92**	**4.80**	**4.73**	**4.64**	**4.56**	**4.51**	**4.45**	**4.41**	**4.36**	**4.33**	**4.31**
10	4.96	4.10	3.71	3.48	3.33	3.22	3.14	3.07	3.02	2.97	2.94	2.91	2.86	2.82	2.77	2.74	2.70	2.67	2.64	2.61	2.59	2.56	2.55	2.54
	10.04	**7.56**	**6.55**	**5.99**	**5.64**	**5.39**	**5.21**	**5.06**	**4.95**	**4.85**	**4.78**	**4.71**	**4.60**	**4.52**	**4.41**	**4.33**	**4.25**	**4.17**	**4.12**	**4.05**	**4.01**	**3.96**	**3.93**	**3.91**
11	4.84	3.98	3.59	3.36	3.20	3.09	3.01	2.95	2.90	2.86	2.82	2.79	2.74	2.70	2.65	2.61	2.57	2.53	2.50	2.47	2.45	2.42	2.41	2.40
	9.65	**7.20**	**6.22**	**5.67**	**5.32**	**5.07**	**4.88**	**4.74**	**4.63**	**4.54**	**4.46**	**4.40**	**4.29**	**4.21**	**4.10**	**4.02**	**3.94**	**3.86**	**3.80**	**3.74**	**3.70**	**3.66**	**3.62**	**3.60**

TABLE A-7. Critical values of the F distribution* (Continued)

df																								
12	4.75 / 9.33	3.88 / 6.93	3.49 / 5.95	3.26 / 5.41	3.11 / 5.06	3.00 / 4.82	2.92 / 4.65	2.85 / 4.50	2.80 / 4.39	2.76 / 4.30	2.72 / 4.22	2.69 / 4.16	2.64 / 4.05	2.60 / 3.98	2.54 / 3.86	2.50 / 3.78	2.46 / 3.70	2.42 / 3.61	2.40 / 3.56	2.36 / 3.49	2.35 / 3.46	2.32 / 3.41	2.31 / 3.38	2.30 / 3.36
13	4.67 / 9.07	3.80 / 6.70	3.41 / 5.74	3.18 / 5.20	3.02 / 4.86	2.92 / 4.62	2.84 / 4.44	2.77 / 4.30	2.72 / 4.19	2.67 / 4.10	2.63 / 4.02	2.60 / 3.96	2.55 / 3.85	2.51 / 3.78	2.46 / 3.67	2.42 / 3.59	2.38 / 3.51	2.34 / 3.42	2.32 / 3.37	2.28 / 3.30	2.26 / 3.27	2.24 / 3.21	2.22 / 3.18	2.21 / 3.16
14	4.60 / 8.86	3.74 / 6.51	3.34 / 5.56	3.11 / 5.03	2.96 / 4.69	2.85 / 4.46	2.77 / 4.28	2.70 / 4.14	2.65 / 4.03	2.60 / 3.94	2.56 / 3.86	2.53 / 3.80	2.48 / 3.70	2.44 / 3.62	2.39 / 3.51	2.35 / 3.43	2.31 / 3.34	2.27 / 3.26	2.24 / 3.21	2.21 / 3.14	2.19 / 3.11	2.16 / 3.06	2.14 / 3.02	2.13 / 3.00
15	4.54 / 8.68	3.68 / 6.36	3.29 / 5.42	3.06 / 4.89	2.90 / 4.56	2.79 / 4.32	2.70 / 4.14	2.64 / 4.00	2.59 / 3.89	2.55 / 3.80	2.51 / 3.73	2.48 / 3.67	2.43 / 3.56	2.39 / 3.48	2.33 / 3.36	2.29 / 3.29	2.25 / 3.20	2.21 / 3.12	2.18 / 3.07	2.15 / 3.00	2.12 / 2.97	2.10 / 2.92	2.08 / 2.89	2.07 / 2.87
16	4.49 / 8.53	3.63 / 6.23	3.24 / 5.29	3.01 / 4.77	2.85 / 4.44	2.74 / 4.20	2.66 / 4.03	2.59 / 3.89	2.54 / 3.78	2.49 / 3.69	2.45 / 3.61	2.42 / 3.55	2.37 / 3.45	2.33 / 3.37	2.28 / 3.25	2.24 / 3.18	2.20 / 3.10	2.16 / 3.01	2.13 / 2.96	2.09 / 2.89	2.07 / 2.86	2.04 / 2.80	2.02 / 2.77	2.01 / 2.75
17	4.45 / 8.40	3.59 / 6.11	3.20 / 5.18	2.96 / 4.67	2.81 / 4.34	2.70 / 4.10	2.62 / 3.93	2.55 / 3.79	2.50 / 3.68	2.45 / 3.59	2.41 / 3.52	2.38 / 3.45	2.33 / 3.35	2.29 / 3.27	2.23 / 3.16	2.19 / 3.08	2.15 / 3.00	2.11 / 2.92	2.08 / 2.86	2.04 / 2.79	2.02 / 2.76	1.99 / 2.70	1.97 / 2.67	1.96 / 2.65
18	4.41 / 8.28	3.55 / 6.01	3.16 / 5.09	2.93 / 4.58	2.77 / 4.25	2.66 / 4.01	2.58 / 3.85	2.51 / 3.71	2.46 / 3.60	2.41 / 3.51	2.37 / 3.44	2.34 / 3.37	2.29 / 3.27	2.25 / 3.19	2.19 / 3.07	2.15 / 3.00	2.11 / 2.91	2.07 / 2.83	2.04 / 2.78	2.00 / 2.71	1.98 / 2.68	1.95 / 2.62	1.93 / 2.59	1.92 / 2.57
19	4.38 / 8.18	3.52 / 5.93	3.13 / 5.01	2.90 / 4.50	2.74 / 4.17	2.63 / 3.94	2.55 / 3.77	2.48 / 3.63	2.43 / 3.52	2.38 / 3.43	2.34 / 3.36	2.31 / 3.30	2.26 / 3.19	2.21 / 3.12	2.15 / 3.00	2.11 / 2.92	2.07 / 2.84	2.02 / 2.76	2.00 / 2.70	1.96 / 2.63	1.94 / 2.60	1.91 / 2.54	1.90 / 2.51	1.88 / 2.49
20	4.35 / 8.10	3.49 / 5.85	3.10 / 4.94	2.87 / 4.43	2.71 / 4.10	2.60 / 3.87	2.52 / 3.71	2.45 / 3.56	2.40 / 3.45	2.35 / 3.37	2.31 / 3.30	2.28 / 3.23	2.23 / 3.13	2.18 / 3.05	2.12 / 2.94	2.08 / 2.86	2.04 / 2.77	1.99 / 2.69	1.96 / 2.63	1.92 / 2.56	1.90 / 2.53	1.87 / 2.47	1.85 / 2.44	1.84 / 2.42
21	4.32 / 8.02	3.47 / 5.78	3.07 / 4.87	2.84 / 4.37	2.68 / 4.04	2.57 / 3.81	2.49 / 3.65	2.42 / 3.51	2.37 / 3.40	2.32 / 3.31	2.28 / 3.24	2.25 / 3.17	2.20 / 3.07	2.15 / 2.99	2.09 / 2.88	2.05 / 2.80	2.00 / 2.72	1.96 / 2.63	1.93 / 2.58	1.89 / 2.51	1.87 / 2.47	1.84 / 2.42	1.82 / 2.38	1.81 / 2.36
22	4.30 / 7.94	3.44 / 5.72	3.05 / 4.82	2.82 / 4.31	2.66 / 3.99	2.55 / 3.76	2.47 / 3.59	2.40 / 3.45	2.35 / 3.35	2.30 / 3.26	2.26 / 3.18	2.23 / 3.12	2.18 / 3.02	2.13 / 2.94	2.07 / 2.83	2.03 / 2.75	1.98 / 2.67	1.93 / 2.58	1.91 / 2.53	1.87 / 2.46	1.84 / 2.42	1.81 / 2.37	1.80 / 2.33	1.78 / 2.31
23	4.28 / 7.88	3.42 / 5.66	3.03 / 4.76	2.80 / 4.26	2.64 / 3.94	2.53 / 3.71	2.45 / 3.54	2.38 / 3.41	2.32 / 3.30	2.28 / 3.21	2.24 / 3.14	2.20 / 3.07	2.14 / 2.97	2.10 / 2.89	2.04 / 2.78	2.00 / 2.70	1.96 / 2.62	1.91 / 2.53	1.88 / 2.48	1.84 / 2.41	1.82 / 2.37	1.79 / 2.32	1.77 / 2.28	1.76 / 2.26
24	4.26 / 7.82	3.40 / 5.61	3.01 / 4.72	2.78 / 4.22	2.62 / 3.90	2.51 / 3.67	2.43 / 3.50	2.36 / 3.36	2.30 / 3.25	2.26 / 3.17	2.22 / 3.09	2.18 / 3.03	2.13 / 2.93	2.09 / 2.85	2.02 / 2.74	1.98 / 2.66	1.94 / 2.58	1.89 / 2.49	1.86 / 2.44	1.82 / 2.36	1.80 / 2.33	1.76 / 2.27	1.74 / 2.23	1.73 / 2.21

TABLE A-7. Critical values of the F distribution* (Continued)

Degrees of freedom for greater mean square

Degrees of freedom for lesser mean square	1	2	3	4	5	6	7	8	9	10	11	12	14	16	20	24	30	40	50	75	100	200	500	∞
25	4.24 / 7.77	3.38 / 5.57	2.99 / 4.68	2.76 / 4.18	2.60 / 3.86	2.49 / 3.63	2.41 / 3.46	2.34 / 3.32	2.28 / 3.21	2.24 / 3.13	2.20 / 3.05	2.16 / 2.99	2.11 / 2.89	2.06 / 2.81	2.00 / 2.70	1.96 / 2.62	1.92 / 2.54	1.87 / 2.45	1.84 / 2.40	1.80 / 2.32	1.77 / 2.29	1.74 / 2.23	1.72 / 2.19	1.71 / 2.17
26	4.22 / 7.72	3.37 / 5.53	2.98 / 4.64	2.74 / 4.14	2.59 / 3.82	2.47 / 3.59	2.39 / 3.42	2.32 / 3.29	2.27 / 3.17	2.22 / 3.09	2.18 / 3.02	2.15 / 2.96	2.10 / 2.86	2.05 / 2.77	1.99 / 2.66	1.95 / 2.58	1.90 / 2.50	1.85 / 2.41	1.82 / 2.36	1.78 / 2.28	1.76 / 2.25	1.72 / 2.19	1.70 / 2.15	1.69 / 2.13
27	4.21 / 7.68	3.35 / 5.49	2.96 / 4.60	2.73 / 4.11	2.57 / 3.79	2.46 / 3.56	2.37 / 3.39	2.30 / 3.26	2.25 / 3.14	2.20 / 3.06	2.16 / 2.98	2.13 / 2.93	2.08 / 2.83	2.03 / 2.74	1.97 / 2.63	1.93 / 2.55	1.88 / 2.47	1.84 / 2.38	1.80 / 2.33	1.76 / 2.25	1.74 / 2.21	1.71 / 2.16	1.68 / 2.12	1.67 / 2.10
28	4.20 / 7.64	3.34 / 5.45	2.95 / 4.57	2.71 / 4.07	2.56 / 3.76	2.44 / 3.53	2.36 / 3.36	2.29 / 3.23	2.24 / 3.11	2.19 / 3.03	2.15 / 2.95	2.12 / 2.90	2.06 / 2.80	2.02 / 2.71	1.96 / 2.60	1.91 / 2.52	1.87 / 2.44	1.81 / 2.35	1.78 / 2.30	1.75 / 2.22	1.72 / 2.18	1.69 / 2.13	1.67 / 2.09	1.65 / 2.06
29	4.18 / 7.60	3.33 / 5.42	2.93 / 4.54	2.70 / 4.04	2.54 / 3.73	2.43 / 3.50	2.35 / 3.33	2.28 / 3.20	2.22 / 3.08	2.18 / 3.00	2.14 / 2.92	2.10 / 2.87	2.05 / 2.77	2.00 / 2.68	1.94 / 2.57	1.90 / 2.49	1.85 / 2.41	1.80 / 2.32	1.77 / 2.27	1.73 / 2.19	1.71 / 2.15	1.68 / 2.10	1.65 / 2.06	1.64 / 2.03
30	4.17 / 7.56	3.32 / 5.39	2.92 / 4.51	2.69 / 4.02	2.53 / 3.70	2.42 / 3.47	2.34 / 3.30	2.27 / 3.17	2.21 / 3.06	2.16 / 2.98	2.12 / 2.90	2.09 / 2.84	2.04 / 2.74	1.99 / 2.66	1.93 / 2.55	1.89 / 2.47	1.84 / 2.38	1.79 / 2.29	1.76 / 2.24	1.72 / 2.16	1.69 / 2.13	1.66 / 2.07	1.64 / 2.03	1.62 / 2.01
32	4.15 / 7.50	3.30 / 5.34	2.90 / 4.46	2.67 / 3.97	2.51 / 3.66	2.40 / 3.42	2.32 / 3.25	2.25 / 3.12	2.19 / 3.01	2.14 / 2.94	2.10 / 2.86	2.07 / 2.80	2.02 / 2.70	1.97 / 2.62	1.91 / 2.51	1.86 / 2.42	1.82 / 2.34	1.76 / 2.25	1.74 / 2.20	1.69 / 2.12	1.67 / 2.08	1.64 / 2.02	1.61 / 1.98	1.59 / 1.96
34	4.13 / 7.44	3.28 / 5.29	2.88 / 4.42	2.65 / 3.93	2.49 / 3.61	2.38 / 3.38	2.30 / 3.21	2.23 / 3.08	2.17 / 2.97	2.12 / 2.89	2.08 / 2.82	2.05 / 2.76	2.00 / 2.66	1.95 / 2.58	1.89 / 2.47	1.84 / 2.38	1.80 / 2.30	1.74 / 2.21	1.71 / 2.15	1.67 / 2.08	1.64 / 2.04	1.61 / 1.98	1.59 / 1.94	1.57 / 1.91
36	4.11 / 7.39	3.26 / 5.25	2.86 / 4.38	2.63 / 3.89	2.48 / 3.58	2.36 / 3.35	2.28 / 3.18	2.21 / 3.04	2.15 / 2.94	2.10 / 2.86	2.06 / 2.78	2.03 / 2.72	1.98 / 2.62	1.93 / 2.54	1.87 / 2.43	1.82 / 2.35	1.78 / 2.26	1.72 / 2.17	1.69 / 2.12	1.65 / 2.04	1.62 / 2.00	1.59 / 1.94	1.56 / 1.90	1.55 / 1.87
38	4.10 / 7.35	3.25 / 5.21	2.85 / 4.34	2.62 / 3.86	2.46 / 3.54	2.35 / 3.32	2.26 / 3.15	2.19 / 3.02	2.14 / 2.91	2.09 / 2.82	2.05 / 2.75	2.02 / 2.69	1.96 / 2.59	1.92 / 2.51	1.85 / 2.40	1.80 / 2.32	1.76 / 2.22	1.71 / 2.14	1.67 / 2.08	1.63 / 2.00	1.60 / 1.97	1.57 / 1.90	1.54 / 1.86	1.53 / 1.84
40	4.08 / 7.31	3.23 / 5.18	2.84 / 4.31	2.61 / 3.83	2.45 / 3.51	2.34 / 3.29	2.25 / 3.12	2.18 / 2.99	2.12 / 2.88	2.07 / 2.80	2.04 / 2.73	2.00 / 2.66	1.95 / 2.56	1.90 / 2.49	1.84 / 2.37	1.79 / 2.29	1.74 / 2.20	1.69 / 2.11	1.66 / 2.05	1.61 / 1.97	1.59 / 1.94	1.55 / 1.88	1.53 / 1.84	1.51 / 1.81
42	4.07 / 7.27	3.22 / 5.15	2.83 / 4.29	2.59 / 3.80	2.44 / 3.49	2.32 / 3.26	2.24 / 3.10	2.17 / 2.96	2.11 / 2.86	2.06 / 2.77	2.02 / 2.70	1.99 / 2.64	1.94 / 2.54	1.89 / 2.46	1.82 / 2.35	1.78 / 2.26	1.73 / 2.17	1.68 / 2.08	1.64 / 2.02	1.60 / 1.94	1.57 / 1.91	1.54 / 1.85	1.51 / 1.80	1.49 / 1.78
44	4.06 / 7.24	3.21 / 5.12	2.82 / 4.26	2.58 / 3.78	2.43 / 3.46	2.31 / 3.24	2.23 / 3.07	2.16 / 2.94	2.10 / 2.84	2.05 / 2.75	2.01 / 2.68	1.98 / 2.62	1.92 / 2.52	1.88 / 2.44	1.81 / 2.32	1.76 / 2.24	1.72 / 2.15	1.66 / 2.06	1.63 / 2.00	1.58 / 1.92	1.56 / 1.88	1.52 / 1.82	1.50 / 1.78	1.48 / 1.75

TABLE A-7. Critical values of the F distribution* (Continued)

df																								
46	4.05 / 7.21	3.20 / 5.10	2.81 / 4.24	2.57 / 3.76	2.42 / 3.44	2.30 / 3.22	2.22 / 3.05	2.14 / 2.92	2.09 / 2.82	2.04 / 2.73	2.00 / 2.66	1.97 / 2.60	1.91 / 2.50	1.87 / 2.42	1.80 / 2.30	1.75 / 2.22	1.71 / 2.13	1.65 / 2.04	1.62 / 1.98	1.57 / 1.90	1.54 / 1.86	1.51 / 1.80	1.48 / 1.76	1.46 / 1.72
48	4.04 / 7.19	3.19 / 5.08	2.80 / 4.22	2.56 / 3.74	2.41 / 3.42	2.30 / 3.20	2.21 / 3.04	2.14 / 2.90	2.08 / 2.80	2.03 / 2.71	1.99 / 2.64	1.96 / 2.58	1.90 / 2.48	1.86 / 2.40	1.79 / 2.28	1.74 / 2.20	1.70 / 2.11	1.64 / 2.02	1.61 / 1.96	1.56 / 1.88	1.53 / 1.84	1.50 / 1.78	1.47 / 1.73	1.45 / 1.70
50	4.03 / 7.17	3.18 / 5.06	2.79 / 4.20	2.56 / 3.72	2.40 / 3.41	2.29 / 3.18	2.20 / 3.02	2.13 / 2.88	2.07 / 2.78	2.02 / 2.70	1.98 / 2.62	1.95 / 2.56	1.90 / 2.46	1.85 / 2.39	1.78 / 2.26	1.74 / 2.18	1.69 / 2.10	1.63 / 2.00	1.60 / 1.94	1.55 / 1.86	1.52 / 1.82	1.48 / 1.76	1.46 / 1.71	1.44 / 1.68
55	4.02 / 7.12	3.17 / 5.01	2.78 / 4.16	2.54 / 3.68	2.38 / 3.37	2.27 / 3.15	2.18 / 2.98	2.11 / 2.85	2.05 / 2.75	2.00 / 2.66	1.97 / 2.59	1.93 / 2.53	1.88 / 2.43	1.83 / 2.35	1.76 / 2.23	1.72 / 2.15	1.67 / 2.06	1.61 / 1.96	1.58 / 1.90	1.52 / 1.82	1.50 / 1.78	1.46 / 1.71	1.43 / 1.66	1.41 / 1.64
60	4.00 / 7.08	3.15 / 4.98	2.76 / 4.13	2.52 / 3.65	2.37 / 3.34	2.25 / 3.12	2.17 / 2.95	2.10 / 2.82	2.04 / 2.72	1.99 / 2.63	1.95 / 2.56	1.92 / 2.50	1.86 / 2.40	1.81 / 2.32	1.75 / 2.20	1.70 / 2.12	1.65 / 2.03	1.59 / 1.93	1.56 / 1.87	1.50 / 1.79	1.48 / 1.74	1.44 / 1.68	1.41 / 1.63	1.39 / 1.60
65	3.99 / 7.04	3.14 / 4.95	2.75 / 4.10	2.51 / 3.62	2.36 / 3.31	2.24 / 3.09	2.15 / 2.93	2.08 / 2.79	2.02 / 2.70	1.98 / 2.61	1.94 / 2.54	1.90 / 2.47	1.85 / 2.37	1.80 / 2.30	1.73 / 2.18	1.68 / 2.09	1.63 / 2.00	1.57 / 1.90	1.54 / 1.84	1.49 / 1.76	1.46 / 1.71	1.42 / 1.64	1.39 / 1.60	1.37 / 1.56
70	3.98 / 7.01	3.13 / 4.92	2.74 / 4.08	2.50 / 3.60	2.35 / 3.29	2.23 / 3.07	2.14 / 2.91	2.07 / 2.77	2.01 / 2.67	1.97 / 2.59	1.93 / 2.51	1.89 / 2.45	1.84 / 2.35	1.79 / 2.28	1.72 / 2.15	1.67 / 2.07	1.62 / 1.98	1.56 / 1.88	1.53 / 1.82	1.47 / 1.74	1.45 / 1.69	1.40 / 1.62	1.37 / 1.56	1.35 / 1.53
80	3.96 / 6.96	3.11 / 4.88	2.72 / 4.04	2.48 / 3.56	2.33 / 3.25	2.21 / 3.04	2.12 / 2.87	2.05 / 2.74	1.99 / 2.64	1.95 / 2.55	1.91 / 2.48	1.88 / 2.41	1.82 / 2.32	1.77 / 2.24	1.70 / 2.11	1.65 / 2.03	1.60 / 1.94	1.54 / 1.84	1.51 / 1.78	1.45 / 1.70	1.42 / 1.65	1.38 / 1.57	1.35 / 1.52	1.32 / 1.49
100	3.94 / 6.90	3.09 / 4.82	2.70 / 3.98	2.46 / 3.51	2.30 / 3.20	2.19 / 2.99	2.10 / 2.82	2.03 / 2.69	1.97 / 2.59	1.92 / 2.51	1.88 / 2.43	1.85 / 2.36	1.79 / 2.26	1.75 / 2.19	1.68 / 2.06	1.63 / 1.98	1.57 / 1.89	1.51 / 1.79	1.48 / 1.73	1.42 / 1.64	1.39 / 1.59	1.34 / 1.51	1.30 / 1.46	1.28 / 1.43
125	3.92 / 6.84	3.07 / 4.78	2.68 / 3.94	2.44 / 3.47	2.29 / 3.17	2.17 / 2.95	2.08 / 2.79	2.01 / 2.65	1.95 / 2.56	1.90 / 2.47	1.86 / 2.40	1.83 / 2.33	1.77 / 2.23	1.72 / 2.15	1.65 / 2.03	1.60 / 1.94	1.55 / 1.85	1.49 / 1.75	1.45 / 1.68	1.39 / 1.59	1.36 / 1.54	1.31 / 1.46	1.27 / 1.40	1.25 / 1.37
150	3.91 / 6.81	3.06 / 4.75	2.67 / 3.91	2.43 / 3.44	2.27 / 3.14	2.16 / 2.92	2.07 / 2.76	2.00 / 2.62	1.94 / 2.53	1.89 / 2.44	1.85 / 2.37	1.82 / 2.30	1.76 / 2.20	1.71 / 2.12	1.64 / 2.00	1.59 / 1.91	1.54 / 1.83	1.47 / 1.72	1.44 / 1.66	1.37 / 1.56	1.34 / 1.51	1.29 / 1.43	1.25 / 1.37	1.22 / 1.33
200	3.89 / 6.76	3.04 / 4.71	2.65 / 3.88	2.41 / 3.41	2.26 / 3.11	2.14 / 2.90	2.05 / 2.73	1.98 / 2.60	1.92 / 2.50	1.87 / 2.41	1.83 / 2.34	1.80 / 2.28	1.74 / 2.17	1.69 / 2.09	1.62 / 1.97	1.57 / 1.88	1.52 / 1.79	1.45 / 1.69	1.42 / 1.62	1.35 / 1.53	1.32 / 1.48	1.26 / 1.39	1.22 / 1.33	1.19 / 1.28
400	3.86 / 6.70	3.02 / 4.66	2.62 / 3.83	2.39 / 3.36	2.23 / 3.06	2.12 / 2.85	2.03 / 2.69	1.96 / 2.55	1.90 / 2.46	1.85 / 2.37	1.81 / 2.29	1.78 / 2.23	1.72 / 2.12	1.67 / 2.04	1.60 / 1.92	1.54 / 1.84	1.49 / 1.74	1.42 / 1.64	1.38 / 1.57	1.32 / 1.47	1.28 / 1.42	1.22 / 1.32	1.16 / 1.24	1.13 / 1.19
1000	3.85 / 6.66	3.00 / 4.62	2.61 / 3.80	2.38 / 3.34	2.22 / 3.04	2.10 / 2.82	2.02 / 2.66	1.95 / 2.53	1.89 / 2.43	1.84 / 2.34	1.80 / 2.26	1.76 / 2.20	1.70 / 2.09	1.65 / 2.01	1.58 / 1.89	1.53 / 1.81	1.47 / 1.71	1.41 / 1.61	1.36 / 1.54	1.30 / 1.44	1.26 / 1.38	1.19 / 1.28	1.13 / 1.19	1.08 / 1.11
∞	3.84 / 6.64	2.99 / 4.60	2.60 / 3.78	2.37 / 3.32	2.21 / 3.02	2.09 / 2.80	2.01 / 2.64	1.94 / 2.51	1.88 / 2.41	1.83 / 2.32	1.79 / 2.24	1.75 / 2.18	1.69 / 2.07	1.64 / 1.99	1.57 / 1.87	1.52 / 1.79	1.46 / 1.69	1.40 / 1.59	1.35 / 1.52	1.28 / 1.41	1.24 / 1.36	1.17 / 1.25	1.11 / 1.15	1.00 / 1.00

*Reprinted by permission, from G. W. Snedecor and William G. Cochran, *Statistical methods*, 7th ed., ©1980, Iowa State University Press, Ames, Iowa.

TABLE A-8. Critical values for the sign test*

The entries in this table are the critical values for the number of the least frequent sign for a two-tailed test at α for the binomial $p = 0.5$.

n	0.01	0.05	0.10	0.25	n	0.01	0.05	0.10	0.25
1					51	15	18	19	20
2					52	16	18	19	21
3				0	53	16	18	20	21
4				0	54	17	19	20	22
5			0	0	55	17	19	20	22
6		0	0	1	56	17	20	21	23
7		0	0	1	57	18	20	21	23
8	0	0	1	1	58	18	21	22	24
9	0	1	1	2	59	19	21	22	24
10	0	1	1	2	60	19	21	23	25
11	0	1	2	3	61	20	22	23	25
12	1	2	2	3	62	20	22	24	25
13	1	2	3	3	63	20	23	24	26
14	1	2	3	4	64	21	23	24	26
15	2	3	3	4	65	21	24	25	27
16	2	3	4	5	66	22	24	25	27
17	2	4	4	5	67	22	25	26	28
18	3	4	5	6	68	22	25	26	28
19	3	4	5	6	69	23	25	27	29
20	3	5	5	6	70	23	26	27	29
21	4	5	6	7	71	24	26	28	30
22	4	5	6	7	72	24	27	28	30
23	4	6	7	8	73	25	27	28	31
24	5	6	7	8	74	25	28	29	31
25	5	7	7	9	75	25	28	29	32
26	6	7	8	9	76	26	28	30	32
27	6	7	8	10	77	26	29	30	32
28	6	8	9	10	78	27	29	31	33
29	7	8	9	10	79	27	30	31	33
30	7	9	10	11	80	28	30	32	34
31	7	9	10	11	81	28	31	32	34
32	8	9	10	12	82	28	31	33	35
33	8	10	11	12	83	29	32	33	35
34	9	10	11	13	84	29	32	33	36
35	9	11	12	13	85	30	32	34	36
36	9	11	12	14	86	30	33	34	37
37	10	12	13	14	87	31	33	35	37
38	10	12	13	14	88	31	34	35	38
39	11	12	13	15	89	31	34	36	38
40	11	13	14	15	90	32	35	36	39
41	11	13	14	16	91	32	35	37	39
42	12	14	15	16	92	33	36	37	39
43	12	14	15	17	93	33	36	38	40
44	13	15	16	17	94	34	37	38	40
45	13	15	16	18	95	34	37	38	41
46	13	15	16	18	96	34	37	39	41
47	14	16	17	19	97	35	38	39	42
48	14	16	17	19	98	35	38	40	42
49	15	17	18	19	99	36	39	40	43
50	15	17	18	20	100	36	39	41	43

*For specific details on the use of this table, see page 515.

Reprinted from Wilfred J. Dixon and Frank J. Massey, Jr., *Introduction to Statistical Analysis,* 3rd ed., McGraw-Hill Book Company, New York, 1969, p. 509.

TABLE A-9. Critical values for total number of runs (U)

The larger of n_1 and n_2

The smaller of n_1 and n_2	5	6	7	8	9	10	11	12	13	14	15	16	17	18	19	20
2								2/6	2/6	2/6	2/6	2/6	2/6	2/6	2/6	2/6
3		2/8	2/8	2/8	2/8	2/8	2/8	2/8	2/8	2/8	3/8	3/8	3/8	3/8	3/8	3/8
4	2/9	2/9	2/10	3/10	3/10	3/10	3/10	3/10	3/10	3/10	3/10	4/10	4/10	4/10	4/10	4/10
5	2/10	3/10	3/11	3/11	3/12	3/12	4/12	4/12	4/12	4/12	4/12	4/12	4/12	5/12	5/12	5/12
6		3/11	3/12	3/12	4/13	4/13	4/13	4/13	5/14	5/14	5/14	5/14	5/14	5/14	6/14	6/14
7			3/13	4/13	4/14	5/14	5/14	5/14	5/15	5/15	6/15	6/16	6/16	6/16	6/16	6/16
8				4/14	5/14	5/15	5/15	6/16	6/16	6/16	6/16	6/17	7/17	7/17	7/17	7/17
9					5/15	5/16	6/16	6/16	6/17	7/17	7/18	7/18	7/18	8/18	8/18	8/18
10						6/16	6/17	7/17	7/18	7/18	7/18	8/19	8/19	8/19	8/20	9/20
11							7/17	7/18	7/19	8/19	8/19	8/20	9/20	9/20	9/21	9/21
12								7/19	8/19	8/20	8/20	9/21	9/21	9/21	10/22	10/22
13									8/20	9/20	9/21	9/21	10/22	10/22	10/23	10/23
14										9/21	9/22	10/22	10/23	10/23	11/23	11/24
15											10/22	10/23	11/23	11/24	11/24	12/25
16												11/23	11/24	11/25	12/25	12/25
17													11/25	12/25	12/26	13/26
18														12/26	13/26	13/27
19															13/27	13/27
20																14/28

From C. Eisenhart and F. Swed, "Tables for testing randomness of grouping in a sequence of alternatives," *The Annals of Statistics,* Vol. 14 (1943), pp. 66–87. Reprinted by permission.

TABLE A-10a.   Quantiles of the Komogorov test statistic

One-sided test	p = 0.90	0.95	0.975	0.99	0.995
Two-sided test	p = 0.80	0.90	0.95	0.98	0.99
n = 1	.900	.950	.975	.990	.995
2	.684	.776	.842	.900	.929
3	.565	.636	.708	.785	.829
4	.493	.565	.624	.689	.734
5	.447	.509	.563	.627	.669
6	.410	.468	.519	.577	.617
7	.381	.436	.483	.538	.576
8	.358	.410	.454	.507	.542
9	.339	.387	.430	.480	.513
10	.323	.369	.409	.457	.489
11	.308	.352	.391	.437	.468
12	.296	.338	.375	.419	.449
13	.285	.325	.361	.404	.432
14	.275	.314	.349	.390	.418
15	.266	.304	.338	.377	.404
16	.258	.295	.327	.366	.392
17	.250	.286	.318	.355	.381
18	.244	.279	.309	.346	.371
19	.237	.271	.301	.337	.361
20	.232	.265	.294	.329	.352
21	.226	.259	.287	.321	.344
22	.221	.253	.281	.314	.337
23	.216	.247	.275	.307	.330
24	.212	.242	.269	.301	.323
25	.208	.238	.264	.295	.317
26	.204	.233	.259	.290	.311
27	.200	.229	.254	.284	.305
28	.197	.225	.250	.279	.300
29	.193	.221	.246	.275	.295
30	.190	.218	.242	.270	.290
31	.187	.214	.238	.266	.285
32	.184	.211	.234	.262	.281
33	.182	.208	.231	.258	.277
34	.179	.205	.227	.254	.273
35	.177	.202	.224	.251	.269
36	.174	.199	.221	.247	.265
37	.172	.196	.218	.244	.262
38	.170	.194	.215	.241	.258
39	.168	.191	.213	.238	.255
40	.165	.189	.210	.235	.252
Approximation for n > 40:	$\dfrac{1.07}{\sqrt{n}}$	$\dfrac{1.22}{\sqrt{n}}$	$\dfrac{1.36}{\sqrt{n}}$	$\dfrac{1.52}{\sqrt{n}}$	$\dfrac{1.63}{\sqrt{n}}$

Source: L. H. Miller, "Table of Percentage Points of Kolmogorov Statistics," *J. Amer. Statist. Assoc.,* 51 (1956), 111–121.

TABLE A-10b. Quantities of the Smirnov test statistic for two samples of equal size n

One-sided test	$p = 0.90$	0.95	0.975	0.99	0.995
Two-sided test	$p = 0.80$	0.90	0.95	0.98	0.99
$n = 3$	2/3	2/3			
4	3/4	3/4	3/4		
5	3/5	3/5	4/5	4/5	4/5
6	3/6	4/6	4/6	5/6	5/6
7	4/7	4/7	5/7	5/7	5/7
8	4/8	4/8	5/8	5/8	6/8
9	4/9	5/9	5/9	6/9	6/9
10	4/10	5/10	6/10	6/10	7/10
11	5/11	5/11	6/11	7/11	7/11
12	5/12	5/12	6/12	7/12	7/12
13	5/13	6/13	6/13	7/13	8/13
14	5/14	6/14	7/14	7/14	8/14
15	5/15	6/15	7/15	8/15	8/15
16	6/16	6/16	7/16	8/16	9/16
17	6/17	7/17	7/17	8/17	9/17
18	6/18	7/18	8/18	9/18	9/18
19	6/19	7/19	8/19	9/19	9/19
20	6/20	7/20	8/20	9/20	10/20
21	6/21	7/21	8/21	9/21	10/21
22	7/22	8/22	8/22	10/22	10/22
23	7/23	8/23	9/23	10/23	10/23
24	7/24	8/24	9/24	10/24	11/24
25	7/25	8/25	9/25	10/25	11/25
26	7/26	8/26	9/26	10/26	11/26
27	7/27	8/27	9/27	11/27	11/27
28	8/28	9/28	10/28	11/28	12/28
29	8/29	9/29	10/29	11/29	12/29
30	8/30	9/30	10/30	11/30	12/30
31	8/31	9/31	10/31	11/31	12/31
32	8/32	9/32	10/32	12/32	12/32
33	8/33	9/33	11/33	12/33	13/33
34	8/34	10/34	11/34	12/34	13/34
35	8/35	10/35	11/35	12/35	13/35
36	9/36	10/36	11/36	12/36	13/36
37	9/37	10/37	11/37	13/37	13/37
38	9/38	10/38	11/38	13/38	14/38
39	9/39	10/39	11/39	13/39	14/39
40	9/40	10/40	12/40	13/40	14/40
Approximation for $n > 40$:	$\dfrac{1.52}{\sqrt{n}}$	$\dfrac{1.73}{\sqrt{n}}$	$\dfrac{1.92}{\sqrt{n}}$	$\dfrac{2.15}{\sqrt{n}}$	$\dfrac{2.30}{\sqrt{n}}$

Source: Z. W. Birnbaum and R. A. Hall, "Small-Sample Distribution for Multi-Sample Statistics of the Smirnov Type," *Ann. Math. Statist.*, 31 (1960), 710–720.

TABLE A-10b. Quantiles of the Smirnov test statistic for two samples of different size

One-sided test Two-sided test		$p = 0.90$ $p = 0.80$	0.95 0.90	0.975 0.95	0.99 0.98	0.995 0.99
$N_1 = 1$	$N_2 = 9$	17/18				
	10	9/10				
$N_1 = 2$	$N_2 = 3$	5/6				
	4	3/4				
	5	4/5	4/5			
	6	5/6	5/6			
	7	5/7	6/7			
	8	3/4	7/8	7/8		
	9	7/9	8/9	8/9		
	10	7/10	4/5	9/10		
$N_1 = 3$	$N_2 = 4$	3/4	3/4			
	5	2/3	4/5	4/5		
	6	2/3	2/3	5/6		
	7	2/3	5/7	6/7	6/7	
	8	5/8	3/4	3/4	7/8	
	9	2/3	2/3	7/9	8/9	8/9
	10	3/5	7/10	4/5	9/10	9/10
	12	7/12	2/3	3/4	5/6	11/12
$N_1 = 4$	$N_2 = 5$	3/5	3/4	4/5	4/5	
	6	7/12	2/3	3/4	5/6	5/6
	7	17/28	5/7	3/4	6/7	6/7
	8	5/8	5/8	3/4	7/8	7/8
	9	5/9	2/3	3/4	7/9	8/9
	10	11/20	13/20	7/10	4/5	4/5
	12	7/12	2/3	2/3	3/4	5/6
	16	9/16	5/8	11/16	3/4	13/16
$N_1 = 5$	$N_2 = 6$	3/5	2/3	2/3	5/6	5/6
	7	4/7	23/35	5/7	29/35	6/7
	8	11/20	5/8	27/40	4/5	4/5
	9	5/9	3/5	31/45	7/9	4/5
	10	1/2	3/5	7/10	7/10	4/5
	15	8/15	3/5	2/3	11/15	11/15
	20	1/2	11/20	3/5	7/10	3/4
$N_1 = 6$	$N_2 = 7$	23/42	4/7	29/42	5/7	5/6
	8	1/2	7/12	2/3	3/4	3/4
	9	1/2	5/9	2/3	13/18	7/9
	10	1/2	17/30	19/30	7/10	11/15
	12	1/2	7/12	7/12	2/3	3/4
	18	4/9	5/9	11/18	2/3	13/18
	24	11/24	1/2	7/12	5/8	2/3
$N_1 = 7$	$N_2 = 8$	27/56	33/56	5/8	41/56	3/4
	9	31/63	5/9	40/63	5/7	47/63
	10	33/70	39/70	43/70	7/10	5/7
	14	3/7	1/2	4/7	9/14	5/7
	28	3/7	13/28	15/28	17/28	9/14
$N_1 = 8$	$N_2 = 9$	4/9	13/24	5/8	2/3	3/4
	10	19/40	21/40	23/40	27/40	7/10
	12	11/24	1/2	7/12	5/8	2/3
	16	7/16	1/2	9/16	5/8	5/8
	32	13/32	7/16	1/2	9/16	19/32

TABLE A-10b. Quantiles of the Smirnov test statistic for two samples of different size (Continued)

One-sided test Two-sided test		$p = 0.90$ $p = 0.80$	0.95 0.90	0.975 0.95	0.99 0.98	0.995 0.99
$N_1 = 9$	$N_2 = 10$	7/15	1/2	26/45	2/3	31/45
	12	4/9	1/2	5/9	11/18	2/3
	15	19/45	22/45	8/15	3/5	29/45
	18	7/18	4/9	1/2	5/9	11/18
	36	13/36	5/12	17/36	19/36	5/9
$N_1 = 10$	$N_2 = 15$	2/5	7/15	1/2	17/30	19/30
	20	2/5	9/20	1/2	11/20	3/5
	40	7/20	2/5	9/20	1/2	
$N_1 = 12$	$N_2 = 15$	23/60	9/20	1/2	11/20	7/12
	16	3/8	7/16	23/48	13/24	7/12
	18	13/36	5/12	17/36	19/36	5/9
	20	11/30	5/12	7/15	31/60	17/30
$N_1 = 15$	$N_2 = 20$	7/20	2/5	13/30	29/60	31/60
$N_1 = 16$	$N_2 = 20$	27/80	31/80	17/40	19/40	41/80

Large-sample approximation:

$$1.07 \sqrt{\frac{m+n}{mn}} \quad 1.22 \sqrt{\frac{m+n}{mn}} \quad 1.36 \sqrt{\frac{m+n}{mn}} \quad 1.52 \sqrt{\frac{m+n}{mn}} \quad 1.63 \sqrt{\frac{m+n}{mn}}$$

Source: Frank J. Massey, Jr., "Distribution Table for the Deviation Between Two Sample Cumulatives," *Ann. Math. Statist.,* 23 (1952), 435--441. This table incorporates the corrections reported in Louis S. Davis, "Table Errata 266," *Math. of Computation,* 12 (1958), 262–263.

TABLE A-11. Critical values of T in the Wilcoxon paired-difference test

One sided	Two-sided	n = 5	n = 6	n = 7	n = 8	n = 9	n = 10
P = .05	P = .10	1	2	4	6	8	11
P = .025	P = .05		1	2	4	6	8
P = .01	P = .02			0	2	3	5
P = .005	P = .01				0	2	3

One-sided	Two-sided	n = 11	n = 12	n = 13	n = 14	n = 15	n = 16
P = .05	P = .10	14	17	21	26	30	36
P = .025	P = .05	11	14	17	21	25	30
P = .01	P = .02	7	10	13	16	20	24
P = .005	P = .01	5	7	10	13	16	19

One-sided	Two-sided	n = 17	n = 18	n = 19	n = 20	n = 21	n = 22
P = .05	P = .10	41	47	54	60	68	75
P = .025	P = .05	35	40	46	52	59	66
P = .01	P = .02	28	33	38	43	49	56
P = .005	P = .01	23	28	32	37	43	49

One-sided	Two-sided	n = 23	n = 24	n = 25	n = 26	n = 27	n = 28
P = .05	P = .10	83	92	101	110	120	130
P = .025	P = .05	73	81	90	98	107	117
P = .01	P = .02	62	69	77	85	93	102
P = .005	P = .01	55	68	68	76	84	92

One-sided	Two-sided	n = 29	n = 30	n = 31	n = 32	n = 33	n = 34
P = .05	P = .10	141	152	163	175	188	201
P = .025	P = .05	127	137	148	159	171	183
P = .01	P = .02	111	120	130	141	151	162
P = .005	P = .01	100	109	118	128	138	149

One-sided	Two-sided	n = 35	n = 36	n = 37	n = 38	n = 39	
P = .05	P = .10	214	228	242	256	271	
P = .025	P = .05	195	208	222	235	250	
P = .01	P = .02	174	186	198	211	224	
P = .005	P = .01	160	171	183	195	208	

One-sided	Two-sided	n = 40	n = 41	n = 42	n = 43	n = 44	n = 45
P = .05	P = .10	287	303	319	336	353	371
P = .025	P = .05	264	279	295	311	327	344
P = .01	P = .02	238	252	267	281	297	313
P = .005	P = .01	221	234	248	262	277	292

One-sided	Two-sided	n = 46	n = 47	n = 48	n = 49	n = 50	
P = .05	P = .10	389	408	427	446	466	
P = .025	P = .05	361	379	397	415	434	
P = .01	P = .02	329	345	362	380	398	
P = .005	P = .01	307	323	339	356	373	

From ''Some Rapid Approximate Statistical Procedures'' (1964), 28, F. Wilcoxon and R. A. Wilcox. Reported with the permission of Lederle Laboratories, a division of American Cyanamid Company.

TABLE A-12. Critical values of U in the Mann-Whitney test

In the first table the entries are the critical values of U for a one-tailed test at 0.025 or for a two-tailed test at 0.05; in the second, for a one-tailed test at 0.05 or for a two-tailed test at 0.10.

$n_2 \backslash n_1$	1	2	3	4	5	6	7	8	9	10	11	12	13	14	15	16	17	18	19	20
1																				
2								0	0	0	0	1	1	1	1	1	2	2	2	2
3					0	1	1	2	2	3	3	4	4	5	5	6	6	7	7	8
4				0	1	2	3	4	4	5	6	7	8	9	10	11	11	12	13	13
5			0	1	2	3	5	6	7	8	9	11	12	13	14	15	17	18	19	20
6			1	2	3	5	6	8	10	11	13	14	16	17	19	21	22	24	25	27
7			1	3	5	6	8	10	12	14	16	18	20	22	24	26	28	30	32	34
8		0	2	4	6	8	10	13	15	17	19	22	24	26	29	31	34	36	38	41
9		0	2	4	7	10	12	15	17	20	23	26	28	31	34	37	39	42	45	48
10		0	3	5	8	11	14	17	20	23	26	29	33	36	39	42	45	48	52	55
11		0	3	6	9	13	16	19	23	26	30	33	37	40	44	47	51	55	58	62
12		1	4	7	11	14	18	22	26	29	33	37	41	45	49	53	57	61	65	69
13		1	4	8	12	16	20	24	28	33	37	41	45	50	54	59	63	67	72	76
14		1	5	9	13	17	22	26	31	36	40	45	50	55	59	64	69	74	78	83
15		1	5	10	14	19	24	29	34	39	44	49	54	59	64	70	75	80	85	90
16		1	6	11	15	21	26	31	37	42	47	53	59	64	70	75	81	86	92	98
17		2	6	11	17	22	28	34	39	45	51	57	63	69	75	81	87	93	99	105
18		2	7	12	18	24	30	36	42	48	55	61	67	74	80	86	93	99	106	112
19		2	7	13	19	25	32	38	45	52	58	65	72	78	85	92	99	106	113	119
20		2	8	13	20	27	34	41	48	55	62	69	76	83	90	98	105	112	119	127

$n_2 \backslash n_1$	1	2	3	4	5	6	7	8	9	10	11	12	13	14	15	16	17	18	19	20
1																			0	0
2					0	0	0	1	1	1	1	2	2	2	3	3	3	4	4	4
3			0	0	1	2	2	3	3	4	5	5	6	7	7	8	9	9	10	11
4			0	1	2	3	4	5	6	7	8	9	10	11	12	14	15	16	17	18
5		0	1	2	4	5	6	8	9	11	12	13	15	16	18	19	20	22	23	25
6		0	2	3	5	7	8	10	12	14	16	17	19	21	23	25	26	28	30	32
7		0	2	4	6	8	11	13	15	17	19	21	24	26	28	30	33	35	37	39
8		1	3	5	8	10	13	15	18	20	23	26	28	31	33	36	39	41	44	47
9		1	3	6	9	12	15	18	21	24	27	30	33	36	39	42	45	48	51	54
10		1	4	7	11	14	17	20	24	27	31	34	37	41	44	48	51	55	58	62
11		1	5	8	12	16	19	23	27	31	34	38	42	46	50	54	57	61	65	69
12		2	5	9	13	17	21	26	30	34	38	42	47	51	55	60	64	68	72	77
13		2	6	10	15	19	24	28	33	37	42	47	51	56	61	65	70	75	80	84
14		2	7	11	16	21	26	31	36	41	46	51	56	61	66	71	77	82	87	92
15		3	7	12	18	23	28	33	39	44	50	55	61	66	72	77	83	88	94	100
16		3	8	14	19	25	30	36	42	48	54	60	65	71	77	83	89	95	101	107
17		3	9	15	20	26	33	39	45	51	57	64	70	77	83	89	96	102	109	115
18		4	9	16	22	28	35	41	48	55	61	68	75	82	88	95	102	109	116	123
19	0	4	10	17	23	30	37	44	51	58	65	72	80	87	94	101	109	116	123	130
20	0	4	11	18	25	32	39	47	54	62	69	77	84	92	100	107	115	123	130	138

Reproduced from the *Bulletin of the Institute of Educational Research* at Indiana University, Vol. 1, No. 2; with the permission of the author and the publisher. D. Auble developed the extended tables.

TABLE A-13. Critical values for the Hartley test of homogeneity of variance

The Hartley F_{max} Test for Homogeneity of Variances

$df = n - 1$	α	\multicolumn{11}{c}{$k = $ number of variances}
		2	3	4	5	6	7	8	9	10	11	12
4	.05	9.60	15.5	20.6	25.2	29.5	33.6	37.5	41.4	44.6	48.0	51.4
	.01	23.2	37.	49.	59.	69.	79.	89.	97.	106.	113.	120.
5	.05	7.15	10.8	13.7	16.3	18.7	20.8	22.9	24.7	26.5	28.2	29.9
	.01	14.9	22.	28.	33.	38.	42.	46.	50.	54.	57.	60.
6	.05	5.82	8.38	10.4	12.1	13.7	15.0	16.3	17.5	18.6	19.7	20.7
	.01	11.1	15.5	19.1	22.	25.	27.	30.	32.	34.	36.	37.
7	.05	4.99	6.94	8.44	9.70	10.8	11.8	12.7	13.5	14.3	15.1	15.8
	.01	8.89	12.1	14.5	16.5	18.4	20.	22.	23.	24.	26.	27.
8	.05	4.43	6.00	7.18	8.12	9.03	9.78	10.5	11.1	11.7	12.2	12.7
	.01	7.50	9.9	11.7	13.2	14.5	15.8	16.9	17.9	18.9	19.8	21.
9	.05	4.03	5.34	6.31	7.11	7.80	8.41	8.95	9.45	9.91	10.3	10.7
	.01	6.54	8.5	9.9	11.1	12.1	13.1	13.9	14.7	15.3	16.0	16.6
10	.05	3.72	4.85	5.67	6.34	6.92	7.42	7.87	8.28	8.66	9.01	9.34
	.01	5.85	7.4	8.6	9.6	10.4	11.1	11.8	12.4	12.9	13.4	13.9
12	.05	3.28	4.16	4.79	5.30	5.72	6.09	6.42	6.72	7.00	7.25	7.48
	.01	4.91	6.1	6.9	7.6	8.2	8.7	9.1	9.5	9.9	10.2	10.6
15	.05	2.86	3.54	4.01	4.37	4.68	4.95	5.19	5.40	5.59	5.77	5.93
	.01	4.07	4.9	5.5	6.0	6.4	6.7	7.1	7.3	7.5	7.8	8.0
20	.05	2.46	2.95	3.29	3.54	3.76	3.94	4.10	4.24	4.37	4.49	4.59
	.01	3.32	3.8	4.3	4.6	4.9	5.1	5.3	5.5	5.6	5.8	5.9
30	.05	2.07	2.40	2.61	2.78	2.91	3.02	3.12	3.21	3.29	3.36	3.39
	.01	2.63	3.0	3.3	3.4	3.6	3.7	3.8	3.9	4.0	4.1	4.2
60	.05	1.67	1.85	1.96	2.04	2.11	2.17	2.22	2.26	2.30	2.33	2.36
	.01	1.96	2.2	2.3	2.4	2.4	2.5	2.5	2.6	2.6	2.7	2.7
∞	.05	1.00	1.00	1.00	1.00	1.00	1.00	1.00	1.00	1.00	1.00	1.00
	.01	1.00	1.00	1.00	1.00	1.00	1.00	1.00	1.00	1.00	1.00	1.00

TABLE A-14. Critical values of the Cochran test of homogeneity of variance

The Cochran Test for Homogeneity of Variances (.01 Level)

	\multicolumn{11}{c}{k = number of variances}
n	2	3	4	5	6	7	8	9	10	12	15
5	.959	.834	.721	.633	.564	.508	.463	.425	.393	.343	.288
6	.937	.793	.676	.588	.520	.466	.423	.387	.357	.310	.259
7	.917	.761	.641	.553	.487	.435	.393	.359	.331	.286	.239
8	.899	.734	.613	.526	.461	.411	.370	.338	.311	.268	.223
9	.882	.711	.590	.504	.440	.391	.352	.321	.295	.254	.210
10	.867	.691	.570	.485	.423	.375	.337	.307	.281	.242	.200
11	.855	.675	.554	.470	.408	.361	.325	.295	.270	.232	.191
12	.843	.660	.539	.456	.396	.350	.314	.285	.261	.223	.184
13	.831	.647	.527	.445	.385	.340	.305	.276	.253	.216	.178
14	.821	.635	.515	.434	.376	.331	.297	.269	.246	.210	.173
15	.812	.624	.505	.425	.367	.323	.289	.262	.240	.205	.168
16	.803	.615	.496	.417	.360	.317	.283	.256	.234	.200	.164
17	.795	.606	.488	.409	.353	.311	.278	.251	.230	.196	.161
18	.788	.598	.481	.403	.347	.305	.272	.246	.225	.191	.157
19	.781	.591	.474	.396	.341	.300	.267	.242	.220	.188	.154
20	.775	.584	.468	.391	.336	.295	.263	.238	.217	.185	.151
21	.769	.578	.462	.385	.331	.291	.259	.234	.213	.182	.149
22	.763	.572	.457	.381	.327	.287	.255	.231	.210	.179	.147
23	.758	.566	.452	.376	.323	.283	.252	.227	.207	.176	.144
24	.753	.561	.447	.372	.319	.279	.249	.224	.205	.174	.142
25	.748	.556	.443	.368	.315	.276	.246	.222	.202	.172	.140
26	.744	.552	.439	.364	.312	.273	.243	.219	.200	.170	.139
27	.739	.548	.435	.361	.309	.270	.241	.217	.197	.168	.137
28	.735	.544	.431	.358	.306	.268	.238	.215	.195	.166	.136
29	.732	.540	.428	.355	.303	.265	.236	.212	.193	.164	.134
30	.728	.536	.424	.352	.301	.263	.234	.210	.192	.163	.133
32	.721	.530	.418	.346	.296	.258	.230	.207	.188	.160	.130
34	.715	.523	.413	.341	.291	.255	.226	.203	.185	.157	.128
36	.709	.518	.408	.337	.287	.251	.223	.200	.182	.154	.126
38	.704	.513	.403	.333	.284	.248	.220	.198	.180	.152	.124
40	.699	.508	.399	.329	.281	.245	.217	.195	.177	.150	.122
42	.694	.504	.395	.326	.277	.242	.215	.193	.175	.148	.121
44	.690	.500	.392	.323	.275	.239	.212	.191	.173	.147	.119
46	.686	.496	.388	.320	.272	.237	.210	.189	.171	.145	.118
48	.682	.492	.385	.317	.270	.235	.208	.187	.170	.143	.117
50	.679	.489	.382	.314	.267	.234	.206	.185	.168	.142	.116
145	.606	.423	.325	.264	.223	.193	.170	.152	.138	.116	.093
∞	.500	.333	.250	.200	.143	.125	.111	.100	.083	.067	.050

TABLE A-14. Critical values of the Cochran test of homogeneity of variance (*Continued*)

The Cochran Test for Homogeneity of Variances (.05 Level) (*Continued*)

n	2	3	4	5	6	7	8	9	10	12	15
					$k =$ number of variances						
5	.906	.746	.629	.544	.480	.431	.391	.358	.331	.288	.242
6	.877	.707	.590	.507	.445	.397	.360	.329	.303	.262	.220
7	.853	.677	.560	.478	.418	.373	.336	.307	.282	.244	.203
8	.833	.653	.537	.456	.398	.354	.319	.290	.267	.230	.191
9	.816	.633	.518	.439	.382	.338	:304	.277	.254	.219	.182
10	.801	.617	.502	.424	.368	.326	.293	.266	.244	.210	.174
11	.789	.603	.489	.412	.357	.315	.283	.257	.235	.202	.167
12	.777	.590	.477	.401	.347	.306	.275	.249	.228	.195	.161
13	.767	.580	.467	.392	.339	.299	.267	.242	.222	.190	.157
14	.757	.570	.458	.384	.331	.292	.261	.237	.216	.185	.153
15	.749	.561	.450	.377	.325	.286	.256	.231	.211	.181	.149
16	.741	.554	.443	.370	.319	.280	.251	.227	.207	.177	.146
17	.734	.547	.437	.365	.314	.276	.246	.223	.203	.174	.143
18	.728	.540	.431	.359	.309	.271	.242	.219	.200	.170	.140
19	.722	.534	.425	.354	.304	.267	.238	.215	.197	.168	.138
20	.717	.529	.421	.350	.300	.264	.235	.212	.194	.165	.135
21	.712	.524	.416	.346	.297	.260	.232	.209	.191	.163	.133
22	.707	.519	.412	.342	.293	.257	.229	.207	.188	.160	.132
23	.702	.515	.408	.339	.290	.254	.226	.204	.186	.158	.130
24	.698	.511	.404	.335	.287	.251	.224	.202	.184	.157	.128
25	.694	.507	.401	.332	.284	.249	.222	.200	.182	.155	.127
26	.691	.504	.398	.330	.282	.247	.219	.199	.180	.153	.125
27	.687	.500	.395	.327	.279	.244	.217	.196	.178	.152	.124
28	.684	.497	.392	.324	.277	.242	.215	.194	.177	.150	.123
29	.681	.494	.389	.322	.275	.240	.214	.193	.175	.149	.122
30	.678	.491	.387	.320	.273	.238	.212	.191	.174	.148	.121
32	.672	.486	.382	.315	.269	.235	.209	.188	.171	.145	.119
34	.667	.481	.378	.312	.266	.232	.206	.185	.169	.143	.117
36	.662	.477	.374	.308	.263	.229	.203	.183	.166	.141	.115
38	.658	.473	.370	.305	.260	.227	.201	.181	.164	.139	.114
40	.654	.469	.367	.302	.257	.224	.199	.179	.163	.138	.112
42	.651	.466	.364	.299	.255	.222	.197	.177	.161	.136	.111
44	.647	.463	.361	.297	.253	.220	.195	.175	.159	.135	.110
46	.644	.460	.359	.295	.251	.218	.193	.174	.158	.134	.109
48	.641	.457	.357	.293	.249	.216	.192	.172	.156	.132	.108
50	.638	.454	.354	.290	.247	.215	.190	.171	.155	.131	.107
145	.581	.403	.309	.251	.212	.183	.162	.145	.131	.110	.089
∞	.500	.333	.250	.167	.143	.125	.111	.100	.083	.067	.050

This table was derived by John T. Roscoe, William R. Veitch, and Donaldson G. Woods.

TABLE A-15. Random numbers

86992	51815	32944	94883	64228	46332	94596	04279	88828	09233	63866	51602
58116	95642	82774	54532	18256	19574	76051	12236	16137	13641	82818	74449
10401	62369	06313	70879	15127	62569	50684	62900	60671	52519	97427	92713
92469	78381	72651	72335	16815	31054	90263	81038	94427	82100	40938	26181
39271	43366	67050	24279	60229	62736	07700	14103	98387	55926	84838	07476
78948	34602	33675	72618	11401	64159	61148	23584	14040	52468	15340	68484
09260	43120	29840	92038	12841	33689	44895	67865	97694	62147	66681	48004
45075	98475	91376	02066	28071	92121	58473	89273	06653	94883	60187	96942
16081	16903	51537	42935	86396	34432	14524	70594	91829	98814	91933	92662
58475	88941	60966	58190	42302	41738	90356	29641	42276	31501	02172	53521
48966	83037	35038	20880	53563	02054	23107	88367	38765	17106	25654	28758
42099	48838	96726	87622	35691	54067	51273	28455	26970	94186	36910	97888
95404	66063	16786	49600	95217	84630	99753	26990	93708	10386	01622	93280
30731	89737	61265	42071	00493	22491	10575	37933	67657	09515	97279	10889
74295	55028	83970	11703	87637	95979	06082	79141	38179	31851	83825	96419
53600	46558	25571	78769	04095	17467	54276	20810	67280	63488	43970	26234
09185	38612	82646	51957	77073	60050	95108	76285	70803	53313	12963	94714
23297	92521	95037	93314	53311	89869	73173	04689	52388	14643	99421	35796
79624	21723	00085	00922	77499	87817	46583	14120	46229	29992	36398	43702
81289	92018	16589	53396	52072	21450	97642	66817	47598	10858	65539	71244
77288	93012	31992	46500	00455	30237	64290	08935	68745	04291	00502	10466
41314	24712	48591	17394	31306	97182	31898	43678	72535	72118	87867	14328
42018	43726	57433	48442	63557	37655	99617	71185	70503	50671	01144	45329
38259	39902	52271	35655	80575	32830	35317	23639	67594	44331	10111	41037
76160	50897	97508	85679	88607	38156	47657	41123	73141	06547	26607	28649

TABLE A-15. Random numbers (Continued)

11276	88017	02125	91547	23572	54663	73422	53371	35690	53177	28631	32142
91304	91748	07856	63448	96615	63374	26074	75539	25730	52525	43959	18277
45223	86507	29760	15641	84933	96973	98723	28789	94965	09778	43503	93595
20547	52143	76325	21553	16562	89085	23911	77231	09529	44288	86958	88276
52131	25414	39005	65766	59599	66424	13389	27927	58731	36304	26339	24280
97840	18459	34161	93935	01919	57659	32172	15600	94499	51415	47903	03543
66853	26892	22398	11872	71676	13257	15191	93898	04795	95604	88956	95803
75490	52105	26667	92866	33005	91170	20140	63283	67953	12867	73778	52816
55519	90051	55736	70579	34298	91535	22031	22052	53263	37628	77880	55885
17808	98641	50300	47246	13409	61393	31609	82690	57318	78511	82845	57896
44751	59258	21247	48728	12558	54579	32852	64724	50367	87125	10823	52003
13963	48849	41505	83577	46978	80679	52791	28395	02067	49578	95531	35704
71790	32482	83002	37059	09233	47251	08703	96575	32523	98973	77850	71069
81624	46647	69811	63844	80525	81288	44719	39017	88548	75996	20701	19294
02381	15678	11869	36881	07344	70523	90022	05843	39936	52446	70685	64361
03275	97894	56667	90017	24049	49535	76607	19567	11185	06529	49642	38303
97786	21658	67898	23173	93408	75321	93206	67447	93611	46925	63355	62253
72420	27487	59215	75943	96405	08873	57482	80381	76957	89711	08797	77528
22796	79031	36291	64566	30216	39222	69608	95629	51191	03818	20142	99857
20599	78174	81074	72710	74939	94358	98173	55281	58025	92970	80284	41447
36146	08151	00427	26232	26519	69468	76852	19542	13204	90016	44048	38284
94221	24101	93274	25665	95111	85027	46556	07385	00932	64543	10048	14671
83220	41312	76932	91741	18032	49183	16839	76100	90828	94709	71798	73009
07956	43661	80055	40935	21044	82243	89463	15722	43971	37815	72813	65806
20264	98945	71312	92847	42238	36228	97543	24144	16981	97917	96591	28346

Appendix B: Selected Statistics from USNEWS File

Sample	Mean	Median	Standard Deviation	Range
1	15332.000	14768	2702.458	7765
2	16808.143	15577	1325.745	3665
3	15233.143	16224	2631.145	6877
4	15135.857	15238	1639.052	5087
5	15660.143	15456	3053.627	9853
6	14280.286	14291	2240.489	5872
7	15465.286	15917	2180.970	6150
8	13559.857	13412	1634.478	5013
9	13984.429	13658	2423.710	7749
10	15631.714	16298	2411.589	7184
11	14025.857	13459	2242.998	6710
12	14619.857	14818	1174.834	5074
13	15820.857	15918	1278.901	3694
14	15299.000	15261	1399.060	4463
15	14940.429	14228	2811.617	9079
16	15575.143	14410	1864.080	4272
17	14917.429	15261	1880.537	4838
18	15865.571	15917	1929.053	4923
19	17480.429	17782	1983.918	5305
20	15139.000	15918	2274.648	6921
21	15489.857	15226	2530.293	7688
22	14332.429	14524	2577.183	6973
23	14907.411	15784	1707.411	5003
24	15030.286	14524	2249.709	5612
25	15952.571	16547	1942.040	5060
26	15237.000	15848	2672.519	8032
27	15806.429	16577	2044.955	5481
28	16412.857	16622	3073.620	9853
29	14469.714	14132	2237.289	6655
30	15794.571	15963	1539.764	4842
31	15458.143	15681	3137.546	10041
32	15666.000	16178	2395.593	5775
33	16457.286	17890	2344.266	5612
34	15262.286	15456	1653.510	4817

Sample	Mean	Median	Standard Deviation	Range
35	16328.857	16224	1651.421	4605
36	14397.857	14298	2564.835	7457
37	13147.000	12708	1816.327	5717
38	15967.000	16349	2427.270	7087
39	14945.286	15784	2679.405	8022
40	15558.143	16349	2956.444	7063
41	14924.571	14298	1627.590	4021
42	14697.571	15963	2434.418	6340
43	14523.143	15226	1993.139	5812
44	14309.571	14132	2694.679	6878
45	15783.857	15918	2516.347	7080
46	15520.714	15053	1958.201	5281
47	15340.714	14702	2117.280	6096
48	14272.286	14410	1699.728	4906
49	13413.286	13790	2346.248	6012
50	15028.571	14410	2504.595	7670
51	14015.857	13002	2734.304	7064
52	14926.000	14142	2790.248	8295
53	15169.286	15053	1801.056	5481
54	14693.000	15784	2249.902	6159
55	14164.714	14132	1984.362	5812
56	16276.429	15456	1863.421	4897
57	15808.857	16337	2354.713	7728
58	14518.000	14524	2146.624	6217
59	14264.429	14524	2721.712	6615
60	16189.571	16547	2034.212	5281
61	14077.714	14291	3165.446	8025
62	15657.429	15073	1839.747	4608
63	15015.000	15226	2673.949	8146
64	15074.571	14524	1641.438	4930
65	15252.714	15261	2932.792	8295
66	14507.571	14524	2685.048	8403
67	14428.000	13979	2752.104	8402
68	16074.571	16298	1667.408	4232
69	15585.714	14818	1965.775	4603
70	14897.857	15456	3184.916	8022
71	13116.286	13658	2093.801	5841
72	14459.286	13790	2338.078	6615

Sample	Mean	Median	Standard Deviation	Range
73	14210.143	14410	1947.188	5520
74	14483.000	14291	2407.821	7101
75	15479.286	15918	2401.920	6455
76	15239.857	15918	2447.337	6734
77	13637.571	13658	2139.081	5610
78	15184.714	15053	2332.358	7350
79	15572.000	15784	2411.542	7121
80	14784.571	14228	2702.659	7471
81	15354.000	15918	1834.299	5395
82	15834.000	17139	2811.299	7157
83	15708.714	15784	1136.198	3532
84	14898.000	14083	2586.738	6198
85	13837.000	13979	1694.477	4730
86	13513.857	13351	3092.743	7269
87	16210.571	16024	1931.929	5079
88	14701.571	14410	2303.106	6591
89	15109.714	15238	2251.235	6594
90	15740.429	15917	1900.340	5281
91	15229.286	14410	2570.560	7507
92	15542.143	15917	2635.933	8025
93	15373.000	15261	2873.914	9660
94	15297.857	14410	2367.449	6436
95	15882.571	15963	1956.701	5941
96	15122.429	14768	2730.711	7208
97	15442.286	14132	3533.788	9103
98	16160.143	16547	2402.859	7184
99	14513.429	14818	2772.286	8008
100	15656.571	15053	2299.291	5869

*Selected statistics of 100 samples of size 7 from the values of x5 in the USNEWS file.

Appendix C: Data Files

TABLE C-1. Auto data file*

Car	Basic Price (X1)	Displacement cc (X2)	Horsepower (X3)	Curb Weight lb. (X4)	MPG (Y)
Alfa Romeo 2000	$11,195	1962	111	2550	22
Alfa Romeo Sports Sedan	$ 9,695	1962	111	2815	22
Alfa Romeo Sprint	$10,495	1962	111	2660	21
AMC Spirit & AMX	$ 3,899	4229	110	2630	17
AMC Pacer	$ 4,699	4983	125	3205	14
AMC Concord	$ 4,049	1984	80	2945	20
Audi Fox	$ 6,295	1588	78	2070	28
Audi 5000	$ 8,995	2144	103	2860	19
Avanti	$17,670	5735	185	3500	14
BMW 320i	$ 9,735	1990	110	2530	19
BMW 528i	$15,505	2788	169	3720	17
BMW 633CSi	$26,770	3210	177	3430	12
BMW 733i	$23,575	3210	177	3770	12
Buick Isuzu Opel	$ 4,075	1817	80	2160	26
Cadillac Seville & DL	$15,646	5735	125	4380	21
Chevy Chevette	$ 3,794	1598	74	2040	23
Chevy Monza	$ 3,617	2474	90	2720	24
Chevy Nova	$ 3,955	5735	165	3295	16
Chevy Camaro	$ 4,677	4998	130	3400	16
Corvette	$10,220	5735	195	3520	16
Datsun 210	$ 3,700	1397	65	1995	35
Datsun 310	$ 4,800	1397	65	2015	28
Datsun 200–SX	$ 5,100	1952	92	2365	21
Datsun 510	$ 4,300	1952	92	2240	23
Datsun 810	$ 6,200	2393	120	2755	21
Datsun 280ZX	$11,599	2753	135	2935	18
Dodge Colt Hatchback & Plymouth Champ	$ 4,217	1410	70	1795	34

TABLE C-1. Auto data file* (Continued)

Car	Basic Price (X1)	Displacement cc (X2)	Horsepower (X3)	Curb Weight lb. (X4)	MPG (Y)
Dodge Colt	$ 3,813	1597	77	2040	30
Dodge Colt Challenger	$ 6,167	2555	105	2520	21
Dodge Omni/Plymouth Horizon	$ 4,122	1716	77	2195	25
Dodge Aspen/ Plymouth Volare	$ 3,968	5900	195	3205	16
Ferrari 308 GTE & GTS	$37,348	2926	205	3225	11
Ferrari 308 GT4	$35,345	2926	205	3200	12
Fiat Strada	$ 4,500	1498	75	2100	30
Fiat Xl/9	$ 6,290	1498	67	2100	26
Fiat Spider 2000	$ 7,090	1995	86	2310	18
Fiat Brava	$ 5,290	1998	86	2490	20
Ford Fiesta	$ 4,198	1599	66	1760	28
Ford Pinto & Mercury Bobcat	$ 3,199	2294	88	2425	22
Ford Mustang & Mercury Capri	$ 4,071	2294	130	2515	21
Ford Fairmont/ Mercury Zephyr	$ 3,710	4950	140	2615	16
Ford Granada/ Mercury Monarch	$ 4,342	4098	97	3205	17
Honda Civic	$ 3,649	1488	63	1790	35
Honda Accord	$ 5,799	1751	72	2110	28
Jaguar XJ–S	$25,000	5343	244	3830	10
Jaguar XJ6L & XJ12L	$19,000	4235	176	3980	14
Lamborghini Silhouette	$40,000	2996	260	2750	11.5
Lamborghini Countach S	$80,000	3929	325	3170	11
Lancia Beta Coupe	$10,040	1995	87	2715	21
Lancia Beta Sedan	$ 8,551	1995	87	2790	21
Lotus Eclat	$28,000	1973	156	2270	24
Lotus Espirit II	$27,000	1973	156	2320	24
Lotus Elite	$29,046	1973	156	2450	22

TABLE C-1. Auto data file* (Continued)

Car	Basic Price (X1)	Displacement cc (X2)	Horsepower (X3)	Curb Weight lb. (X4)	MPG (Y)
Maserati Merak/SS	$31,000	2965	182	3185	14
Maserati Bore	$39,927	4931	315	3540	12
Maserati Khamsin	$42,587	4931	315	3800	12
Mazda GLC	$ 3,895	1415	65	1995	30
Mazda RX-7	$ 6,395	1146	100	2350	17
Mercedes–Benz 240D	$20,775	2746	142	3560	14
Mercedes–Benz 300TD	$25,000	2998	77	3780	23
Mercedes–Benz 280CE	$22,481	2998	77	3475	23
Mercedes–Benz 280SE	$25,000	2998	110	3850	24
Mercedez–Benz 450 SL	$34,760	4520	180	3795	12
MG Midget	$ 5,200	1493	50	1850	23
MGB	$ 6,550	1798	67	2415	16
Peugeot 504	$ 7,922	1971	88	2905	22
Peugeot 504	$11,969	2851	133	3210	16
Plymouth Arrow	$ 4,434	1995	93	2480	28
Plymouth Sapporo	$ 6,166	2555	105	2615	26
Pontiac Firebird	$ 4,825	6596	185	3500	14
Porsche 924	$13,950	1983	110	2445	20
Porsche 911 SC	$20,775	2994	117	2425	17
Porsche 928	$29,775	4474	218	3285	16
Porsche Turbo	$28,500	3299	261	2855	15
Renault Le Car	$ 3,950	1289	60	1819	26
Renault 17 Cordini	$ 7,495	1947	95	2440	20
Rolls–Royce Silver Shadow	$55,000	6750	240	4930	11
Rolls–Royce Corniche	$85,000	6750	240	5310	11
Rolls–Royce Camargue	$110,000	6750	240	5175	11

TABLE C-1. Auto data file* (Continued)

Car ·	Basic Price (X1)	Displacement cc (X2)	Horsepower (X3)	Curb Weight lb. (X4)	MPG (Y)
Saab 99 GL & GLE	$ 7,400	1985	115	2750	22
Saab EMS	$ 8,300	1985	115	2600	22
Saab 99 Turbo	$10,500	1985	135	2670	20
Saab 900	$ 8,000	1985	115	2800	22
Subaru DL & GF	$ 3,699	1595	67	1955	30
Subaru 4WD & BRAT	$ 5,300	1595	67	2225	26
Toyota Corolla	$ 3,748	1588	75	2290	24
Toyota Corona	$ 5,719	2189	90	2565	18
Toyota Celica	$ 6,129	2189	90	2530	18
Toyota Cressida	$ 8,279	2563	108	2695	18
Triumph Spitfire	$ 5,795	1493	53	1830	22
Triumph TR7	$ 7,300	1998	86	2470	19
TVR Taimar	$15,900	2994	142	2335	20
Volkswagen Beatle Conv.	$ 6,245	1584	48	2130	20
Volkswagen Rabbit	$ 4,499	1457	71	1780	25
Volkswagen Scirocco	$ 6,545	1588	78	1950	24
Volkswagen Dasher	$ 6,650	1588	78	2160	23
Volvo 240 Series	$ 6,985	2127	107	2845	19
Volvo 260 Series	$11,495	2664	125	3130	17

*This data was collected from issues of *Road and Track* magazine.

TABLE C-2. Comparative price study*

		Kroger	Hy-Vee	Eagle	Niemann	Gamage	W. Pierce	Aldi	Seven-Eleven
1. Cabbage	H:	$N.A.	$ .19	$ .19	$ .19	$ .19	$N.A.	$N.A.	$N.A.
	N:	.19	N.A.	N.A.	N.A.	N.A.	.25	N.A.	N.A.
Orange	H:	N.A.	1.69	1.67	2.78	1.45	N.A.	N.A.	N.A.
	N:	1.39	2.84	N.A.	N.A.	N.A.	3.56	N.A.	N.A.
2. Dairy: milk	H:	1.93	1.89	1.93	1.93	1.98	2.13	N.A.	N.A.
	N:	N.A.	2.06	N.A.	2.06	1.93	2.23	N.A.	1.93
3. Round steak	H:	2.48	2.18	2.09	2.39	2.39	N.A.	N.A.	N.A.
	N:	N.A.	N.A.	N.A.	N.A.	N.A.	$2.29	N.A.	N.A.
4. Soup: tomato	H:	.22	.19	.21	N.A.	.25	N.A.	.23	N.A.
	N:	.23	.24	.24	.31	.28	.29	N.A.	.37
Vegetable	H:	.23	.26	.26	N.A.	.25	N.A.	.23	N.A.
	N:	.27	.28	.29	.30	.31	.33	N.A.	.39
5. Cranberry	H:	.39	.39	.41	N.A.	N.A.	N.A.	N.A.	N.A.
	N:	.45	.51	.44	.35	.55	.54	N.A.	.85
6. Peas	H:	.39	.32	.33	.27	.49	N.A.	.25	N.A.
	N:	.34	.38	.38	.36	.29	.37	.35	.65
Tomato	H:	.25	.38	.34	N.A.	.49	N.A.	N.A.	N.A.
	N:	.56	.53	.52	.39	.59	.41	.35	.65
7. Beef stew	H:	.66	N.A.	1.25	N.A.	N.A.	N.A.	N.A.	N.A.
	N:	.91	.96	1.46	.79	1.06	.99	.99	1.37
Spam type	H:	1.32	N.A.	N.A.	N.A.	N.A.	N.A.	N.A.	N.A.
	N:	1.53	1.53	1.08	1.65	1.93	1.99	.99	2.60
8. Ravioli	H:	.74	N.A.	.57	N.A.	N.A.	N.A.	N.A.	N.A.
	N:	1.19	N.A.	.62	N.A.	.84	N.A.	N.A.	N.A.
9. Mayonnaise	H:	.79	N.A.	N.A.	N.A.	N.A.	N.A.	N.A.	N.A.
	N:	1.13	.88	.87	.94	$1.03	.98	.60	1.29
Sugar	H:	1.69	1.64	1.64	N.A.	N.A.	N.A.	N.A.	N.A.
	N:	2.09	1.69	1.70	1.64	2.13	1.99	2.13	2.49
10. Sauce: tomato	H:	.65	.33	.36	.35	.52	N.A.	.19	N.A.
	N:	.60	.42	.42	.49	.56	.51	N.A.	.74
11. Moist foil pack	H:	N.A.	N.A.	N.A.	N.A.	.53	N.A.	1.99	N.A.
	N:	.68	.76	.76	.65	.69	.92	N.A.	1.19

TABLE C-2. Comparative price study* (Continued)

		Kroger	Hy-Vee	Eagle	Niemann	Gamage	W. Pierce	Aldi	Seven-Eleven
12. Peanut butter	H:	.70	.88	.98	.70	1.06	N.A.	.79	N.A.
	N:	1.05	1.03	1.16	.70	1.20	1.19	N.A.	1.72
13. Brownies	H:	.69	.73	1.06	.69	.91	N.A.	N.A.	N.A.
	N:	1.32	.99	1.06	1.36	1.06	1.04	N.A.	1.35
14. Grease remover	H:	.79	N.A.	N.A.	N.A.	N.A.	N.A.	N.A.	N.A.
	N:	1.19	1.13	.99	1.34	1.01	1.29	.99	1.29
15. Boxed pizza kit	H:	N.A.	N.A.	.77	N.A.	N.A.	N.A.	N.A.	N.A.
	N:	1.17	1.18	1.18	1.30	1.22	1.29	.89	1.69
16. Salt	H:	.17	.14	.23	N.A.	.18	N.A.	.17	N.A.
	N:	.24	.15	.26	.15	.19	.16	N.A.	.24
Garlic	H:	3.23	N.A.	N.A.	N.A.	N.A.	N.A.	N.A.	N.A.
	N:	N.A.	3.58	1.38	2.19	3.40	N.A.	N.A.	N.A.
17. Dietetic food	H:	.90	1.02	1.08	N.A.	N.A.	N.A.	1.02	N.A.
	N:	1.39	1.69	1.50	1.38	1.50	1.44	N.A.	2.19
18. Beef	H:	N.A.	N.A.	N.A.	N.A.	N.A.	N.A.	N.A.	N.A.
	N:	1.58	2.68	3.14	N.A.	3.38	3.68	N.A.	N.A.
19. Cookies	H:	.85	N.A.	N.A.	N.A.	N.A.	N.A.	.99	N.A.
	N:	1.29	1.22	1.14	1.10	1.76	.89	N.A.	1.25
20. Paper plates	H:	.32	N.A.	N.A.	N.A.	N.A.	N.A.	.21	N.A.
	N:	.21	1.33	.25	.21	2.09	2.00	N.A.	1.09
Toilet paper	H:	N.A.	.68	.81	N.A.	N.A.	N.A.	.59	N.A.
	N:	.95	.99	.91	1.03	1.25	1.29	N.A.	N.A.
21. Unsweetened cold cereal	H:	.74	.70	.68	N.A.	.84	N.A.	.69	N.A.
	N:	.95	.79	.84	.97	.99	1.52	1.09	1.45
22. Corn	H:	.33	.39	.36	.51	N.A.	N.A.	N.A.	N.A.
	N:	.47	.48	.78	.51	.77	.53	N.A.	N.A.
Green beans	H:	.49	.47	.41	.51	N.A.	N.A.	N.A.	N.A.
	N:	.54	.56	.78	.68	.55	.59	N.A.	N.A.
Angel food cake	H:	N.A.	N.A.	1.79	N.A.	N.A.	N.A.	N.A.	N.A.
	N:	N.A.	N.A.	N.A.	N.A.	N.A.	N.A.	N.A.	2.90

TABLE C-2. Comparative price study* (Continued)

		Kroger	Hy-Vee	Eagle	Niemann	Gamage	W. Pierce	Aldi	Seven-Eleven
23. Canned fruit drink	H:	.47	.38	.69	.31	.41	N.A.	.29	N.A.
	N:	.55	.40	.58	.41	.51	.96	N.A.	1.09
Boxed tea	H:	1.43	1.67	1.67	1.59	1.73	N.A.	N.A.	N.A.
	N:	2.38	2.39	2.39	2.13	2.79	2.77	1.40	3.73
24. Peanuts	H:	1.19	1.33	1.59	N.A.	N.A.	N.A.	.89	N.A.
	N:	1.59	1.69	1.66	1.76	1.85	1.69	N.A.	6.35
Ding-Dongs	H:	1.19	N.A.	N.A.	N.A.	N.A.	N.A.	.99	N.A.
	N:	1.39	1.69	1.66	1.69	1.69	1.69	N.A.	.41
25. Cheese spread	H:	1.49	N.A.	N.A.	N.A.	N.A.	N.A.	N.A.	N.A.
	N:	1.84	2.49	N.A.	2.19	2.07	1.98	2.19	N.A.
26. Shampoo	H:	.59	N.A.	N.A.	N.A.	N.A.	N.A.	N.A.	N.A.
	N:	1.19	1.33	.99	1.34	1.73	2.58	.69	1.85

H: House brand   N: Name brand   NA: Not Available

*This data file was obtained in 1980 when completing a servey of grocery stores in Macomb, Illinois.

TABLE C-3. Educational data file*

```
VARIABLE LIST    L,AGE,SEX,YRS,DEGREE,MAJOR1,MAJOR2,MAJOR3,MAJOR4
                 HRS,ASSIST1,ASSIST2,ASSIST3,ASSIST4,ASSIST5,ASSIST6,PRLS
                 PRAD,PRC,ITEM1 TO ITEM14
INPUT FORMAT     FIXED(F5,0,4F2,0,4F3,0,7F2,0,3F3,0,/,5X,14F2,0)
IF               (ASSIST1 EQ 1) PDCAT=1
IF               (ASSIST2 EQ 2) PDCAT=2
IF               (ASSIST3 EQ 3) PDCAT=3
IF               (ASSIST4 EQ 4) PDCAT=4
IF               (ASSIST5 EQ 5) PDCAT=5
IF               (ASSIST6 EQ 6) PDCAT=6
VAR LABELS       SEX SEX OF RESPONDEE/
                 YRS NUMBER OF YEARS IN TRAINING PROFESSION/
                 DEGREE HIGHEST LEVEL OF EDUCATION/
                 MAJOR1 ASSOCIATE MAJOR/
                 AGE AGE OF RESPONDEE/
                 MAJOR2 BA-BS MAJOR/
                 MAJOR3 MASTER-S MAJOR/
                 MAJOR4 DOCTORATE MAJOR/
                 HRS HOURS PER WEEK IN PROF. DEVELOPMENT/
                 ASSIST1 COLLEGE OR UNIVERSITY/
                 ASSIST2 COMMUNITY COLLEGE/
                 ASSIST3 PROFESSIONAL ORGANIZATION/
                 ASSIST4 IN-HOUSE PROGRAM-INTERNAL STAFF/
                 ASSIST5 IN-HOUSE PROGRAM-EXTERNAL CONSULTANT/
                 ASSIST6 OTHER/
                 PRLS % LEARNING SPECIALIST TIME/
                 PRAD % ADMINISTRATOR TIME/
                 PRC % CONSULTANT TIME/
                 ITEM1 TO ITEM14 PDNA LEVELS OF NEED FOR THE ITEM/
                 PDCAT PROFESSIONAL DEVELOPMENT CATEGORY
VALUE LABELS     AGE (1)21-32 (2)33-44 (3)45-56 (4)57 & OVER/
                 SEX (1)MALE (2)FEMALE/
                 YRS (1)0-4 (2)5-10 (3)11-24 (4)25 OR MORE/
                 DEGREE (1) HIGH SCHOOL
                 (2) ASSOCIATE DEGREE
                 (3) COLLEGE, NO DEGREE
                 (4)BA OR BS DEGREE
                 (5)BA OR BS DEGREE PLUS GRAD HOURS
                 (6)MASTER-S OR EQUIVALENT
                 (7)MASTER-S PLUS GRAD HOURS
                 (8)DOCTORATE/
                 MAJOR1 TO MAJOR4 (1)IND. RELATIONS-PERSONNEL
                 (2)BUSINESS
                 (3)EDUCATION
                 (4)ENGINEERING
                 (5)HEALTH-PSYCHOLOGY
                 (6)LIBERAL ARTS
                 (7)SCIENCES
                 (8)JOURNALISM-COMM,
                 (9)HRD-O,D,
                 (10)OTHER/
                 HRS (1)0 (2)1-5 (3)6-10 (4)10 OR MORE/
                 ITEM1 TO ITEM14 (1)VERY LITTLE NEED
                 (2) LITTLE NEED
                 (3)NEUTRAL
                 (4)SOME NEED
                 (5)NEEDED
                 (6)GREATLY NEEDED/
```

TABLE C-3. Educational data file* (Continued)

```
                    POCAT (1) COLLEGE OR UNIVERSITY
                    (2) COMMUNITY COLLEGE
                    (3) PROFESSIONAL ORGANIZATION
                    (4) IN-HOUSE PROGRAM-INTERNAL STAFF
                    (5) IN-HOUSE PROGRAM-EXTERNAL CONSULTANT
                    (6) OTHER
MISSING VALUES AGE,SEX,DEGREE,HRS
                    PRLS,PRAD,POCAT
                    PRC (0)
RECODE              PRLS(98,5 THRU 99,5=100)
RECODE              PRAD(98,5 THRU 99,5=100)
RECODE              PRC(98,5 THRU 99,5=100)
N OF CASES          137
```

```
00095 3 1 4 7 00 06 03 00 3 1 2 3 4 5 0 20 30 50
00100 1 1 1 4 3 1 1 2 4 3 3 1 2 4
00105 1 2 1 7 00 03 09 00 2 0 0 0 4 0 0 80 10 10
00110 0 0 1 2 0 0 1 2 5 0 2 0 5 5
00115 1 2 1 4 00 02 00 00 2 0 0 0 0 5 0 60 30 10
00120 5 5 5 4 5 4 5 4 5 5 5 5 5 4
00125 2 1 2 3 11 00 00 00 3 0 0 3 4 5 6 20 50 30
00130 0 0 0 2 0 0 0 0 3 1 0 0 0 0
00135 1 2 1 1 11 00 00 00 3 0 0 3 4 5 6 80 20 00
00140 3 4 4 1 5 2 2 0 4 4 3 4 1 0
00145 2 1 3 3 11 00 00 00 2 1 0 3 4 5 0 30 60 10
00150 1 3 3 2 4 1 3 5 3 3 4 5 5 4
00155 2 1 2 6 00 02 09 00 3 0 0 3 4 0 0 50 30 20
00160 3 2 1 1 4 0 3 2 1 4 2 1 3 1
00165 1 2 1 7 00 06 06 00 2 1 0 3 4 5 6 30 40 30
00170 1 3 5 3 1 0 0 0 3 5 0 1 0 0
00175 3 1 3 5 00 02 02 00 2 0 0 3 0 0 0 40 50 10
00180 1 3 4 2 2 3 3 1 4 5 1 4 4 4
00185 2 1 2 5 00 01 00 00 4 1 0 3 4 5 0 10 70 20
00190 1 1 1 1 0 0 1 0 3 1 1 0 1 1
00195 3 1 3 5 00 00 02 00 2 0 0 0 4 5 0 20 70 10
00200 0 5 0 2 0 1 4 1 3 2 1 5 1 1
00205 3 1 2 6 00 06 06 00 3 0 0 0 4 5 0 40 20 40
00210 5 2 2 4 3 1 3 3 4 3 3 2 2 2
00215 2 1 3 7 00 00 00 00 2 1 0 3 4 0 0 00 00 99
00220 2 3 4 1 4 5 3 5 4 3 3 4 5 4
00225 3 1 3 6 00 06 01 00 2 0 0 0 0 0 0 80 20 00
00230 1 2 2 2 2 1 2 5 3 4 4 3 4 2
00235 3 1 2 6 00 10 10 00 2 0 0 3 4 5 0 00 75 25
00240 5 5 4 4 3 3 3 4 5 4 3 4 3 4
00245 3 1 4 8 00 02 02 03 2 0 0 3 0 0 0 10 05 85
00250 1 1 1 3 2 2 4 1 5 4 2 1 1 1
00255 3 1 3 6 00 01 02 00 2 0 0 3 4 5 6 30 35 35
00260 1 5 2 4 5 0 0 1 5 2 1 1 4 0
00265 1 2 2 6 00 06 06 00 2 0 0 3 0 0 6 05 45 50
00270 0 0 1 0 3 0 1 0 4 4 0 2 0 0
00275 4 1 4 3 11 00 00 00 3 0 0 3 0 0 0 10 60 30
00280 0 3 3 5 0 0 3 2 3 1 0 0 2 1
00285 1 1 1 6 00 06 06 00 2 1 0 3 0 0 0 40 20 40
00290 1 0 4 3 2 0 3 3 4 4 2 0 5 2 *
00295 1 1 1 6 00 08 08 00 2 0 0 3 0 0 0 00 00 99
```

TABLE C-3. Educational data file* (Continued)

```
00300 1 3 3 3 5 1 1 1 3 2 3 4 3 4
00305 3 1 2 5 00 02 00 00 2 0 0 3 4 5 0 10 80 10
00310 3 3 3 2 3 4 4 2 4 1 4 4 3 2
00315 1 1 1 5 00 05 00 00 2 0 0 3 4 5 0 30 50 20
00320 4 1 4 1 4 0 1 1 4 3 1 4 1 1
00325 2 2 3 5 00 06 02 00 2 0 0 3 4 5 0 01 80 19
00330 0 0 5 2 1 0 0 5 3 4 1 0 5 0
00335 3 2 2 1 11 00 00 00 3 1 0 3 0 5 6 60 20 20
00340 4 2 4 2 2 3 5 4 5 4 3 5 5 5
00345 2 2 2 3 11 00 00 00 2 1 0 3 0 0 0 40 30 30
00350 3 3 4 2 3 2 4 3 2 3 3 3 4 4
00355 1 2 1 6 00 08 08 00 3 0 0 3 0 5 6 25 50 25
00360 2 3 3 4 3 3 4 3 4 4 2 4 3 3
00365 1 1 2 7 00 06 09 00 3 1 0 3 0 0 0 10 80 10
00370 0 0 3 4 3 0 3 0 0 0 0 0 0 0
00375 3 1 3 7 00 02 02 00 2 1 0 3 0 0 0 00 99 00
00380 2 2 2 4 2 2 2 2 5 2 2 2 2 2
00385 1 2 5 00 02 00 00 2 0 0 0 0 5 0 00 50 50
00390 2 2 4 4 3 1 3 2 3 3 2 2 1 1
00395 1 2 1 5 00 06 00 00 2 0 0 3 0 0 0 20 30 50
00400 2 2 3 3 1 2 4 0 4 4 0 5 1 0
00405 2 1 1 6 00 02 02 00 2 0 0 3 0 0 0 20 30 50
00410 2 3 4 3 4 5 1 4 4 2 2 5 1 1
00415 3 2 2 5 05 05 00 00 4 1 2 3 4 5 0 00 00 60
00420 4 3 3 2 3 1 5 0 2 3 2 4 3 2
00425 4 1 4 7 00 03 03 00 0 1 0 3 0 0 0 20 00 00
00430 2 2 2 4 3 2 1 1 3 1 1 4 1 1
00435 1 2 1 5 00 03 00 00 2 1 0 3 4 5 0 75 05 20
00440 4 4 2 4 2 1 1 3 4 4 1 3 4 2
00445 3 1 2 4 00 10 00 00 2 0 0 3 4 5 0 00 90 10
00450 3 2 4 2 1 1 2 2 1 2 1 3 2 2
00455 2 1 3 5 00 03 03 00 3 0 0 3 4 5 0 20 60 20
00460 3 3 4 4 4 3 2 4 3 5 5 5 5 3
00465 1 2 1 6 00 06 06 00 2 0 0 3 4 5 6 00 60 40
00470 4 4 5 2 4 3 5 2 4 3 4 4 5 2
00475 3 2 3 5 06 06 00 00 4 1 0 3 4 0 0 00 75 25
00480 4 2 1 1 2 1 2 2 3 4 1 1 2 2
00485 1 2 1 1 00 00 09 00 7 0 2 3 0 5 0 20 70 10
00490 2 4 2 4 3 1 5 1 4 1 5 2 3 1
00495 2 1 1 7 00 04 06 00 2 0 0 3 0 0 6 60 40 00
00500 2 2 3 5 2 1 1 3 4 5 4 5 5 5
00505 2 1 2 6 06 04 04 00 2 0 0 3 0 0 0 99 00 00
00510 5 2 0 0 3 3 1 1 4 2 3 2 1 2
00515 2 2 1 5 00 06 02 00 4 1 0 0 0 0 0 00 99 00
00520 1 4 4 1 3 1 2 3 3 2 2 3 3 2
00525 2 1 3 7 00 06 03 00 3 1 0 3 0 0 0 10 50 40
00530 2 4 4 4 3 2 2 3 3 4 5 4 4 4
00535 2 1 2 7 00 06 03 00 2 0 0 3 0 0 0 40 20 40
00540 2 2 3 2 2 1 5 4 3 4 3 3 5 3
00545 1 1 1 4 00 07 00 00 4 0 0 3 4 0 6 20 40 40
00550 5 4 3 5 5 4 3 3 4 1 5 5 5 2
00555 3 2 2 5 06 03 00 00 3 0 0 3 0 0 0 00 70 30
00560 1 0 1 3 1 0 1 1 4 3 2 1 1 1
00565 2 1 2 6 00 01 01 00 2 1 0 3 4 5 0 20 20 60
00570 2 4 5 4 5 2 4 4 3 1 2 4 4 3
00575 1 2 2 3 00 01 00 00 2 1 0 3 4 0 6 05 75 20
00580 2 3 4 4 3 0 1 2 3 3 1 1 2 1
```

TABLE C-3. Educational data file* (Continued)

```
00585 2 1 2 7 07 05 06 00 2 1 2 0 0 0 0 30 50 20
00590 0 1 3 4 3 0 3 1 2 1 0 0 1 0
00595 2 2 2 7 00 03 03 00 2 0 0 3 4 0 0 30 50 20
00600 2 2 2 4 2 2 4 2 2 3 4 3 4 4
00605 1 2 2 7 00 06 06 00 3 1 0 3 0 0 6 40 45 15
00610 4 0 0 3 2 0 3 4 5 3 5 2 5 2
00615 2 2 2 5 00 02 00 00 2 0 0 3 4 5 0 00 95 05
00620 0 0 3 0 0 1 4 5 3 4 2 0 5 0
00625 4 1 2 4 00 07 00 00 2 0 0 3 4 5 0 20 80 00
00630 0 0 0 0 0 0 0 4 3 5 2 0 2 0
00635 3 1 3 5 00 03 03 00 3 1 0 3 0 0 6 00 00 40
00640 0 2 4 4 5 0 4 2 4 4 3 3 4 1
00645 3 1 1 5 00 06 00 00 3 0 2 3 4 0 0 30 70 00
00650 3 1 3 4 2 4 2 1 4 4 1 3 3 0
00655 1 2 2 6 00 03 03 00 4 0 0 3 4 5 0 00 90 10
00660 3 5 5 3 3 1 2 2 3 5 3 3 4 2
00665 2 1 3 5 00 08 00 00 2 0 0 3 4 5 0 50 25 25
00670 3 1 0 2 3 0 0 0 4 2 0 0 0 0
00675 3 1 2 2 02 00 00 00 2 0 0 0 0 0 6 90 10 00
00680 5 4 2 3 2 5 1 1 4 1 2 4 3 0
00685 3 2 3 7 00 06 03 00 4 0 0 3 4 0 6 20 50 30
00690 0 0 3 0 0 0 4 3 4 0 0 0 0 0
00695 3 1 2 7 00 06 06 00 2 0 0 3 4 0 0 10 55 35
00700 1 2 1 3 1 0 2 3 5 4 1 1 2 2
00705 2 1 1 6 00 05 03 00 2 1 0 3 4 0 0 75 20 05
00710 2 3 4 3 4 0 4 4 3 1 5 2 3 1
00715 2 2 1 8 00 06 06 06 2 0 0 3 0 0 0 30 40 30
00720 1 2 4 4 2 1 2 1 4 2 3 3 2 1
00725 4 1 4 5 00 04 00 00 3 0 0 3 4 5 0 00 00 99
00730 0 1 1 3 1 1 3 1 4 1 4 1 1 1
00735 1 2 1 1 11 00 00 00 2 0 0 3 0 0 0 40 50 10
00740 2 2 3 4 1 1 3 3 3 2 2 1 4 4
00745 1 2 1 4 00 06 00 00 2 0 0 3 4 0 0 80 00 20
00750 2 3 2 2 3 1 5 3 2 4 1 1 2 4
00755 2 2 3 4 00 08 00 00 3 1 0 3 0 0 0 10 50 40
00760 0 1 1 2 2 0 4 2 5 5 5 1 5 5
00765 2 1 3 7 06 06 06 00 2 0 0 3 0 0 6 50 10 40
00770 0 2 3 3 1 0 1 0 4 2 0 0 2 4
00775 3 1 3 7 00 05 03 00 2 0 0 3 0 0 0 15 25 60
00780 3 4 4 2 2 3 4 4 3 3 5 4 3 1
00785 3 1 3 7 00 05 03 00 2 0 0 3 0 0 0 15 25 60
00790 1 4 3 0 0 0 5 4 3 2 3 4 2 1
00795 3 1 3 3 11 00 00 00 4 1 0 3 0 5 0 00 10 60
00800 5 5 5 1 3 0 5 1 5 5 0 5 0 4
00805 2 1 1 5 00 03 00 00 4 0 0 0 5 0 50 30 20
00810 4 4 2 2 2 2 2 4 4 4 4 4 3 3
00815 4 1 3 4 00 03 00 00 2 0 0 3 0 0 0 00 50 50
00820 3 5 4 1 3 0 0 1 4 3 4 1 5 5
00825 2 1 3 5 00 03 00 00 2 1 0 3 4 0 0 15 50 35
00830 3 2 5 4 2 1 2 1 3 4 4 3 1 1
00835 3 1 3 6 00 06 05 00 2 0 2 0 4 0 0 50 35 15
00840 2 3 4 4 5 2 1 1 3 2 5 3 5 1
00845 3 2 2 6 00 02 09 00 3 1 2 3 0 0 0 05 90 05
00850 2 5 3 5 5 5 5 5 5 3 3 5 5 3
00855 1 1 2 6 00 05 09 00 2 0 0 3 0 0 0 30 50 20
00860 0 0 2 2 2 0 2 0 1 3 0 0 0 0
00865 4 1 4 5 00 04 00 00 2 0 0 3 4 5 0 00 00 00
00870 4 2 1 5 4 0 3 2 3 4 1 5 1 0
```

TABLE C-3. Educational data file* (Continued)

```
00875 2 1 2 6 00 05 06 00 3 0 0 3 0 0 6 30 50 20
00880 0 2 2 0 2 0 0 2 0 4 4 1 4 3
00885 2 1 3 7 00 03 03 00 3 1 0 0 4 5 0 10 80 10
00890 1 2 1 3 3 0 1 1 5 5 1 1 2 1
00895 3 1 3 7 00 03 08 00 2 0 0 3 4 5 0 15 60 25
00900 4 4 1 4 2 1 3 1 3 1 1 3 2 1
00905 2 1 1 5 00 02 00 00 2 0 0 0 0 0 0 00 10 60
00910 1 4 3 3 2 0 3 1 4 2 2 2 2 2
00915 2 1 1 4 00 02 00 00 2 1 0 3 4 5 0 20 50 30
00920 3 4 2 4 4 4 4 5 4 2 2 2 5 3
00925 3 1 2 4 00 02 00 00 2 0 0 3 4 0 0 00 80 20
00930 3 2 3 2 4 3 1 1 2 2 2 3 4 3
00935 1 2 2 5 04 06 00 00 2 0 0 3 4 5 0 30 40 30
00940 0 3 3 2 4 3 4 3 4 1 5 5 5 5
00945 1 2 1 4 00 03 00 00 3 1 0 3 0 5 0 20 00 20
00950 1 1 3 4 2 3 3 1 4 2 5 1 3 2
00955 2 1 3 4 00 60 00 00 2 0 2 0 0 0 6 20 00 20
00960 0 1 0 2 3 2 3 2 3 1 1 0 1 0
00965 2 1 1 1 11 00 00 00 2 0 0 0 4 5 0 25 50 25
00970 1 2 4 4 3 1 4 1 3 2 1 2 0 2
00975 3 1 2 5 00 04 00 00 3 1 0 0 0 5 0 40 40 20
00980 1 2 3 2 2 0 3 1 5 3 3 1 5 5
00985 3 1 3 6 00 02 05 00 2 1 0 3 0 0 0 45 40 15
00990 2 3 4 5 3 2 2 1 2 5 2 2 1 1
00995 2 1 3 7 00 06 03 00 2 0 0 3 0 0 0 30 50 20
01000 1 1 0 1 0 0 0 0 2 2 0 1 0 0
01005 3 1 1 4 00 08 00 00 1 0 0 3 0 0 0 25 50 25
01010 2 2 0 1 2 0 1 2 4 3 3 3 5 1
01015 2 2 3 7 00 06 05 05 4 0 0 3 0 0 6 50 10 40
01020 1 1 1 3 1 1 1 1 1 1 1 1 2 2
01025 2 1 2 5 00 05 09 00 2 1 0 3 4 5 0 00 50 00
01030 1 1 1 5 2 1 5 3 3 2 4 2 4 3
01035 3 1 3 7 00 05 01 01 2 0 0 3 0 5 0 20 30 50
01040 2 5 4 4 5 1 4 3 3 2 3 4 2 4
01045 2 1 2 4 00 02 00 00 2 0 0 0 0 5 0 00 00 05
01050 1 2 3 3 2 0 1 3 5 4 3 2 4 4
01055 2 1 3 8 00 07 03 03 2 0 0 3 0 5 0 70 30 00
01060 3 1 2 1 2 1 1 2 3 5 3 1 3 4
01065 2 2 2 6 00 00 03 00 2 0 0 3 4 0 0 25 00 75
01070 5 3 2 4 1 0 0 2 0 1 5 2 4 2
01075 2 2 2 7 05 05 03 00 4 1 0 3 0 0 0 00 00 10
01080 0 2 5 3 2 1 1 0 2 1 5 1 1 2
01085 2 1 1 5 00 02 00 00 2 0 0 0 4 5 0 00 99 00
01090 3 2 3 1 3 2 1 3 2 3 1 4 5 3
01095 2 1 3 4 00 05 00 00 2 0 0 3 0 0 0 20 70 10
01100 2 3 3 3 4 2 5 2 4 3 5 5 5 5
01105 3 1 3 5 00 03 00 00 2 1 0 3 4 5 0 00 70 30
01110 4 3 4 4 3 4 3 3 3 3 1 4 3 2
01115 1 2 1 4 00 05 00 00 2 1 0 3 4 0 0 20 30 50
01120 3 2 1 5 4 1 5 0 4 0 3 3 0 0
01125 2 1 1 5 00 10 00 00 3 1 0 3 0 5 0 05 70 25
01130 1 1 2 2 3 3 2 1 2 2 3 3 2 2
01135 2 1 3 4 00 00 00 00 2 1 0 0 0 0 0 20 50 30
01140 1 1 1 1 4 1 1 1 1 1 1 1 1 1
01145 3 1 3 6 00 06 06 00 2 0 0 3 0 0 6 00 20 50
01150 3 2 3 2 2 1 4 2 4 4 2 3 2 2
01155 2 1 3 5 00 05 00 00 2 0 0 3 1 5 0 30 10 60
01160 0 0 0 0 0 0 5 3 0 0 0 0 0 0
```

TABLE C-3. Educational data file* (Continued)

```
01165 4 1 3 5 00 04 00 00 2 0 0 0 4 0 0 40 40 20
01170 5 4 5 5 5 3 4 4 0 0 0 0 0 0
01175 3 2 1 6 00 06 06 00 2 0 0 0 0 0 0 00 00 00
01180 4 2 4 3 4 2 3 5 5 4 5 4 3 1
01185 2 1 3 3 00 00 00 00 2 1 2 0 0 0 0 0 20 50 30
01190 5 4 4 1 2 4 3 4 3 3 4 4 3 3
01195 3 1 2 2 04 00 00 00 3 1 0 3 4 0 0 80 10 10
01200 1 1 3 1 1 1 1 2 1 1 1 2 1 1
01205 1 1 2 4 00 06 00 00 2 0 0 3 4 0 0 00 40 60
01210 2 2 0 3 3 4 0 0 3 0 3 3 0 0
01215 4 1 1 3 00 00 00 00 3 0 0 3 4 5 0 70 20 10
01220 0 0 0 0 0 0 0 0 3 0 0 0 0 0
01225 3 1 3 6 00 00 02 00 4 0 0 3 4 5 0 75 00 25
01230 1 0 3 4 4 3 3 1 2 0 3 3 3 2
01235 2 2 1 5 00 03 00 00 2 0 2 3 0 5 0 45 45 00
01240 0 0 0 0 0 0 0 0 0 0 0 0 0
01245 2 1 2 6 00 04 04 00 2 0 0 0 0 0 6 00 30 70
01250 1 3 5 2 4 1 3 2 4 2 3 1 4 2
01255 2 1 3 7 00 00 06 00 2 0 0 3 0 5 0 25 50 25
01260 0 4 3 4 4 4 5 1 3 1 4 3 3 1
01265 2 1 2 6 00 02 02 00 2 0 0 3 0 5 0 00 20 10
01270 5 5 5 3 2 5 5 1 4 5 3 5 5 4
01275 1 1 1 5 00 06 00 00 2 0 0 3 4 0 0 20 50 30
01280 1 3 1 0 0 0 3 3 1 0 2 0 0
01285 1 1 2 6 00 06 03 00 2 1 0 0 4 5 0 20 30 50
01290 3 2 3 1 4 2 2 3 4 4 2 3 3 2
01295 2 1 3 4 00 06 00 00 3 0 0 3 0 0 0 00 30 70
01300 1 4 3 5 5 1 4 4 5 4 5 1 5 1
01305 4 2 3 6 00 06 06 00 2 0 0 3 4 5 6 70 10 30
01310 0 0 0 0 0 0 0 0 0 0 0 0 0 0
01315 2 1 2 8 00 05 05 05 2 0 0 0 0 0 0 30 20 50
01320 3 3 4 3 3 0 1 1 1 1 2 2 1 3
01325 2 1 3 6 00 06 06 00 4 1 0 3 0 0 0 20 40 40
01330 0 1 2 0 3 0 3 2 4 3 0 1 0 0
01335 3 1 4 3 00 00 00 00 2 1 0 3 0 0 0 10 10 80
01340 2 3 4 3 1 0 1 0 5 2 2 3 1 1
01345 2 1 1 6 06 06 03 00 2 0 0 3 4 5 0 05 05 90
01350 4 2 1 2 4 0 1 2 2 3 1 3 2 0
01355 2 2 2 5 00 05 00 90 3 0 0 3 4 0 0 40 20 40
01360 4 2 2 3 3 2 2 1 1 1 1 4 3 1
01365 2 1 2 7 00 06 03 00 2 0 0 3 0 0 0 10 70 20
01370 3 1 1 4 1 3 2 4 4 4 1 2 4 4
01375 1 2 2 5 00 06 02 00 3 1 2 3 4 0 0 05 15 80
01380 1 4 4 3 2 4 2 1 4 5 5 3 1 2
01385 2 2 1 3 00 00 00 00 2 0 2 3 4 0 0 00 00 00
01390 0 0 0 2 3 0 0 3 4 3 3 2 4 2
01395 4 1 4 5 00 04 00 00 3 0 2 0 4 0 0 25 75 00
01400 3 0 2 1 2 0 3 4 4 4 2 3 5 1
01405 2 2 3 6 00 06 03 00 2 0 0 3 0 0 0 35 35 30
01410 4 3 3 4 3 1 2 5 4 5 3 3 5 5
01415 2 1 3 8 00 06 05 05 2 0 0 3 0 0 0 10 60 30
01420 0 0 1 3 1 0 2 2 5 0 2 0 3 1
01425 2 1 3 6 00 02 02 00 2 0 0 3 4 0 0 20 60 20
01430 4 4 5 2 3 4 4 4 3 5 5 5 5 4
```

TABLE C-3. Educational data file* (Continued)

```
01435 1 1 1 6 00 03 03 00 2 0 0 3 4 5 0 80 20 00
01440 3 2 3 2 1 1 2 2 3 3 2 2 2 2
01445 3 2 3 5 00 06 00 00 3 1 0 3 0 0 0 00 50 50
01450 2 2 2 2 2 2 2 3 5 3 3 2 2 1
01455 3 1 3 6 00 03 03 00 2 0 0 3 4 0 0 35 60 05
01460 0 1 2 5 2 2 4 1 4 3 5 1 1 2
```

*This data file was obtained from: Ms. Jean Hauser, Director, Project CREATION, LaSalle-Peru, Illinois. More information concerning the purpose and content of the study can be obtained by writing Ms. Hauser directly.

Appendix D: SPSS Data Files

A SELECTED STATISTICAL BIBLIOGRAPHY

Descriptive Statistics and Probability

Clark, C.T., and L.L. Schkade. *Statistical Methods for Business Decisions.* Cincinnati: South-Western Publishing, 1969.

Draper, N., and H. Smith. *Applied Regression Analysis.* New York: John Wiley, 1966.

Guilford, J.P. *Fundamental Statistics in Psychology and Education.* New York: McGraw-Hill, 1965.

Johnson, R.R. *Elementary Statistics.* N. Scituate, MA: Duxbury Press, 1973.

Kelly, F.J., D.L. Beggs, K.A. McNeil, T. Eichelberger, and J. Lyon. *Multiple Regression Approach.* Carbondale, IL: Southern Illinois Press, 1969.

Mendenhall, William, and James E. Reinmuth. *Statistics for Management and Economics*, 3rd ed. N. Scituate, MA: Duxbury Press, 1978.

Tsokos, C.P. *Probability Distributions: An Introduction to Probability Theory with Applications.* Belmont, CA: Duxbury Press, 1972.

Statistical Inference

Bottenberg, R.A., and R.E. Christal. *An Iterative Technique for Clustering Criteria Which Retains Optimum Predictive Efficiency.* Lackland Air Force Base, Texas: Personnel Laboratory, Wreight Air Development Division, March 1961 (WADD-IN-61-3, ASTIA Document AS-261 615).

Dixon, W.J., and F.J. Massey, Jr. *Introduction to Statistical Analysis*, 3rd ed. New York: McGraw-Hill, 1969.

Harnett, D.L., and J.L. Murphy. *Introductory Statistical Analysis*. Reading, MA: Addison-Wesley, 1975.

Huntsberger, D.V. *Elements of Statistical Inference*, 3rd ed. Boston: Allyn and Bacon, 1971.

Johnson, R.R. *Elementary Statistics*. N. Scituate, MA: Duxbury Press, 1973.

Mendenhall, W. *An Introduction to Linear Models and the Design and Analysis of Experiments*. Belmont, CA: Wadsworth, 1967.

Neter, J., and W. Wasserman. *Applied Linear Statistical Models*. Homewood, IL: Richard D. Irwin, 1974.

Ward, J.H., Jr., and E. Jennings. *Introduction to Linear Models*. Englewood Cliffs, NJ: Prentice-Hall, 1973.

Wonnacott, T.H., and R.J. Wonnacott. *Introductory Statistics. for Business and Economics,* 2nd ed. New York: John Wiley and Sons, 1977.

Analysis of Variance

Dunn, O.J., and V.A. Clark. *Applied Statistics: Analysis of Variance and Regression*. New York: John Wiley, 1974.

Guenther, William C. *Analysis of Variance*. Englewood Cliffs, NJ: Prentice-Hall, 1964.

Mendenhall, W. *An Introduction to Linear Models and the Design and Analysis of Experiments*. Belmont, CA: Wadsworth, 1968.

Neter, J., and W. Wasserman. *Applied Linear Statistical Models*. Homewood, IL: Richard D. Irwin, 1974.

Romano, A. *Applied Statistics for Science and Industry*. Boston: Allyn and Bacon, 1977.

Scheffé, Henry. *The Analysis of Variance*. New York: John Wiley, 1959.

Experimental Design

Campbell, D.T., and J.C. Stanley. *Experimental and Quasi-Experimental Designs for Research*. Chicago: Rand McNally, 1963.

Cochran, W.G., and G.M. Cox. *Experimental Designs*, 2nd ed. New York: John Wiley, 1957.

Davies, O.L., ed. *The Design and Analysis of Industrial Experiments*. New York: Hafner Publishing, 1956.

Kerlinger, F.N. *Foundations of Behavioral Research*. New York: Holt, Rinehart and Winston, 1965.

Winer, B.J. *Statistical Principles in Experimental Design*, 2nd ed. New York: McGraw-Hill, 1971.

Nonparametric Statistics

Conover, W.J. *Practical Nonparametric Statistics*. New York: John Wiley, 1971.

Daniel, W.W. *Applied Nonparametric Statistics*. Boston: Houghton Mifflin, 1978.

Fisher, F.E. *Fundamental Statistical Concepts*. New York: Canfield Press, 1973.

Gibbons, J.D. *Nonparametric Methods for Quantitative Analysis*. New York: Holt, Rinehart and Winston, 1976.

Noether, G.E. *Introduction to Statistics: A Nonparametric Approach*, 2nd ed. Boston: Houghton Mifflin, 1976.

Siegel, S. *Nonparametric Statistics for the Behavioral Sciences*. New York: McGraw-Hill, 1956.

Walsh, James E. *Handbook of Nonparametric Statistics*, 3 vols. New York: Holt, Rinehart and Winston, 1953.

Survey Sampling

Cochran, W.G. *Sampling Techniques*. New York: John Wiley, 1963.

Kish, L. *Survey Sampling*. New York: John Wiley, 1965.

Mendenhall, W., L. Ott, and R.L. Scheaffer. *Elementary Survey Sampling*. Belmont, CA: Wadsworth, 1971.

Sudman, S. *Applied Sampling*. New York: Academic Press, 1976.

Statistical Packages

Moore, Richard W. *Introduction to the Use of Computer Packages for Statistical Analyses*. Englewood Cliffs, NJ: Prentice-Hall, 1978.

Ramsey, James B., and Gerald L. Musgrave. *APL-STAT: A-DO-IT-YOURSELF Guide to Computational Statistics*. Belmont, CA: Lifetime Learning Publications, 1981.

Statistical Applications Hardware

Hollingworth, Dennis. *Minicomputers: A Review of Current Technology, Systems, and Applications*. Rand Report R-1279. Santa Monica, CA: The Rand Corporation, July 1973.

Kenney, Donald P. *Minicomputers*. New York: AMACOM, 1973.

The MATYC Journal: Special Calculator Issue. Fall 1978, Vol. 12, No. 3. The issue is devoted to calculators and mini/microcomputer systems.

Mini-Micro Systems. November–December 1977. The issue is devoted to the fundamentals of microcomputer systems.

Scientific American. September 1977. The issue is devoted to microelectronics and microcomputer systems.

Zaffarano, Joan. "Calculators: The Programmables and Prepros." *Administrative Management*, August 1973, pp. 40–48.

Fun Reading

Campbell, S.K., *Flaws and Fallacies in Statistical Thinking*. Englewood Cliffs, NJ: Prentice-Hall, 1974.

David, F.N. *Games, Gods, and Gambling*. London: Charles Griffin and Company, 1962.

Epstein, R.A. *The Theory of Gambling and Statistical Logic*. New York: Academic Press, 1967.

Huff, D. *How to Lie with Statistics*. New York: W.W. Norton, 1954.

Pearson, E.S., and M.G. Kendall. *Studies in the History of Statistics and Probability*. Darien, CT: Hafner Publishing Company, 1970.

Sellers, G.R. *Elementary Statistics*. Philadelphia: W.B. Saunders Co., 1977.

Zeisel, H. *Say It with Figures*. New York: Harper & Row, 1957.

SOLUTIONS TO SELECTED
ODD-NUMBERED EXERCISES

Chapter 1

1. a. Ratio   **d.** Ratio   **f.** Ordinal

     b. Nominal   **e.** Ratio   **g.** Nominal

     c. Ordinal

5. a. Population

     b. Sample

     c. Sample, population

Chapter 3

3. ⟋ 17185   26700

Chapter 4

1. Mean = 7; median = 6; mode = 6.

5. 18.14

9. Range = 9; variance = 11.96; standard deviation = 3.46.

11. Sample; range = 3.65; variance = 1.73; standard deviation = 1.315.

13. $\%^2$

15. No.

17. **a.** 8/9

    b. 13.079, 24.125

19. Skewness = 0.958; kurtosis = 1.622.

21. CONDESCRIPTIVE; X3

    STATISTICS; 7, 8

23. GET FILE; USNEWS

    FREQUENCIES; GENERAL = ALL

    STATISTICS; ALL

    OPTIONS; 7

25. RECODE; X7 (LO THRU $-10.00 = -15.00$) (-9.99 THRU $0.00 = -5.00$)

    (0.01 THRU 10.00 = 5.00) (10.01 THRU 20.00 = 15.00)

    (20.01 THRU 30.00 = 25.00) (30.01 THRU HI = 35.00)

27. SELECT IF; (X1 EQ X2)

29. VALUE LABELS: X5 (10000) 11000 AND BELOW

    (12000) 11001 TO 13000 (14000) 13001 TO 15000

    (16000) 15001 TO 17000 (18000) 17001 TO 19000

    (20000) ABOVE 19000

Chapter 5

1. **a.** $S = \{1, 2, 3, 4, 5, 6\}$
   c. $S = \{0, 1, 2, 3, \ldots\}$
   e. S is the set of all groups of 5 people in the class.

3. **a.** 1/52   **d.** 3/52
   b. 1/4   **e.** 11/26
   c. 3/13

5. **a.** 1/6   **c.** 1/2
   b. 1/36   **d.** 1/9

7. **a.** It is hot and rains.
   c. It does not rain.
   e. It is not hot or does not rain.

9. **a.**

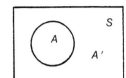

c.

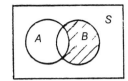

e.

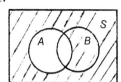

11. **a** . 0.75

    b . 0.70

    c . 0.30

    d. 0.15

    e. 0.90

13. **a** . 0

    b . 0.78

    c. 0.33

    d. 0.55

15. 15/16

17. **a** . 2401

    b . 840

19. **a** . 252

    b . 210

21. 5040

23. 712842

25. **a** . 0.171

    b . 0.486

    c . 0

    d. 0

    e. 0.208

    f. 0.893

29. **a** . 0.625

    b . 0.833

31. **a** . 0

    b . 417

    c . 1

33. **a** . A and B are independent.

    c . A and C are not independent.

37. Yes; yes; yes.

41. $f(x) = \begin{cases} 0.324, x=4 \\ 0.459, x=6 \\ 0.162, x=8 \\ 0.054, x=1000 \\ 0, \text{all other } x \end{cases}$

43. **b** . $E(X) = 7/3$; $\text{Var}(X) = 5/9$.

45. $E(X) = 59.346$; $\text{Var}(X) = 50506.6$.

47. $E(X) = 712$; $\text{Var}(X) = 35/12$.
$E(Y) = 7$; $\text{Var}(Y) = 35/3$.
$E(Z) = 116$; $\text{Var}(Z) = 5/36$.
$E(X + Y + Z) = 32/3$.

49. **a** . 35   **d.** 1

  b . 6   **e.** -35

  c . 15/4   **f.** 4

Chapter 6

3. **a** . Y is b (10, 1/2)   **d.** 0.377

  b . $E(Y) = 2.5$ **e.** 0.161

  c . 0.205   **f.** 0.568

9. **a** . 0.3179

  b . 0.1372

  c . $E(X) = 3.6$; $\text{Var}(X) = 2.653$; $\sigma(X) = 1.629$.

15. 1.5; 18.

17. 0.8187

25. **a** . 1/5

  b . $f(x) = \begin{cases} 1/50000, \; 50000 \le x \le 100000 \\ 0, \text{ otherwise} \end{cases}$

  c . 2/5

  d . 75,000; 208,333,333.3; 14433.76

27. **a** . 1/4 minute

  b . 0.0183

  c . 0.632

31. **a** . 1.24   **e.** -1.23

  b . 1.91   **f.** 0.06

  c . 1.22   **g.** 1.00

  d . -0.27 **h.** 1.12

35. **a.** 79.012   **e.** 66.016

  b. 73.236   **f.** 74.984

  c. 60.392   **g.** 7.752

  d. 67.308

37. **a.** 0.7123

  b. 0.6632

  c. 125.12

39. 0.9007; 0.8899

41. 0.8571

43. 0.6488

45. **a.** 0.8812

  b. 0.7960

  c. 0.0036

  d. 0.6628

Chapter 7

3. **a.** X is $b(74, 0.10)$; $E(X) = 7.4$; $\text{Var}(X) = 6.66$.

   b. 0.1547

5. **a.** 0.5762   **c.** 0.9918

   b. 0.0548   **d.** 10.85

7. **a.** 0.6778

   b. 0.8384

   c. 0.9438

9. **a.** $\bar{X}$ is approximately $N(\mu, 304.06)$.

   b. 0.4495

   c. $n \geq 801$

11. **a.** 0.9298

    b. 0.2327

13. **a.** $N(\mu, 381)$

    b. 0.8098

    c. $n \geq 1257$

    d. No.

17. **a.** p is approximately $N(0.10, 0.00018)$.

    b. 0.8638

    c. 0.1314

    d. 0.1020

19. **a.** 0.8098

    b. 0.0336

Chapter 8

3. The sample mean.

7. **a.** Let M be the sample maximum.

   b. $E(M) = \dfrac{5}{6} N - \dfrac{1}{2} + \dfrac{5}{12N} - \dfrac{1}{12N^3}$

13. 1.28, 1.645, 1.96, 2.33, 2.58

15. **a.** [412.55, 439.45]

    b. [408.71, 443.29]

    c. [398.89, 453.11]

17. [3.745, 7.301], [3.404, 7.642], [3.004, 8.042]

19. [1991.6, 2762.7]

21. [0.051, 0.177], [0.040, 0.188], [0.016, 0.212]

23. [0.093, 0.313]

27. **a.** Do not reject H_o.

    b. Reject H_o.

29. Do not reject H_o.

Chapter 9

1. **a.** $\bar{X}^* = 310.30$

    b. 0.8023, 0.5000, 0.2266, 0.0606

    d. $\bar{X}^* = 314.59;\ \beta = 0.4721$

    e. Reject H_o; Type I.

    f. $z = 1.92 > 1.645;\ 0.0274 < 0.05 = \alpha.$

5. **a.** $H_o: \mu \leq 5$
   $H_a: \mu > 5$
   $\bar{X}^* = 6.779$

    b. $\beta = 0.8810$

    d. Do not reject H_o; Type II.

7. **a.** $H_o: \mu \geq 10$
   $H_a: \mu < 10$
   $\bar{X}^* = 8.931$

    b. $\beta = 0.5438$

    e. Reject H_o.

9. 2.58, 2.33, 1.96, 1.645

11. **a.** $H_o: \mu = 5$
    $H_a: \mu \neq 5$
    The critical values are ± 2.052.

    b. $t = 1.32$; do not reject H_o.

    c. GET FILE; USNEWS
    BREAKDOWN; X3 BY X1

13. **a.** The t distribution.

    b. Do not reject H_o.

15. **a.** $\bar{X} \pm t\ \dfrac{s}{\sqrt{n}}$

    b. [13874.5, 15684.3]

    c. GET FILE; USNEWS
    BREAKDOWN; X5 BY X1

17. It cannot be so claimed.

19. $[-0.56, 11.51]$

21. **a.** [3.54, 8.96]

    b. [6.12, 20.93]

23. Yes.

27. Do not reject H_o.

Chapter 10

1. 4.30, 2.76, 0.344, 0.212

3. **a.** Yes.
   b. Yes.

5. That the populations were normally distributed.

7. $t = -1.82$; do not reject H_o.
9. **a.** $t = -2.72$; reject H_o.
   b. Reject H_o.

11. **b.** No.
    c. No.
    d. Yes.
    e. T-TEST; GROUPS = X1(1, 2)/VARIABLES = X5

13. Do not reject H_o.

15. Reject H_o.

17. **a.** $t = 1.00$; no.
    b. Do not reject H_o.

19. **a.** Yes; the correlation is positive.
    b. Yes.
    d. T-TEST; PAIRS = X2 WITH X4

23. Reject H_o.

Chapter 11

1. 0.081

3. No.

5. $X^2 = 9.22$; reject H_o.

7. $X^2 = 34.66$; reject H_o.

11. Reject H_o.

15. $X^2 = 7.035$; do not reject H_o.

17. $X^2 = 55.991$; reject H_o.

Chapter 12

1. **b.** $R = -0.7077$
   c. Yes; inverse.

3. **b.** $R = -0.7313$
   c. $t = -3.391$; yes.

7. **a.** SCATTERGRAM; X3 WITH X4
STATISTICS; 1, 3
   b. $R = -0.37438$
   c. Yes.

11. Yes.

13. **a.** $R = 0$
    b. $R = 0.9050$
    c. $R = -0.9213$

17. **a.** $M_1 = 79.833$
    b. $M_0 = 46.875$
    c. $p = 6/14; q = 8/14$
    d. $R_{pbi} = 0.7516$

19. $R_{bi} = -0.8342$

27. **a.** $R_{tet} = 0.068$

35. $R_{HP,MPG.W} = -0.5326; t = -1.888;$ yes.

37. $R_{X4,X7.X5} = 0.5818$

Chapter 13

5. **a.** $\bar{Y}_x = 31.09 - 0.0859X$
   b. 18.63
   c. 3.8517
   d. Yes.
   e. [23.96, 38.22]
   f. Decrease mileage by 2.15.
   g. [15.74, 21.52]

9. **a.** GET FILE; USNEWS
SCATTERGRAM; X3 WITH X5
STATISTICS; ALL
   b. At $\alpha = 5\%$, but not at $\alpha = 1\%$.
   c. $\bar{X3}_{x5} = 2.812 + 0.00019 \cdot X5$
   d. $R^2 = 0.07443$
   e. [0.00003, 0.00035]
   g. 6.33

11. **a.** [0.00026, 0.00282]
    c. [-1.216, 1.867]

15. [9.57, 27.69]

17. **a.** $\bar{X}_Y = 248.8 - 62289\,Y$
    b. $X = 361.9 + 11.64\,Y$

19. **a.** Yes; positive linear.

    b. $\bar{Y}_x = 0.989 + 0.0961\,X$; 3.680

    c. [0.0793, 0.1129]; yes.

    d. [−0.0029, 2.000]

21. $\beta_{z_1} = 0.142$; $\beta_{z_2} = 0.450$.

23. **a.** $\bar{X}_1 = 2895.417$; $s_{X_1} = 730.661$.
      $\bar{X}_2 = 118.500$; $s_{X_2} = 45.862$.
      $\bar{Y} = 20.917$; $s_Y = 5.384$.

    b.
$$\begin{bmatrix} 1.000 & -0.708 & -0.731 \\ -0.708 & 1.000 & 0.613 \\ -0.731 & 0.613 & 1.000 \end{bmatrix}$$

    c.
$$\begin{bmatrix} 1.000 & 0.613 \\ 0.613 & 1.000 \end{bmatrix} \times \begin{bmatrix} \beta_{z_1} \\ \beta_{z_2} \end{bmatrix} = \begin{bmatrix} -0.708 \\ -0.731 \end{bmatrix}$$

    d. $\beta_{z_1} = -0.416$; $\beta_{z_2} = -0.476$

    e. $Y' = 22.883$

    f. $Y = -0.00294\,X_1 - 0.0537\,X_2 + 35.803$

    g. $Y' = 22.815$

29. **a.** Consider $Y = AX^B$.

    b. $Y = 0.9661\,X^{1.9976}$

    c. 11.80

31. **b.** Consider $Y' = b_0 + b_1 X + b_2 X^2$.

Chapter 14

1. **b.** fixed

    c. $F = \dfrac{114.43}{78.12} = 1.465$; do not reject H_o.

3. **a.** $F = \dfrac{140.06}{11.39} = 12.30$; reject H_o.

    b. $B = 0.260$; do not reject H_o.

5. $F_{max} = 2.97$; $C = 0.443$. In both tests, do not reject H_o.

7. $B = 2.99$; do not reject H_o.

13. In the COMPUTE command, use the function SQRT.

15. ONEWAY;X7 BY X1(1, 3)
    STATISTICS:ALL

19. $\dbinom{k}{2} = \dfrac{k(k-1)}{2}$

21. Significant differences are shown between the means of these pairs of groups: 1 and 2; 1 and 4; 1 and 5; 2 and 3; and 3 and 4.

23. The differences in the means of groups 1 and 2, and groups 1 and 3, are significant.

25. $F_{row} = \dfrac{2394.02}{39.34} = 60.85.$

$F_{col} = \dfrac{773.16}{39.34} = 19.65.$

$F_{int} = \dfrac{48.12}{39.34} = 1.22.$

31. $F_{row} = \dfrac{300.84}{30.83} = 9.76.$

$F_{col} = \dfrac{91.00}{30.83} = 2.95.$

33. $F_{row} = \dfrac{2.06}{0.038} = 54.21.$

$F_{col} = \dfrac{0.024}{0.038} = 0.632.$

$F_{tr} = \dfrac{0.35}{0.038} = 9.21.$

35. $F_{tr} = \dfrac{1656.8}{109.1} = 15.19.$

$F_{diff} = \dfrac{21.2}{109.1} = 0.194.$

Chapter 15

3. **a.** No.
   b. Yes.

7. **a.** t-test for related samples.
   b. ANOVA.

Chapter 16

1. $X^2 = 3.20$; do not reject H_o.

3. $X^2 = 39.06$; reject H_o.

5. All the E_i are 10.

9. $X^2 = 39.63$; reject H_o.

11. $X^2 = 17.86$; do not reject H_o.

13. $X^2 = 4.65$; do not reject $H_{...}$

17. **a.**

Hand	Probability
All different	0.350
One pair	0.525
Two pair	0.052
Three of a kind	0.070
Four of a kind	0.003

   b. $X^2 = 1.73$; do not reject H_o.

19. **a.** $U = 10$; do not reject H_o.

  b. $U = 10$; do not reject H_o

  c. $U = 14$; do not reject H_o.

21. $U = 18$; no.

25. $U = 27$; do not reject H_o.

27. $U = 24$; do not reject H_o.

29. NPAR TESTS; RUNS $(0.5) = X/$

31. **a.** NPAR TESTS; RUNS (MEAN) $= X4/$
  Do not reject H_o.

  b. NPAR TESTS; RUNS (MEDIAN) $= X4/$
  Do not reject H_o.

35. $U = 32$; reject H_o.

37. $U = 36$; reject H_o.

39. $D = 0.2865$; do not reject H_o.

41. NPAR TESTS; K-S (NORMAL, 100, 20) $= X/$

43. NPAR TESTS; K-S (NORMAL, 9, 10) $= X2/$

45. $D = 0.3661$; reject H_o.

47. NPAR TESTS; K-S $= X$ BY Y $(1,2)/$

49. NPAR TESTS; K-S $= X7$ BY X1 $(1,3)/$

51. $R_{38} = 427$, $R_{45} = 203$; reject H_0.

53. NPAR TESTS; M-W $= X$ BY Y $(1,2)/$

55. NPAR TESTS; M-W $= X5$ BY X1 $(1,3)/$

57. $T^+ = 6.5$, $T^- = 38.5$; reject H_o.

59. NPAR TESTS; WILCOXON $= X$ WITH Y/

61. NPAR TESTS; WILCOXON $= X4$ WITH X7/

63. $H = 4.35$; do not reject H_o.

65. NPAR TESTS; K-W $= X$ BY Y $(1,4)/$

67. NPAR TESTS; K-W $= X6$ BY X1 $(1,3)/$

69. $R_s = 0.30$

71. NONPAR CORR; X WITH Y

73. NONPAR CORR; X2 WITH X4

INDEX

030-22